The Physiology of Growth

The Physiology of Growth

RICHARD J. GOSS

Division of Biology and Medicine
Brown University
Providence, Rhode Island

ACADEMIC PRESS   New York   San Francisco   London   1978

A Subsidiary of Harcourt Brace Jovanovich, Publishers

ACADEMIC PRESS, INC.
111 Fifth Avenue, New York, New York 10003

United Kingdom Edition published by
ACADEMIC PRESS, INC. (LONDON) LTD.
24/28 Oval Road, London NW1 7DX

Library of Congress Cataloging in Publication Data

Goss, Richard, J., 1978
    The physiology of growth.

    Includes bibliographies and index.
    1. Growth. I. Title.   [DNLM: 1. Growth.
2. Regeneration. 3. Physiology.   QT104.3 G677p]
QP84.G63   599'.03'1   78-51239
ISBN 0-12-293055-X

PRINTED IN THE UNITED STATES OF AMERICA

Contents

Preface

Recent years have witnessed an upsurge of interest in problems of growth regulation. Few organs and tissues of the body have escaped the probing experiments of researchers, and although unanswered questions still abound, there has been no dearth of breakthroughs. The time is ripe, therefore, to evaluate what has been learned about the mechanisms and controls of growth and to redefine some of the unsolved problems.

When I used to teach histology, the challenge was to make what was in danger of becoming merely a descriptive course in microscopic anatomy something which would arouse the interests of students—and sustain year after year the enthusiasm of the instructor. The task has been an enjoyable one pursued with mounting interest for more than two decades, a period coinciding with the exponential increase in our knowledge of how tissues and organs of the body grow. In an effort never to go into class without having learned something different about the subject under discussion, formerly uninteresting parts of the body took on new life as exciting experiments were uncovered. Salivary glands, once dutifully alluded to in my course, have in recent years been shown to be responsive to a fascinating variety of physiological stimuli. And the straightforward histology of blood vessels, for example, became more meaningful in light of discoveries about how capillary growth is stimulated and atherosclerotic plaques originate. So it went, until my files became as hypertrophic as some of the organs they covered.

This book is intended for the physiologist willing to concede that development is a physiological process, and for the developmental biologist concerned with the functional dimensions of growth. It will be of equal interest to students of histology, students who want to go beyond the static constraints of descriptive microscopic anatomy to probe the mysteries of how structure is shaped by function. The reader is assumed already to have a basic knowledge of anatomy and histology, coupled with a healthy curiosity about why things are the way they are. The approach is both developmental and comparative because organs of the body are the end result of ontogeny and phylogeny. Even more important are the experimental, and sometimes pathological, approaches upon which much of our knowledge of how growth is controlled depends. I have always found questions more intriguing than answers. Accordingly, that which is known is all the more appetizing when flavored with the challenge of what remains to be discovered.

Each chapter focuses on one or more related organs or tissues. No attempt has been made to be comprehensive, nor to document each and every item that is covered. To become bogged down in too much detail would detract from the main theme, namely, the physiological interplay of factors affecting the growth of organs and tissues. Thus, the text is sprinkled with numbers designed to pinpoint key references in the literature, with priority given to those that are recent, in English, and published in accessible journals.

The author has benefited from the intellectual nourishment of students, too numerous to mention, who have crossed my path at Brown University. Without the challenge of younger minds one is not so motivated to keep up with the advancing front of research, nor to explore those interesting diverticula out of which unforeseen discoveries sometimes develop. We have learned much from each other over the years, if only to appreciate the limits of our own understanding.

No one person has been more indispensable to the writing of this book than Lois T. Brex. Part secretary, part librarian, part research assistant, she has coordinated my efforts with a loyalty which sought no thanks. Her dedication cannot easily be compensated, nor can the efforts of the members of her family who were pressed into service from time to time.

Others have also played important roles, not the least of whom is my artist, Susan Tilberry. Without her graphic talents, and her patience with the author's criticisms, the illustrations could not have done justice to the text. Her drawings speak for themselves. I am also indebted to Grazina Kulawas for her flawless typing of the final manuscript.

The length of the reference lists testifies to the essential role a library plays in writing a book. The Brown University Science Library is staffed with professionals who willingly rise to the challenge of tracking down even the most elusive references. For their unselfish cooperation I am ever grateful.

The etymological trivia which introduce each chapter are derived from many sources. My colleague, Professor George Erikson, has been especially stimulating along these lines. Authorities on the derivation of words have been "The Story Behind the Word" by Harry Wain (C. C. Thomas, 1958) and "The Origin of Medical Terms" by H. A. Skinner (Williams and Wilkins, 1961), both of which make delightful reading for anyone who finds the phylogeny of language as fascinating as that which it describes.

Richard J. Goss

1

Introduction

Mitosis: Gr., *mitos*, thread

Development has two dimensions. The qualitative relates to problems of cellular differentiation. The quantitative one is growth and size determination. It is the latter aspect of development which is the primary subject of the present account.

There are many kinds of growth. Once the developing embryo has matured into an adult, growth does not stop. The physiological turnover of living material is one of life's most basic processes, an activity that balances the rate of synthesis to that of degradation (6). Compensatory growth is exhibited by nearly every organ and tissue of the body in response to overwork. There are no tissues (except for teeth) incapable of repairing injuries to themselves, and in some forms considerable feats of regeneration may occur (9). Growth may be pathological, as when morphological increases are not accompanied by commensurate improvements in functional efficiency. Indeed, reduced physiological activity is associated with atrophy and dystrophy, as well as with the depreciations of aging. These and other manifestations of growth demand to be explained, preferably by a unifying hypothesis. In our present state of ignorance such is clearly more than can reasonably be expected. It is this challenge, however, which has made the resolution of growth's unsolved problems the important goal it has become.

THE STRATEGY OF GROWTH

By classifying different modes of growth, it is possible to gain valuable insight into its nature. One of the first and most successful attempts to do this was Guilio Bizzozero's classification in 1894 (*1*) of tissues according to their

1

mitotic potentials. Although the role of chromosomes in inheritance was not recognized at that time, the process of mitosis was one which had intrigued microscopists for some years. Bizzozero recognized three basic patterns of cellular proliferation in tissues which have since been referred to as renewing, expanding, and static (Fig. 1). According to this interpretation, some tissues of the body (e.g., epidermis, blood cells) are in a lifelong state of renewal at the cellular level, a renewal in which the rate of loss is carefully balanced by the rate of replacement. It is typical of renewing tissues that their germinative compartments are spatially distinct from the differentiated ones. Sometimes the two may be in juxtaposition, as in the epidermis where the mature cells directly overlie the stem cells in the basal layer, or in the intestinal mucosa with the crypts of Lieberkühn giving rise to nondividing cells lining the villi. In hemopoiesis, on the other hand, the differentiated blood cells in the circulation may become far removed from their sites of origin in the marrow and lymphatic organs. Another important attribute of renewing tissues is that the differentiated cells have limited life spans, considerably shorter than that of the organism of which they are a part. Finally, such cells in the fully differentiated state are no longer capable of proliferation.

Expanding organs differ from renewing tissues in several important respects. One is that they do not possess a growth zone, the germinative compartment coinciding with the differentiated one. Another is the poten-

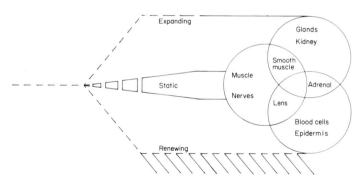

Fig. 1. Diagram of the pathways by which Bizzozero's three types of tissues develop from embryo to adult (*1, 5*). Broken lines signify mitotic proliferation. All tissues are expanding in the embryo (left), gradually ceasing to divide (top) as maturity is approached. The cells of static tissues (middle) hypertrophy when they lose mitotic competence. Renewing tissues (bottom) retain a line of proliferative stem cells from which nondividing descendants differentiate. On the right are indicated examples of the three categories of adult tissues, including some with overlapping attributes. From Goss, R. J., *Science* **153**, 1615–1620, 30 September 1966. Copyright 1966 by the American Association for the Advancement of Science.

tially indefinite longevity of such cells which, in the absence of turnover, are not subject to programmed demise. Perhaps the most important characteristic of expanding organs, however, is the fact that differentiation is not incompatible with mitosis. Virtually every cell in such organs is capable of dividing, and does so frequently during the course of development. The pattern of proliferation under these circumstances is diffuse. Once maturity is attained and there is no longer a need for the organ to keep pace with the overall growth of the body, cell division becomes unnecessary. Nevertheless, even in the adult the fully differentiated cells of expanding organs retain their potential for proliferation, a potential that is readily elicited by injury, reduction in mass, or the stimulus of heightened work loads. Many of the endocrine and exocrine organs of the body fall into this category.

Mitotically static tissues comprise nerves and muscles. In both cases, the fully differentiated cell is so specialized as to have abandoned its capacity for division in early developmental stages. Neuroblasts are used up prenatally in most mammals, and the last myoblast differentiates not long afterward. The absence of cell division is correlated with two important features of mitotically static tissues: the lack of a growth zone and the potentially lifelong survival of such cells. In other respects, however, they have something in common with renewing tissues in that the fully differentiated cells in both cases are incapable of mitosis. The stem cells of renewing tissues are spatially restricted but persist through life, while in static tissues they are temporally confined to prenatal stages of development.

It is evident that in some cells differentiation is accompanied by loss of mitotic potential, while in others it is not. The cells of renewing and static tissues are included in the former category, those of expanding organs in the latter. Nondividing differentiated cells possess specific end products that are retained in the cytoplasm. These include keratin, hemoglobin, and myosin. Further, the physiological activity of such cells tends to be of a more physical nature, e.g., contraction, conduction, or mechanical protection. The cells of expanding organs, on the other hand, synthesize their end products for export. These may take the form of hormones or of various exocrine secretions, e.g., tears, milk, bile, or digestive enzymes. The functions of such cells are more chemical in nature. Although these generalizations are not without their exceptions, the conclusion is inescapable that the disposition of cell-specific end products is not without its effects on mitotic potentials, a possibility that may be explained in terms of negative feedback. Thus, the dissipation of end products leaves a cell's mitotic competence uninhibited, while their intracellular retention mitigates against cell division. A similar principle is seen with respect to cartilage and bone, in which the absence of mitotic activity in chondrocytes and osteocytes may be attributed to the close proximity to their end products in the surrounding matrix.

However useful Bizzozero's classification has been, not all organs and tissues of the body fit neatly into his three categories. Smooth muscle cells, for example, share characteristics with both static and expanding tissues. Like nondividing striated muscle fibers, they retain within their cytoplasm the contractile proteins which are diagnostic of their species, but under appropriate circumstances they are known to be capable of cell division. The adrenal cortex resembles expanding and renewing tissues. All of its cells are capable of proliferation when appropriately stimulated, yet the normal turnover of cells in this organ depends upon a zone of proliferation between the glomerulosa and fasciculata from which cells migrate into other regions of the cortex eventually to disappear in the zona reticularis. Characteristics of renewing and static tissues are combined in cartilage and bone. Since growth of these tissues depends largely upon proliferation in the perichondrium and periosteum, or in the cartilaginous plates of long bones, they have spatially segregated growth zones not unlike those characteristic of renewing tissues. Unlike renewing tissues, however, their cells are not in a constant state of turnover, except insofar as the osteons of solid bone may go through phases of resorption and deposition, a process in which the cells tend to be recycled rather than replaced. Yet the nonmitotic characteristics of fully differentiated osteocytes and chondrocytes are suggestive of mitotically static tissues. The lens of the eye is another hard-to-classify organ. The presence of an equatorial growth zone in the lens epithelium is suggestive of renewing characteristics, as in the nonmitotic nature of the fully differentiated lens fibers. However, the fact that the latter cells survive throughout life without being replaced might be reason to classify them as a static tissue. They may also be considered as expanding tissues inasmuch as their cells proliferate primarily to keep pace with the growth of the body as a whole.

It is the existence of these several tissues which are neither exclusively renewing, expanding, nor static that renders Bizzozero's scheme not entirely satisfactory. Still another shortcoming is that it is based solely upon the capacities of various tissues to undergo cell division. However important mitosis may be, it is by definition limited to the cellular level of organization and is not necessarily the most physiologically meaningful mechanism of tissue growth. It may be more appropriate, therefore, to adopt a classification based not so much on mitotic activity *per se* as on the relative abilities of different tissues to increase their functional capabilities as they grow in mass (5). Augmentation of the population is not necessarily the most efficient way to do so.

Accordingly, it is important to recognize that each organ and tissue of the body is made up of subunits upon which specific physiological activity depends. These functional units may be defined as the smallest irreducible structures still capable of performing the specific physiological activities characteristic of the organ of which they are a part. In many instances, functional units are synonymous with cells, as in the case of blood cells or the less histologically complex endocrine glands. In other tissues, however, functional units may be represented by subcellular structures. The synapse may be considered the functional unit of a neuron, while in striated muscle the sarcomere is the smallest structure still capable of contraction. At higher levels of organization, functional units are multicellular structures represented by such histological entities as the exocrine secretory acini, seminiferous tubules, thyroid or ovarian follicles, renal nephrons, pulmonary alveoli, or intestinal villi. If the growth of an organ or tissue is to be more than merely an increase in mass, it must involve a multiplication of the functional units (hyperplasia) rather than the enlargement of preexisting ones (hypertrophy). The classification of organs on this basis yields considerable insight into the physiological efficiency of growth (Figs. 2 and 3).

In mitotically static tissues growth can occur only by cellular hypertrophy, yet this is achieved in part by hyperplasia of cytoplasmic organelles. In striated muscle, for example, the prodigious enlargement of the fibers is achieved by augmenting the number of myofibrils, thereby increasing the population of sarcomeres. Nevertheless, muscle hypertrophy is limited by the ultimate size beyond which such cells cannot grow. In the case of nerves, if the synapse is regarded as the functional unit the number of such junctions is presumably limited by the constraints of neuronal enlargement. Organs capable of cell division but with functional units at the histological level of organization are also incapable of exercising their maximum potential for physiological growth. The cells of the kidney can multiply but the number of nephrons is fixed early in development. Adult renal growth involves an increase in functional competence only to the extent that preexisting nephrons can enlarge by the addition of more cells to their tubules and glomeruli, a growth in function which falls far short of the hypothetical alternative of adding extra nephrons. A similar situation obtains in the case of the lungs, the pulmonary alveoli of which are not capable of multiplication beyond childhood years. Growth of the lungs is therefore achieved by expansion of the original alveoli to increase the surface area for gas exchange, but this is not so efficient as would have been the production of new alveoli. Likewise, the intestinal villi can increase their dimensions, but not their numbers, during the course of intestinal enlargement after early stages of development. In these examples of organs made up of functional units at the histological level, cell division helps to increase functional abilities to some extent, but it is far from optimal. The ideal situation would be one in which hyperplasia could occur above the cellular level to enable organs to increase

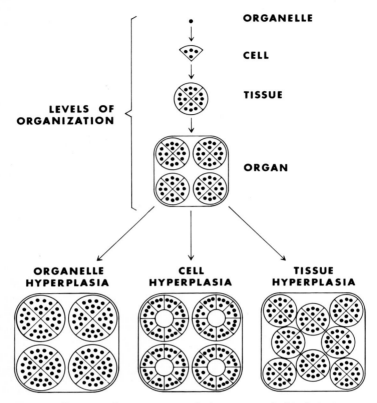

Fig. 2.   Diagram illustrating alternate ways in which organs can double their sizes according to the levels of organization at which hyperplasia takes place (5). From Goss, R. J., *Science* **153,** 1615–1620, 30 September 1966. Copyright 1966 by the American Association for the Advancement of Science.

their population of functional units without limit. The liver is a case in point, for it is an organ which does not lose the capacity to add new tissue units in the form of lobules and hepatic cords. Similarly, other exocrine glands retain their abilities to produce new secretory acini, and the thyroid and ovary can make new follicles in the adult animal.

    If one subscribes to the notion that organisms evolve according to the precepts of functional efficiency and that anything is possible so long as it is within the rules of chemistry and physics, one wonders at the uneven distribution of modes of growth from organ to organ (Fig. 4). Some of the most vitally essential organs of the body (e.g., kidneys, lungs, heart, brain) exhibit the least efficient modes of growth and are singularly lacking in regenerative abilities. Conversely, many glands that are not necessary for survival seem to

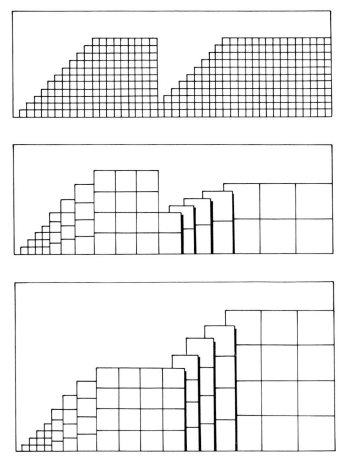

Fig. 3.   Profiles of various patterns of tissue growth based on the proliferation or enlargement of functional units (5). Each square represents n functional units, which increase in numbers (hyperplasia) during early stages of development and in size (hypertrophy) later. Shown at the top is an indeterminate tissue (e.g., liver) which never outgrows the capacity to augment its population of functional units and is therefore capable of complete regeneration. Below are tissues of determinate size (e.g., kidney, heart) which can compensate for reductions in mass (middle) or increased workload (bottom) only by enlarging preexisting functional units. Their regenerative capacities are correspondingly limited. From Goss, R. J., *Science* **153**, 1615–1620, 30 September 1966. Copyright 1966 by the American Association for the Advancement of Science.

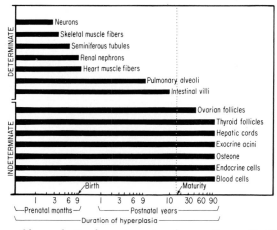

Fig. 4.   Duration of hyperplastic phases in various human tissues (5). Determinate tissues (above) are those which lose the capacity to multiply their functional units before maturity. Indeterminate ones (below) remain hyperplastic throughout life. From Goss, R. J., *Science* **153**, 1615–1620, 30 September 1966. Copyright 1966 by the American Association for the Advancement of Science.

have relatively unlimited capacities for growth. One can only surmise that this apparent paradox has evolved to limit the stature of birds and mammals, if not to insure their eventual demise, as the process of aging results in, or is itself caused by, the attrition of irreplaceable functional units in vital organs. It is worth·noting that some of the lower vertebrates are not subject to such constraints. Their kidneys can augment the nephron population throughout life and their hearts grow by increasing the number of myocardial fibers. Even their central nervous systems may not entirely lose the capacity for neurogenesis. Thus, the patterns of growth described above apply to homeothermic vertebrates, suggesting that growth in higher forms may be a special case of a more general phenomenon. Indeed, since we would seem to be an exception to the rule, one can only contemplate what might have been the selective pressures responsible for these changes during the course of vertebrate evolution.

TACTICS AND PARAMETERS

There are many ways to measure growth, the most direct of which is by weight. Measurements of mass, however, can be deceptive unless one takes into consideration the distinction between fluid accumulation and the addition of living substance *per se*. This possibility can be assessed by measuring the dry weight of an organ after keeping it in an oven until it ceases to lose weight by desiccation. If the ratio of dry to wet weight of an organ is the same in experimentals as in controls, there is reason to believe that changes in mass were not attributable simply to fluid uptake. Relative weights are also important, for animals seldom stop growing during the course of an experiment. This can be corrected by the following formula:

$$\text{Relative weight} = \frac{\text{wet weight of organ}}{\text{body weight}} \times 100$$

Thus, the relative weight of an organ can be expressed as a percentage of the body mass and is a more accurate indication of experimental change than is absolute weight. It is important to bear in mind, however, that even if there is no change in relative weight during the course of an experiment there may still be considerable growth on an absolute scale, as when the organ enlarges at the same rate as the body. A further complication is that relative weight itself may not stay constant over a range of body sizes in a normally developing organism. If it does, growth is referred to as isometric. If not, it is allometric. The equation by which allometry may be expressed is as follows (8):

$$x = by^k$$

The weight of an organ, x, can be expressed as a fraction, b, of the body weight, y. The exponent, k, is a coefficient of growth that takes into account whether the component organ is growing faster or slower than the body as a whole. If k is less than one then the organ is losing relative weight and allometry is negative. In positive allometry, the organ enlarges more rapidly than the rest of the body and k is greater than one. Thus, in the course of development each organ follows a growth curve that is seldom a straight line and that coincides only by accident with that of the body as a whole. The concept of allometry can be applied also to evolutionary relationships. Small species can be compared with large ones just as young animals can be compared with adults. Thus, from the phylogenetic standpoint the relative sizes of organs may increase or decrease with respect to evolving body sizes.

Measurements of organ weights yield information only about the end results of growth. To analyze the process of growth it is necessary to sample tissues and organs at intervals. In a few cases longitudinal studies can be carried out on individual animals where the sampling technique neither affects the growth process nor requires sacrifice of the experimental animal. Blood cells, for example, can be withdrawn in sufficiently small quantities to monitor, without affecting, the course of hemopoiesis. Repeated biopsies of most tissues and organs, however, are not feasible, as a result of which experimental animals must usually be sacrificed in groups on a staggered schedule. In those tissues that grow by cellular proliferation it is useful to record the numbers of dividing cells as a function of postoperative time. This can be achieved by counting mitotic figures in random fields, and expressing the mitotic index as the percentage of dividing cells in the total population at different points in time. Where mitotic counts are low, they can be amplified by pretreatment with colchicine to arrest division in the metaphase and thus accumulate mitoses over a period of 6 to 8 hours prior to sacrifice. Alternatively, it is possible to inject animals with ^{3}H-thymidine and then make autoradiographs of sections of the organ in question. In this procedure the histological sections are mounted on slides which are then dipped in a photographic emulsion to be stored in the dark for a number of weeks. The slides are then developed photographically and the tissues stained to pinpoint the location of silver grains as an indication of where the ^{3}H-thymidine was taken up into synthesized DNA. The resulting labeling index is typically higher than the corresponding mitotic index because cells spend more time making DNA than they do in overt mitosis.

Growth is a sequence of many events and like other developmental processes the earliest reactions tend to be the most important. It is during these initial stages that the biochemical antecedents of hyperplasia and hypertrophy take place. A significant parameter of growth is the change in the rate of synthesis of RNA and proteins, changes that can be studied autoradiographically or by biochemical analysis of tissue homogenates, depending upon how precisely one wishes to localize such reactions. These mechanisms tend to be much the same from one organ to another, except insofar as some tissues rely more on mitosis than others. Equally important, however, are the physiological factors responsible for regulating the rates of these growth mechanisms.

Experimental approaches are called for if the control of growth is to be investigated. Sometimes such experiments are natural ones, as in the comparative study of animal adaptations or the pathological consequences of physiological changes. Man-made experiments are even more revealing, for the nexus between cause and effect is more conveniently discerned under conditions created by the investigator. Experiments may involve such diverse interventions as the injection of drugs or hormones, alterations in diet, exposure to different environments, or a variety of surgical operations. The latter include deletion experiments, whereby part of an organ or tissue is removed to investigate the reactions of what remains, or may involve the transplantation of extra parts. It is relatively simple to augment a tissue such as blood, but very challenging if an organ is to be grafted by microvascular surgery. Of particular interest are transplants made between small and large, or young and old, individuals. When organ and body sizes are mismatched, one can explore to what extent the growth of an organ is intrinsically programmed or is regulated by systemic factors. In addition to extirpation and transplantation, other surgical procedures are sometimes useful to alter the functional activities of an organ, including denervation, ligation, and the rearrangement, displacement, or deviation of parts.

THEORIES OF GROWTH

Any hypothesis formulated to provide a conceptual basis to explain the diversity of growth phenomena must propose the operation of inhibitors, stimulators, or a combination of the two. It must also suggest the origins of such regulators of growth, their modes of delivery to the target organs and the conditions under which they are produced. The nature of these conditions has been a point of contention, some holding that they relate to functional demands, others contending that it is organ mass *per se* which is instrumental in governing the production of growth regulators.

Phenomena of growth are replete with examples of negative feedback whereby an organ is responsible for turning off its own growth when the "right size" is attained. An important hypothesis, therefore, relates to the possibility that organs produce self-inhibitors in proportion to the mass to which they have grown. These self-inhibitors have been variously referred to as antitemplates or chalones, and are postulated to be distributed by diffusion through interstitial spaces or by circulation in the bloodstream (2, 7, 11). According to these models the rate of growth is controlled by inhibitors acting upon the innate tendency of cells to proliferate. The implication of this hypothesis is that each organ is endowed with information as to the eventual dimensions to which it will grow such that as this limit is approached the production of inhibitors rises until a balance is struck between the natural tendency to grow and the suppression by inhibitors.

Superimposed upon this scheme is the theory that organ growth regulation is achieved by a combination of inhibitors and stimulators, the latter from exogenous sources. It is proposed that each organ produces specific end products which are released into the circulation to be monitored elsewhere in the body. Organ-specific growth stimulators are produced in inverse proportion to the concentration of such inhibitory end products. These stimulators are then carried to the target organs where growth is promoted until once again an inhibitory level of end products is produced. It has been suggested that in such feedback loops a single organ in the body may be responsible for the regulation of all others. Candidates for this role include the central nervous system (10) and the lymphatic system (3). All of the above theories have one thing in common, namely, that organ growth is regulated by inhibitors and/or stimulators produced in response to tissue mass without reference to other physiological relationships.

In contrast, the functional demand hypothesis of growth control emphasizes the dual role of physiological regulation in controlling both function and structure (4). This was the earliest hypothesis suggested to explain growth control, for it seemed obvious that functional overload led to hypertrophy and that disuse caused atrophy. Indeed, adaptive growth has been interpreted as a long-range consequence of physiological compensation. In some instances the operation of stimulators is obvious, as in the target organs of tropic hormones. In other examples, growth would seem to be controlled by inhibitory influences operating directly on the organ itself (e. g., suppression of parathyroid growth by elevated $[Ca^{2+}]$, or of the zona glomerulosa by $[Na^+]$). Indirect effects via suppression of stimulators by heightened production of end products from the target organs are found in the classic examples of endocrine control by the anterior pituitary. Many of the visceral organs fall into these categories since the functions they perform serve the body as a whole and are therefore regulated by influences that are systemically distributed. Other parts of the body are subject to more localized regulation. Influences affecting somatic tissues such as muscle, bones, and skin tend to act locally. Surgical depletion or functional overload of such structures is reacted to only by the parts concerned, not by homologous tissues elsewhere in the body. Thus, the mechanical or neural influences on skeletal and muscular tissues are confined to circumscribed anatomical loci, in marked contrast to the tissue-specific growth responses of visceral organs regardless of where they are located.

The roles of tissue mass versus physiological activity in regulating growth are not necessarily mutually exclusive. Indeed, there is compelling evidence to suggest that the true explanation of growth includes some of both theories. It cannot be denied that functional demand plays a significant role in regulating organ size. The evidence in favor of this interpretation is overwhelming. It is equally important to realize that not all phenomena of growth are necessarily developmental consequences of physiological adaptations. Many organs in the prenatal animal are not functional, yet their growth proceeds on a precise schedule suggesting the operation of important regulatory influences. Moreover, disuse atrophy in the adult seldom results in the complete disappearance of the atrophic organ. It would seem, therefore, that in the course of development the parts of the body grow to certain basic dimensions on their own and later come under the influences of functional stimuli. Thus, there would seem to be a genetically inherited predisposition for organs to attain a minimal basic size which is then adjusted in accordance with the functional demands impinging on them. It is for the following chapters to define the spheres of influence within which these two modes of growth operate.

REFERENCES

1. Bizzozero, G. An address on the growth and regeneration of the organism. *Brit. Med. J.* **1**, 728–732 (1894).
2. Bullough, W.S. Mitotic control in adult mammalian tissues. *Biol. Rev.* **50**, 99–127 (1975).
3. Burch, P.R.J., and Burwell, R.G. Self and not-self. A clonal induction approach to immunology. *Q. Rev. Biol.* **40**, 252–279 (1965).
4. Goss, R.J. "Adaptive Growth." Academic Press, New York, 1964.
5. Goss, R.J. Hypertrophy versus hyperplasia. *Science* **153**, 1615–1620 (1966).
6. Goss, R.J. Turnover in cells and tissues. *In* "Adavances in Cell Biology" (D.M. Prescott, L. Goldstein, and E. McConkey, eds.), pp. 233–296. Appleton-Century-Crofts, New York, 1970.
7. Houck, J.C. (ed.) "Chalones." North-Holland Publ., Amsterdam, 1976.
8. Huxley, J. "Problems of Relative Growth." Methuen, London, 1932.
9. McMinn, R.M.H. "Tissue Repair." Academic Press, New York, 1969.
10. Tanner, J.M. Regulation of growth in size in animals. *Nature (London)* **199**, 845–850 (1963).
11. Weiss, P., and Kavanau, J.L. A model of growth and growth control in mathematical terms. *J. Gen. Physiol.* **41**, 1–47 (1957).

2

Renewal of the Epidermis

Squame: L., *squama*, scale

The strategic location of the epidermis—at the interface between the internal tissues of the body and the outside environment—gives it an awesome responsibility. Aside from providing mechanical protection, it is a thermal insulator, a barrier to desiccation, and protector of the body's internal sterility. It is the only tissue (except for the iris of the eye) that is externally visible, and as such it is a costume adorned with colors and appendages in rich variety designed to attract a mate, scare off an enemy, or blend into the background. No other tissue surpasses the epidermis in the variety of specializations into which its cells are capable of differentiating. Whether horns or hooves, plumage or pelage, teeth or taste buds, the seemingly endless configurations in the integument testify to the remarkable versatility of the epidermal cell.

The problem of epidermal growth, like that of all renewing tissues, is to strike a balance between the rates of loss and replacement. Here the distance between the germinative and differentiated compartments is at a minimum, affording the opportunity for feedback regulation of the most direct kind. The kinetics of epidermal proliferation are under local control presumably mediated by influences emanating from the layers of differentiated cells overlying the basal epidermis where cell division goes on. Whether these influences are chemical or physical is a matter for conjecture.

Since epidermal turnover depends upon mitotic activity in its basal layer, on the average each time one of these cells divides one of the daughter cells enters the differentiating compartment, but sometimes neither or both may do so. The factors responsible for dispatching such cells from the germinative compartment in individual instances remain a mystery. A clue, however, may emerge from the discovery that epidermal cells are not randomly ar-

14

ranged, nor is the pattern of their mitotic activity, both spatial and temporal, without order.

TURNOVER AND THE EPIDERMAL PROLIFERATIVE UNIT

The architecture of the epidermis is organized around columns of cells, an arrangement most clearly visualized in the stratum corneum (15). Here the dead flattened keratinocytes, if swollen in alkaline buffer, may be seen to be stacked in orderly columns, called squames (Fig. 5). Measuring approximately 30 μm in diameter, these polygonal cells are interdigitated with those in neighboring squames, suggesting an orderly sequence to their differentiation (41). Each of these columns of keratinocytes is situated above a group of 10–11 basal epidermal cells which make up the epidermal proliferative unit (EPU) (Fig. 6) (54). Within any individual EPU, cell divisions most often occur near the periphery. A daughter cell destined for differentiation moves up into the overlying spinous cell layer. As it does so, it flattens out into an area covering the entire EPU (Fig. 7). It is then escalated upward through the granular layer to the stratum corneum where the cell becomes so clogged with keratin that it can no longer survive. In the mouse epidermis the

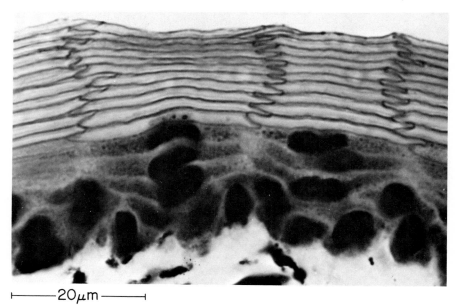

├──────20μm──────┤

Fig. 5. Columnar arrangement of mouse ear epidermal cells is evident following expansion in alkaline buffer (36).

├────── 30μm ──────┤

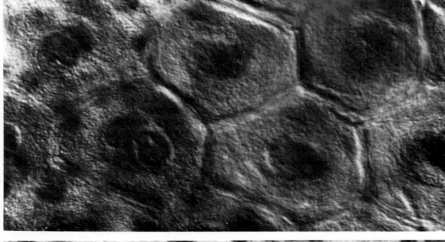

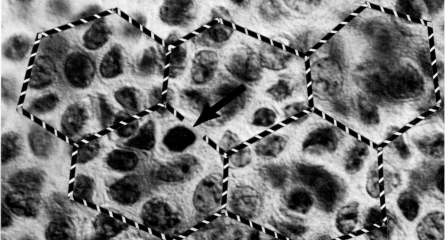

Fig. 6.   Above, surface view of mouse ear epidermis showing hexagonal arrangement of cells in the stratum corneum. Below, the same field is seen at a lower plane of focus with projections of overlying columns outlined. The nuclei of the basal epidermal cells, one of which is in mitosis (arrow), are visible (36).

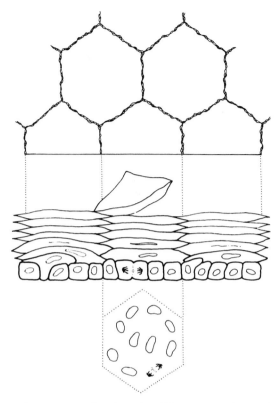

Fig. 7.   Diagrammatic projections of epidermal proliferative units showing how each column of keratinized cells (squames) overlies 10–11 basal cells. Mitotic activity in the basal layer results in the displacement of its cells one at a time into the next outer stratum where they become flattened out as differentiation commences, eventually being escalated to the surface of the squame where sloughing occurs.

spinous layer may be only 1 cell thick, the granular layer 2 and the stratum corneum 4–6 cells in thickness. The peripheral positions of mitoses in the basal layer of the EPU, together with the fact that only one cell usually divides at a time and that the cornified cells above are produced in a nonrandom sequence with respect to those in contiguous EPU's, suggest that the production of epidermal cells may be carefully orchestrated both within and between EPU's (36).

This possibility is the basis of a hypothesis to explain the kinetics of cell turnover in the EPU (55). Instead of all basal cells dividing at random, some of them migrating into the differentiated compartment in a fortuitous sequence, there is no reason to suspect that the 10–11 cells in the basal layer of an EPU progress in an orderly manner through a series of divisions on their way to becoming differentiated. If this were the case, there would have to be a single stem cell in each EPU, a cell perhaps identified with the centrally located Langerhans cell. The divisions of such a cell would be unequal, one of its daughters going on to divide several more times before becoming committed to the differentiated pathway. During this process, the cells might migrate peripherally, if not move in a circular route around the margins of the EPU. The observations that mitoses tend to be peripheral in the EPU, and that only one cell at a time moves up into the suprabasal layer, are consistent with this interpretation. It is further possible that this intra-EPU regulation might be coordinated with inter-EPU controls to account for the geometric precision with which the cells of adjacent squames are interdigitated (36).

Each type of epidermis exhibits its own rate of epidermal turnover ranging from one to several weeks, depending upon anatomical location, age, and species of animal. For example, the rate of proliferation is several times greater in the epidermis on the foot and tail of the mouse than it is in back or ear skin (55). The dwindling rate of proliferation with age is accompanied by decreases in epidermal thickness. It is probably safe to assume that differences in mitotic activity are also reflected by variations in thickness from the paper-thin epidermis of the mouse to the several millimeters or more of epidermis in the whale (23) (Fig. 8), a layer thick enough to be scraped off as "muktuk," a delicacy consumed by Eskimos. It is anticipated that however varied may be its configurations, epidermis wherever it occurs renews itself in a carefully controlled and orderly manner.

Although the turnover of epidermis is in large measure a locally controlled phenomenon, even to the extent of persisting *in vitro*, it is not exempt from exogenous factors. Denervation, for example, tends to depress the rate of epidermal proliferation. Starvation has a similar effect. The daily rhythms of proliferation in the epidermis of rats and mice, with higher rates of DNA synthesis at night and elevated mitotic counts in the morning, would indicate the operation of systemic influences affected by photoperiod and activity rhythms. Mitotic activity in the epidermis is suppressed by adrenaline and cortisone.

One of the most interesting factors to affect epidermal growth is found in extracts of the submaxillary salivary glands of mice and rats (16). This epidermal growth factor (EGF), if injected into newborn animals, accelerates the rate of integumental development (Fig. 9). This is indicated not only by the thickening of the epidermis and increase in its keratinization, but also by the precocious eruption of incisors and the premature opening of the eye lids. In the cornea, EGF stimulates hyperplasia and increases the rate of epidermal migration over wounds when administered topically. Inasmuch as EGF

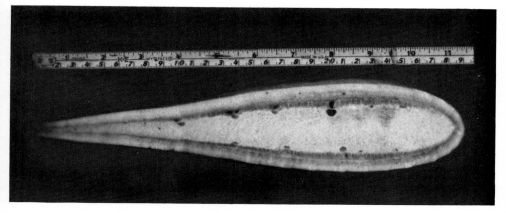

Fig. 8.   Cross section of the fluke of a beluga whale (*Delphinapterus leucas*). The thickness of its epidermis is approximately 1 cm. (Courtesy of Dr. Paul F. Brodie, Bedford Institute of Oceanography, Dartmouth, Nova Scotia).

works as well *in vitro* as *in vivo*, it must act directly on the epidermal cells. However impressive the influences of EGF are on epidermal tissues, the rationale for the production of this and other growth factors by the salivary glands remains obscure.

Recent years have witnessed increased interest in the possibility that epidermal growth may be regulated by feedback inhibition. Much evidence has been forthcoming in support of the existence of epidermal chalones, substances produced by differentiating cells of the epidermis which suppress mitotic activity in the basal layer (*13, 38*). If explants of skin are exposed to aqueous extracts of epidermis, or if such extracts are injected intraperitoneally, there results a reduction in miototic activity and DNA synthesis in the assay tissue. This inhibitory influence has been shown to be specific for the epidermis, but not species-specific. It is believed to act on the cell cycle in the late G_1 or S phase, but there is evidence that proliferation may also be blocked in G_2. The epidermal chalone requires adrenaline as a cofactor, and may act on the cell by regulating cyclic AMP. It is a heat-labile, nondialyzable molecule digested by trypsin.

The chalone hypothesis suggests that, in keeping with experimental evidence, the rate of proliferation in the basal epidermis is inversely proportional to the thickness of the differentiated compartment. If these latter cells are producing an inhibitor which percolates down to the basal layer where it inhibits mitotic activity, one might ask if its role is restricted solely to growth regulation or if it exerts other physiological influences related to the function of the epidermis. In the vast majority of other tissues and organs, growth is

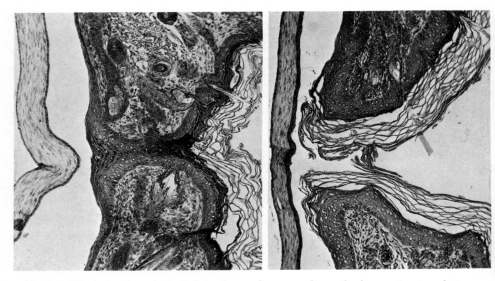

Fig. 9. Effect of epidermal growth factor from male mouse salivary glands on maturation of rat eyelids (*16*). Daily injections of EGF into newborn rats causes precocious keratinization of the epidermis resulting in premature opening of the eyelids at the age of 8 days (right) compared to control (left) in which eyelids do not normally open until about 12 days of age. Reproduced with permission. © 1963 The Williams & Wilkins Co., Baltimore.

known to be regulated primarily by stimulating substances whose principal role is to promote specific physiological activities such that the growth response is an epiphenomenon closely bound up with the demand for increased functions. By extrapolation, the existence of epidermal growth stimulators might be predicted to balance the inhibitory actions of chalones. Much still remains to be explained about the regulatory mechanisms of epidermal renewal, mechanisms which of logical necessity must take into consideration the functions as well as the structure of the epidermis.

One of the most important functions of the epidermis is to act as a barrier to desiccation. Transepidermal water loss is most effectively reduced by the presence of lipids in the epidermis. If this component were reduced, a growth reaction might be anticipated if the regulation of epidermal thickness were linked to the need to act as a barrier to dehydration. Diets deficient in essential fatty acids (EFA) produce such an effect (*40*). They increase the permeability of the skin to water in rats and mice, at the same time stimulating proliferation and increasing the thickness of the epidermis (Fig. 10). That the growth response is causally related to the increased evaporation from the epidermis in such animals is disclosed by the results of experiments in which

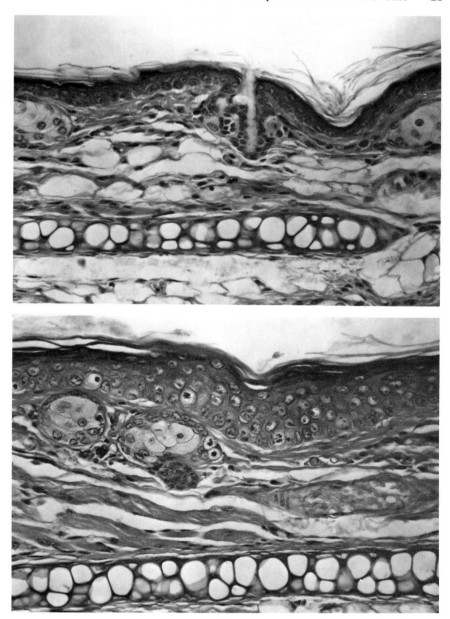

Fig. 10.   Diets deficient in essential fatty acids cause the normally thin mouse ear epidermis (above) to become hyperplastic and thickened (below) (40).

EFA deficient mice were held at high humidity. This drastically reduced the rate of transepidermal water loss and diminished the hyperplastic response of the epidermis to EFA deficient diets.

Still another condition demanding an explanation is psoriasis, a disease characterized by inflammation and thickening of the skin in which the lesion seems to be in the regulatory mechanisms controlling the rate of turnover in the epidermis. The life span of the human epidermal cell normally approximates 4 weeks, but in psoriasis the average cell divides, differentiates, and is sloughed off in less than 1 week. Concomitant with this acceleration in turnover is an expansion of the germinative zone to three basal layers of cells where the incidence of dividing cells may be many times higher than normal (62). Although the overt symptoms of psoriasis are restricted to localized patches of skin, even the uninvolved epidermis exhibits a higher than normal rate of DNA synthesis, though not sufficient to lead to pathological consequences (25). Nevertheless, whatever its etiology might be, it would seem to be more widespread than the regional distribution of symptoms might indicate.

THE WOUND HEALING RESPONSE

The capacity for wound healing is well developed in the epidermis, in keeping with the vulnerability of its exposed location. Like other epithelia, epidermis abhors a discontinuity. The infliction of a wound triggers a prompt centripetal migration of surrounding epidermal cells into the denuded region to resurface the area still exposed after contraction of the wound's margins. Although migrating epidermal cells are not normally found in division, those remaining at the periphery of the wound undergo a burst of proliferation. Once epidermal continuity is reestablished, the wound epithelium proliferates until the original thickness and density has been restored. Indeed, healing may involve an overshoot as the number of cell layers in an abraded area exceeds temporarily the original thickness (Fig. 11) (2). Increased DNA synthesis can be found only hours after injury, with mitotic activity reaching maximal levels a day or two postoperatively.

There are many kinds of epidermal wounds, each one of which results in hyperplasia. Stripping away the outer layers of stratum corneum with an adhesive, for example, decreases the thickness of the epidermis and is followed by hyperplasia in the basal epidermis (53), a response proportional to the number of layers stripped away and increased in the absence of dressings (57). The plucking of nearby hairs also stimulates proliferation in the basal layers of the nearby epidermis (56). However, the intensity of this reaction diminishes with each episode of plucking, disappearing after the seventh

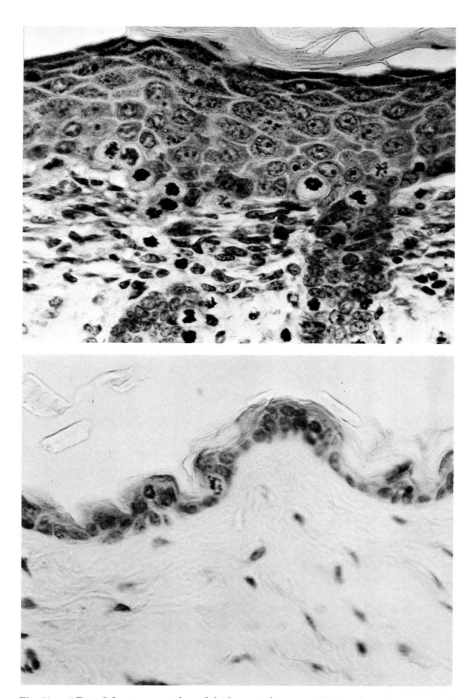

Fig. 11. Effect of abrasion on epidermal thickness in the mouse (2). Five days after removal of the epidermis, the regenerating tissue (above) has overgrown its normal dimensions (below). Colchicine has been used to arrest mitotic figures. Reproduced with permission. © 1976 The Williams & Wilkins Co., Baltimore.

repetition. Even subtle manipulations of the skin such as stretching or pressure are capable of promoting increased DNA synthesis and proliferation of the cells in the basal layer. So basic is the hyperplastic reaction of the epidermis, there are few conditions under which wound healing fails to occur. The application of dressings to a wound decreases the incidence of mitotic activity compared with unoccluded lesions (21). Even unwounded skin will become slightly hypoplastic if protected by a dressing. There seems to be no method, however, by which the proliferation of epidermal cells can be blocked completely.

The migratory response of epidermis to its interrupted continuity does not always occur. In the chick embryo less than 10–12 days of age, the epidermis sometimes fails to heal skin wounds by epidermal migration, although healing proceeds normally *in vitro* (65). In those anatomical sites where the integument is normally interrupted, as in the dental gingiva, there is neither a migratory response nor increased proliferation of the epidermis. A comparable situation is found adjacent to deer antlers in the autumn mating season after the velvet has been shed. The scalp integument on the antler pedicle tolerates this free border until the following spring when the old antler is shed and healing of the exposed surface is finally achieved. How such a fundamental process as wound healing is held in abeyance in these exceptional cases is not known.

THE MORPHOGENESIS OF HAIRS AND FEATHERS

Epidermis can hardly exist in the absence of its dermal substrate. Nowhere is the dermal–epidermal association more important than in the development of such epidermal appendages as hairs and feathers. The production of these structures in the embryo is a cooperative venture between the dermis and epidermis, the latter making up the major part of these keratinized appendages, the former providing nourishment and certain morphogenetic directions.

In the adult feather follicle, for example, it is possible to separate the epidermal and dermal components after plucking, then recombine them in abnormal combinations to see what kind of feather regenerates (63). If the feather papilla is rotated with respect to its epidermal component, the resulting structure is correspondingly disoriented indicating that the axial organization of the feather is under dermal control. On the other hand, if the papilla from a saddle feather is substituted for that in a breast feather follicle, a breast feather is regenerated. This proves that the kind of feather formed in the adult is determined by the origin of the epidermis. Not all integumental structures behave in this way, for the relative ages of the mesodermal and ectodermal components, especially in embryonic systems, play an important role in determining which will dominate the development of the other.

Because of their diminutive dimensions, mesodermal–epidermal recombination experiments are not feasible in hair follicles. This difficulty has been circumvented, however, by taking advantage of the *de novo* production of hairs in the skin of growing deer antlers (Fig. 12). These velvety hairs are quite distinct from those elsewhere on the body. Although their follicles have sebaceous glands, they are not accompanied by arrector pili muscles.

Fig. 12.   Deer in velvet. The antler skin grows hairs from follicles produced at the tips of the branches. At the end of the growing season the velvet is shed, to be renewed when the antlers are replaced in the following spring.

When an old antler is shed, the skin of the subjacent pedicle heals over the raw surface subsequently giving rise to new velvet on the regenerating antler. If the distal centimeter of pedicle skin is circumcised before the old antler is shed and the nearby ear perforated and impaled on the denuded region of the pedicle, its inner epidermis then heals to the remainder of the proximal pedicle skin while the epidermis on the outer surface of the ear remains interrupted until such time as the dead bony antler falls off. Once this happens, the outer ear epidermis migrates over the lesion and, being the only epidermis available, becomes swept up in the process of antler growth. Such antlers are normal in every respect, including the pelage of their velvet derived perforce from ear epidermis which normally produces a very different kind of hair. In this instance (unlike the regenerating feather) it would appear to be the mesodermal tissues which are responsible for dictating the type of hair to be differentiated by the epidermis (26).

REGENERATION OF EPIDERMAL APPENDAGES

The case of the deer antler testifies to the possibility of *de novo* hair follicle formation even in adult mammals. However impressive this special case may be, follicle regeneration under most circumstances is conspicuous by its absence, especially in healing wounds. Yet a number of experimental attempts to induce hair follicle differentiation in wounds have been carried out, some with more success than others. In general, it would appear that the usual absence of *de novo* hair follicle differentiation in healing wounds is associated with the tendency of most wounds to close by contraction, thus markedly reducing the area to be covered by migrating epidermis. When such wound contraction is prevented by insertion of a wire loop to hold the edges of a wound open, new follicles may indeed develop in the wound epidermis where it is in contact with underlying granulation tissues (Fig. 13) (*11*).

Hair follicles are themselves capable of limited regeneration following partial excision. In the case of the rat vibrissa, removal of the upper two-thirds of the hair follicle is followed by its regeneration from the residual proximal portions. If the lower third is excised, however, the papilla cannot be regenerated from its upper parts (*49*). Moreover, transplantation of the denuded papilla along with other kinds of epidermis to subcutaneous sites may result in the reorganization of the hair follicle and the production of hairs by epidermis derived from such diverse regions as the scrotum or lip (*50*). Hair follicles are also capable of regeneration in skin frozen by dry ice. Within several weeks the necrotic hair follicles are reestablished along with their sebaceous glands.

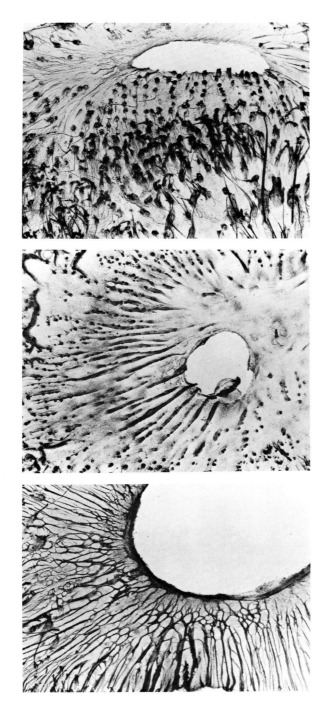

Fig. 13. Neogenesis of hair follicles in wounded rabbit skin made to heal by epidermal migration instead of dermal contraction (11). In these whole mounts of the healing epidermis, radially arranged ridges are seen at 27 days (left) from which hair follicles are budded off by 31 days (middle). After 33 days (right) numerous hairs are seen to have grown from the newly formed follicles.

Hairs, like other parts of the epidermis, are affected in their development by vitamin A. In general, vitamin A deficiency promotes keratinization of the epidermis. Too much vitamin A causes the cells that would otherwise have differentiated into keratinocytes to give rise to mucus-secreting tissues. This "mucoid metaplasia" results not from the transformation of one kind of differentiated cell into another but by the deflection of relatively undifferentiated cells arising in the germinative layer from their potential for keratinization to the mucoid pathway of differentiation. A comparable phenomenon occurs in embryonic mouse skin grown in culture. After a week or so, such tissues normally give rise to hair follicles in which keratinized epidermis is present. Grown in the presence of vitamin A, however, these follicles lose their dermal papillae and sprout lateral buds that develop into glandular structures (29). Although such glands possess alveoli and ducts, their exact nature is uncertain, being different from sebaceous, sweat, or mammary glands.

There are strains of mice that are hairless. Although they normally develop their first coat of hair as infants, this is lost and not replaced, the follicles undergoing regression. The physiological explanation for this failure of normal hair growth could lie at the cellular level or it could be due to systemically distributed factors. To test these alternatives, skin from normal mice has been grafted onto hairless mutants (3). Once such transplants are established, it is possible to clip or pluck the hairs from the graft region to see if they grow back in a hairless host (Fig. 14). The positive results of such experiments suggest that systemic factors are not involved, and that whatever may be responsible for the degeneration of hair follicles in this strain is intrinsic to the cells of the follicles themselves.

One of the most interesting aspects of epidermal appendages is the regulation of growth rates. Human fingernails have been estimated to elongate at the rate of 0.12 mm/day. Hairs grow faster, ranging from 0.27 to 0.44 mm/day, depending on their locations. In the chinchilla, hairs may grow 0.77 mm/day, while those in the rat elongate at the rate of 1.2 mm/day. The horns of a ram, also composed of massive quantities of keratinized epidermis, have been clocked at growth rates of 1.66 mm/day. Perhaps the most rapidly growing epidermal appendages, however, are regenerating feathers. After plucking, the new pin feather may appear above the surface of the skin in less than a week whereupon it plumes out and may attain its full dimensions in only a few weeks. One can only guess at the velocity with which such feathers as those in a peacock's tail must elongate.

It is in the nature of many types of hair to grow intermittently, cycles of activity alternating with periods of arrested growth. Such cycles are commonly associated with molting, which may occur asynchronously as in man or seasonally as in many animals adapted to the temperate zone. In birds, for example, the breeding plumage alternates with the eclipse phase, the molt-

Fig. 14.   To these mice of the hairless strain, grafts of skin from normal animals were made 5 months earlier. The hair was clipped from the grafts and a new generation of pelage was produced at the next hair growth cycle as shown here (3).

ing cycles being coordinated in response to annual changes in the photoperiod. In immature animals such as the rat and mouse, the replacement of one coat of fur with another is synchronized in both time and space. Waves of molting and regrowth, originating at various loci in the body, may be followed as they sweep over the skin from ventral to dorsal sides. A comparable phenomenon occurs in the molting pattern of birds in which equivalent feathers may be lost synchronously on both sides of the body and in a predetermined sequence from one location on the wing to another. These waves of regrowth are intrinsic to the individual participating hair or feather follicles. If dorsal and ventral flaps of rabbit skin are regrafted onto the flank after 90° rotation, the hair growth wave is correspondingly reoriented (22). When it reaches the margins of the operated region it continues to pass horizontally over the ventral flap and then the dorsal one before rejoining the original wave (Fig. 15). Therefore, the hair growth wave is not a domino effect, nor the result of a stimulus migrating through the skin, nor is it interrupted by denervation. It must be concluded, therefore, that each follicle is programmed to replace its hair at a particular time in the maturation of the animal, the hair growth wave being an expression of how these events are related to each other in time and space.

The apparent autonomy of the hair growth cycle should not be taken to indicate that such phenomena are unaffected by systemic factors. For example, if skin grafts are exchanged between animals in different phases of their

Fig. 15. The wave of hair growth which sweeps up the flank of the rabbit cannot be altered by switching positions of skin flaps (22).

hair growth cycles, there is a tendency for the two rhythms to synchronize (18). In such cases the timing of regrowth in the graft may be accelerated if the cycle of the host precedes it. However, hair growth is not delayed in a host whose cycle is normally scheduled after that in the graft.

Hair growth is also affected by the aging process. In old mice the replacement of plucked hairs is delayed in comparison with that in younger animals. If skin is grafted from old to young hosts, however, the hairs plucked from the grafts now regenerate as promptly as those in the skin of the young hosts

Fig. 16. Induction of hair growth next to a wound (1). The back of the mouse was first shaved in order to make it possible to identify activated hairs. Those follicles surrounding the lesion have grown new hairs as indicated by the ring of black fur around the wound.

(30). Such evidence argues in favor of the operation of systemic influences in at least some aspects of hair growth regulation.

Plucking is the standard way to initiate the regrowth of hairs or feathers. Presumably the wound inevitably inflicted by this procedure may act as a stimulus to the onset of new growth, something that normally would not occur until the next molting cycle. The immediate cause of the spontaneous replacement of hairs is even more obscure than that initiated by plucking. Indeed, one of the few clues derives from experiments in which the non-growing hairs of an animal are shaved and a wound inflicted in the skin (Fig. 16). Within a week or so the hair follicles nearby burst into activity giving rise to a ring of new hair growth around the wound which, by mechanisms yet to be explained, is triggered by the proximity to a lesion in the skin (1).

THE UNSOLVED PROBLEM OF TOOTH ERUPTION

Anyone who has looked closely at the jaws of a newborn shark cannot help but be impressed with how difficult it is to determine where the scales leave off and the teeth begin. This is because teeth are in fact modified integumental appendages, and as such, the mechanisms by which they grow and are replaced have much in common with those operating in the cases of hairs and feathers.

Teeth develop by the interaction between ectoderm and mesoderm. These two tissues are responsible for each other's differentiation into ameloblasts and odontoblasts, respectively, and the eventual production of enamel and dentine. The analysis of such interactions, together with studies of dental morphogenesis, have depended heavily upon techniques for growing tooth germs in culture, on the chorioallantoic membrane of the chick embryo, or in the anterior chamber of the eye. Under these conditions, it has been shown that teeth will continue their maturation from embryonic tooth buds or portions of embryonic mandibles. *In vitro* experiments have shown that even halved tooth germs can regulate during subsequent development to give rise to whole teeth, provided the bisection is carried out early in differentiation (24).

The question of whether the type of tooth produced is dictated by the ectoderm or mesoderm can be analyzed by separating these two components and reassembling them in different combinations in the mouse embryo (33, 34). When the enamel organ, an ectodermal derivative, of one kind of tooth and the papilla, of mesodermal origin, of another kind are recombined, the type of tooth produced is determined by the nature of the dental papilla. Even when the dental mesoderm is reassociated with ectoderm from the foot, the latter is induced to develop into the enamel component of a tooth.

Epidermis from the embryonic snout, however, fails to be influenced by the dental mesoderm with which it may be combined. Reciprocal experiments, in which epithelium of dental origin is recombined with mesoderm from the foot or snout, result in the production of keratinized epidermis in the former case and hair follicles in the latter. These experiments confirm that the pathways followed in the differentiation of embryonic ectoderm are dictated by the nature of the underlying mesodermal tissues, a rule, however, that does not necessarily apply to adult integumentary appendages.

The durability of teeth is not without its limitations. Vertebrates have met the need for dental renewal in two ways. Lower vertegrates replace their teeth in endless succession. Mammals have evolved mechanisms for the continuous eruption of certain of their teeth, such as incisors and tusks. Many mammalian teeth, replaced no more than once during infancy, possess little or no capacity for further growth after they have erupted. It is the eventual loss of these teeth which insures that mammals will ultimately die of starvation if they should be so unfortunate as to outlive their dentitions. This is dramatically illustrated in the case of the elephant in which six molars erupt, one after the other, each replacing its ground-down predecessor. When the last pairs of molars wear out, the elephant's days are numbered.

The mechanism by which tooth eruption is achieved has eluded the experimental ingenuity of researchers for many decades. The model system in which to investigate this phenomenon is the continuously erupting incisors of rodents or rabbits (Fig. 17). Other teeth are also famous for their lifelong capacity for growth, not the least of which are the tusks of elephants. Despite the great lengths to which these zoological curiosities may grow, their rate of

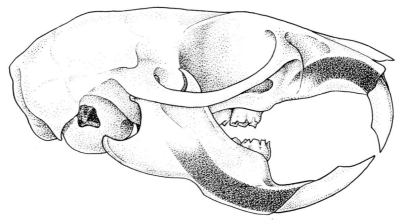

Fig. 17.   Rat skull showing locations of maxillary and mandibular incisors in relation to other structures. Note that the roots of the lower incisors are situated behind the molars.

eruption does not exceed that of the mouse incisor (45). Marine mammals likewise possess continuously growing teeth, perhaps an adaptation to their longevity, if not the possibility for continued somatic growth throughout life. In the toothed whales, for example, annual increments of dentine are visible in longitudinal sections of their teeth, a kind of natural scrimshaw making it possible to determine the ages of whales (Fig. 18).

Continuous eruption of the rat incisor is accompanied by the proliferation of cells in the root of the tooth (47, 60, 61). Here the preameloblasts and preodontoblasts multiply as their descendants migrate distally. When proliferation ceases these cells begin to differentiate. The ameloblasts do so by the elaboration of enamel at the ends in contact with the odontoblasts, their nuclei being displaced to the opposite ends of the cells. Only when their upward movement breaks surface at the gingiva does enamel production cease with the destruction of the ameloblasts. Meanwhile, the odontoblasts lining the pulp cavity follow a similar course of differentiation. Once they have moved upward in the erupting tooth to the point where dentine production commences, their proliferation ceases. Continued growth results in the deposition of layer upon layer of dentine on the inside of the tooth, as a result of which the odontoblasts, while moving distally, are at the same time

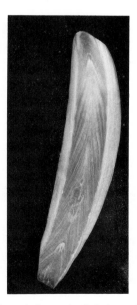

Fig. 18.   Longitudinal section through the tooth of a beluga whale showing the conical laminations of dentine which are gradually worn away distally as they are added at the base. (Courtesy of Dr. Paul F. Brodie, Bedford Institute of Oceanography, Dartmouth, Nova Scotia).

pushed inward. It is this pattern of migration which is responsible for the tapering configuration of the pulp cavity (Fig. 19). Eventually, the point is reached where the concentric production of dentine in a centripetal direction converges to obliterate the pulp cavity itself as well as the layer of odontoblasts which lines its surface. Concomitantly, the nerves and blood vessels supplying the pulp cavity may undergo a comparable turnover (Fig. 20). Originating peripherally, they grow in length with the escalation of tissues in the pulp cavity while being displaced centripetally. (66).

The rate at which the lower incisors of the rat erupt is approximately 0.45 mm/day. The maxillary incisors are shorter and grow at about 70% of this rate. Thus, the entire incisor may be replaced every 6–8 weeks. Theories abound to explain the mechanism by which tooth eruption occurs (6, 39). One possibility is that the proliferation of cells proximally generates the pressure to push the previously developed parts of the tooth upward. However, inhibition of mitosis either pharmacologically or by exposure to radiation does not interrupt the eruption of the tooth. Moreover, resection of the entire root, thus removing all sites of odontoblast and ameloblast proliferation, is followed by continued upward movement of the tooth until it is finally extruded at the surface (9). Clearly, growth pressure from cell proliferation in the root of the tooth cannot account for its eruption.

Another possibility is that blood pressure in the rich network of vessels supplying the pulp cavity might be responsible for dental eruption. This is supported by evidence that denervation of a tooth, resulting in vasodilation, accelerates the rate of eruption (12). Conversely, hypophysectomy, which decreases blood pressure, reduces the eruption rate (35). However, if the blood pressure is reduced pharmacologically there is no detectable effect on the rate of eruption (37).

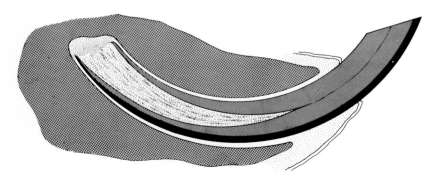

Fig. 19.   Sagittal section through the lower incisor of the rabbit (47). The enamel (black) is limited to the labial side. The tapering configuration of the pulp cavity reflects the progressive deposition of layers of dentine (crosshatched) by the odontoblasts.

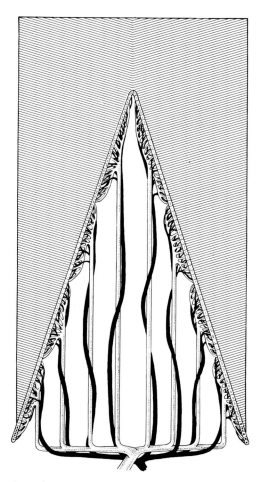

Fig. 20.   Diagram of a rodent incisor as seen in longitudinal section (66). The dentine is crosshatched. The conical configuration of the pulp cavity is the result of the distal migration of odontoblasts (replaced by proliferation below) and the concentric deposition of about 16 μm of dentine per day. Nerves (black) and blood vessels are constantly replaced as they are carried upward and inward during the incessant growth of the tooth.

Experimental techniques which retard or arrest incisor eruption reveal other factors which participate in the regulation of tooth growth. Not surprisingly, the eruption of incisors can be stopped altogether by inserting a pin transversely through both mandible and tooth (19, 44). Although this effectively prevents further eruption, it does not arrest the growth of tissues at the root of the tooth. Eventually, distal abrasion is responsible for wearing down such teeth to the gum line, while continued proliferation in the root causes these tissues, prevented from erupting, to be thrown into folds. A similar reaction occurs following hypophysectomy (59). This inhibits eruption, but the enamel epithelium at the root becomes folded upon itself presumably because its proliferating cells have nowhere else to go. Hypophysectomy does not interfere with the deposition of dentine which, in the absence of eruption, may eventually fill the entire pulp cavity. Whether this is a cause or an effect of the failure of teeth to erupt in the absence of the pituitary is not known, but perhaps it is relevant that incisors also cease to erupt when their pulp cavities are obliterated with dental fillings (19).

The periodontal membrane has long been a leading contender for the role of mediating tooth eruption. The collagen fibers of which it is composed are embedded in the surrounding alveolar bone on one end and the cementum at the base of the tooth on the other. The experimental induction of lathyrism by the administration of β-aminopropionitrile (βAPN) decreases the rate of eruption (42). This agent, like a diet rich in sweet pea (*Lathyrus*) seeds, interferes with the production of cross-linkages between collagen fibrils. In order for the incisor to move with respect to the surrounding mandibular bone the collagen fibers of the periodontal membrane would be expected to be in a constant state of disruption and reformation. Such a turnover is supported by evidence that the highest rate of ^{3}H-proline incorporation in the tooth occurs in the periodontal membrane (46). Nevertheless, it is not clear how noncontractile collagen fibers could exert the kind of pull needed to move the tooth bodily out of its socket unless the fibroblasts involved in collagen synthesis might also provide sufficient force to effect incisor eruption. This possibility is consistent with the discovery of contractile properties of fibroblasts in integumental lesions sufficient to account for the contraction of wounds (see Chap. 3).

Whatever may be the mechanism by which the forces of eruption are generated, abrasion between the upper and lower incisors insures that the rates of attrition balance the rates of eruption. Although the tendency for rodents to gnaw contributes to the attrition of their incisors, consumption of a soft diet does not alter the normal rate of eruption. Indeed, if rats are drugged to make them sleep 18–20 hours a day, the rate at which their incisors erupt increases about 50%, with no alteration in rates of attrition (51). Incisors normally grow more slowly at night when rats are active and eating (14). Therefore eruption is retarded by occlusion but not by attrition *per se*.

The most compelling evidence that occlusion antagonizes eruption comes from experiments in which one incisor is amputated shorter than the other. Under these circumstances, the unimpeded incisor grows at a rate approximately twice that of normal ones (12, 35, 44, 45, 47). Once it catches up with

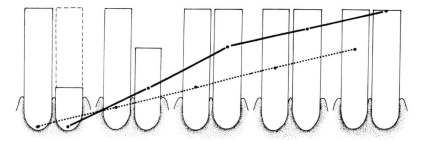

Fig. 21.   The lower incisors (front view) of rats erupt at the rate of about 0.45 mm/day when in normal occlusion with the upper ones, as shown by the distal movement of the marker in the tooth on the left. Amputation of the other incisor results in a doubling of its unimpeded eruption rate until it catches up with its partner, whereupon occlusal pressure again reduces its rate of eruption.

its partner, its rate of eruption again drops to normal levels (Fig. 21). Whether the increase in eruption rate of unimpeded incisors is due to the trauma of amputation, the absence of attrition or the lack of occlusion can be determined by grinding down the lower incisors (55, 65). In the absence of trauma or attrition, the eruption rates of the upper incisors increase, confirming the inhibitory effects of occlusion on dental elongation. The drastic deceleration of eruption in crowned incisors, in which attrition is prevented but occlusal pressures are increased, is also consistent with this interpretation (64), perhaps because of the deleterious effects of increased occlusal trauma on the periodontal membrane (10). Indeed, malocclusion leads to the excessive elongation of opposite incisors prevented from mutual contact, but wheter the resulting tusks continue to erupt at the accelerated rate characteristic of unimpeded incisors remains to be determined.

PATTERNS OF TOOTH REPLACEMENT

The earliest vertebrates solved the problem of dental renewal by replacing entire teeth rather than by the continuous growth of preexisting ones. The difference between mammals and lower vertebrates in this regard is that the deciduous nature of the former's dentition is restricted to one replacement tooth at best, while in lower forms there seems to be an unlimited capacity for producing new teeth throughout life. Indeed, in those salamanders capable of regenerating their lower jaws, substantial proportions of the entire dentition can be replaced. Following amputation of the jaw, the missing mandibular arch is reconstructed, in the course of which the dental lamina extends itself from the stump into the newly formed parts (27). If the dental lamina is removed from the jaw an edentulous regenerate is produced. When short segments of a jaw are resected, the gap is reconstituted by growth from the preexisting dental laminae in both the anterior and 'posterior stumps of larvae, but only from the posterior stumps of adults.

Although fishes, amphibians, and reptiles replace individual teeth from time to time throughout life, there must also be an accommodation to the often continual body growth characteristic of these creatures. This is achieved in part by the production of larger and larger teeth in the course of successive rounds of replacements, and also by augmenting the total number of teeth in the jaw. The differentiation of new tooth germs is an extension of the odontogenic process responsible for the initial development of the original teeth in the immature animal. Although it is not uncommon for new tooth positions to arise between preexisting ones, especially in younger individuals, they are usually added posteriorly (8).

Tooth replacement is not a haphazard process. Like that of other integumental appendages there is a spatial and temporal coordination in the pattern of replacement. The selective advantage in not losing all of one's teeth at once is comparable to the survival value to birds in molting only a few feathers at a time (except in penguins which, like other flightless birds, can afford to replace their plumage wholesale).

The basic pattern in which teeth are replaced has evolved to insure that the loss of old teeth and the development of new ones are staggered along the length of the jaw (Fig. 22). This sequence reflects the order in which the original tooth germs developed in the immature animal (7). At each site where a tooth germ is initiated there will eventually develop a tooth family consisting of a series of tooth generations ranging from the most mature one which may have erupted at the surface to the most recently initiated tooth bud at the back of the line. The number of stages represented at any one time varies from species to species. They are all in a constant state of development, however, insuring a successive replacement of functional teeth at the occlusal end and an unlimited potential for the production of incipient tooth germs at the base. The frequency with which mature teeth are replaced is a function of species, age, and season of the year. In the trout, for example, the generation time from the loss of one tooth to the loss of its successor may be as short as 8 weeks in smaller specimens to twice as long in larger ones (8). In crocodiles, the individual teeth of which may be replaced as infrequently as once a year in older specimens, it has been estimated that each tooth may be replaced up to 50 times during the life span of an animal 4 m long.

What is most intriguing about patterns of tooth replacement is that successive stages of development are to be found in alternate tooth families rather than in adjacent ones. Moreover, successively younger stages of developing

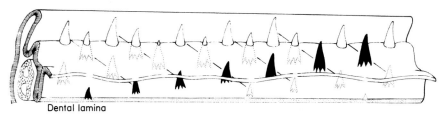

Dental lamina

Fig. 22.   Pattern of tooth replacement as it occurs in the amphibian or reptilian jaw (*17*). Each tooth bud arises in the dental lamina, moving upward during its development in the tooth family. The rows of teeth in successive positions form Zahnreihen, delineated by oblique lines in the diagram. The teeth in every other position are also found to make up a progressive series of developmental stages (drawn in black), reflecting the alternate sequence of tooth bud initiation in the dental lamina from anterior (left) to posterior (right).

teeth usually tend to occur in progressively posterior loci. Thus, as in the original differentiation of tooth germs, the first to be initiated is in an anterior location, the next younger one forming in the second position behind that, and so on. Meanwhile, the same process occurs in alternate tooth positions, as a result of which the development of odd numbered teeth is coordinated in one pattern and that of even numbered ones in another. It is like two relay races being run on the same race course, one in the odd numbered tracks, the other in the even numbered ones, with the added complication that the runners are staggered with respect to each other because they did not all leave the starting line at the same time.

Efforts to explain how all of this is controlled have given rise to several theories of tooth replacement. One of these is based upon the observation that the initiation of tooth development occurs successively in every other tooth position. As the members of alternate tooth families develop, more mature teeth nearer the occlusal surface than are younger ones, they become aligned in diagonal rows of successive stages of tooth development, called Zahnreihen (*17, 43, 52*). If these tooth rows are envisaged in the temporal dimension they can be thought of as waves passing along the jaw. Waves of Zahnreihen activation may pass either posteriorly or anteriorly, depending upon the temporal and spatial kinetics by which the developmental events in each tooth family are related.

It has been suggested that each Zahnreihe represents the passage of a stimulus along the jaw which triggers the onset of tooth differentiation first in the odd numbered positions, then in even numbered ones (*20*). However wavelike the pattern of tooth development may appear to be, it does not necessarily follow that there exists a diffusible substance responsible for initiating the development of each new tooth germ, especially since activation takes place in alternate, not successive, loci. Nevertheless, this hypothesis cannot be ruled out until such time as it is verified or invalidated experimentally, for example, by surgically interrupting the passage of purported growth stimuli along the jaw. Apropos of this, it will be recalled that the progress of hair growth waves follows the original course through the integument even after patches of skin have been grafted back in abnormal orientations (Fig. 15) (22).

Another theory of tooth growth regulation is designed to account for the alternation of Zahnreihen activation (7). According to this hypothesis, the onset of tooth development in one position inhibits that in its nearest neighbor. Consequently, the next tooth to embark on the pathway of differentiation would be in the second position down the line beyond the sphere of inhibitory influences from its predecessor. Odontogenesis would later be initiated in between these two loci only after the suppressive effects had subsided. Although this might explain how tooth development is activated in every other position, it is still predicated on the as yet untested

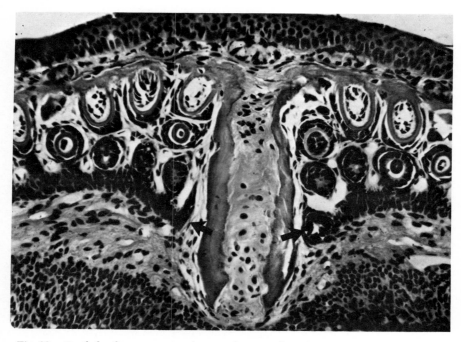

Fig. 23. Tooth families as seen in a horizontal section through the lower jaw of a newt. The mandibular symphysis is in the middle. The several teeth in each family are seen in transverse section, the functional ones on the labial side (above) and progressively less differentiated ones in more lingual positions. The dental lamina from which tooth germs arise is indicated by arrows.

assumption that stimulatory waves travel along the jaw to promote tooth production.

Not unrelated to the mathematical precision with which development is coordinated in the dentition as a whole is the problem of how the relays of tooth buds are timed within each tooth family (Fig. 23). It is conceivable that successive generations of teeth in different stages of development might hold in check the differentiation of their next younger tooth buds. If the functional tooth does in fact suppress the developmental potential of the next one in line, then the extraction of a newly erupted tooth should accelerate the development of its successor just as hair and feather generation is stimulated by plucking. These are not comparable situations, however, since hair and feather follicles do not consist of a series of individual appendages in different stages of development as is seen in tooth families. Whether or not the extraction of a functional tooth promotes the precocious development of those in the rest of its family seems not to have been put to the test in lower vertebrates. If this were to occur, however, it would be expected to disrupt the pattern of tooth replacement in the entire jaw. In view of such potential complications, the more prudent interpretation of tooth replacement patterns might be to assume that the members of each tooth family are marching to their own tunes, and that the precision with which the entire dentition is coordinated is not orchestrated by a single pacemaker, but is the natural outcome of the order in which the positions developed in the first place.

NEUROTROPHIC MAINTENANCE OF TASTE BUDS

Among the skin's most intriguing organs are its taste buds. In the mammal they are mostly found on the walls of crypts surrounding the lingual papillae, their specific sensitivities (sweet, sour, salt, bitter) varying with the position on the tongue. In fishes and amphibians, taste buds may be situated on the outside skin, as in catfish taste barbels.

Like other epidermal structures, taste buds are renewing tissues (4). Labeling studies with ^{3}H-thymidine have shown that the average life span of a taste cell is approximately 10 days, with new cells being added about every 10 hours (Fig. 24). There is evidence, however, that some cells may be more permanent. When pregnant rats are injected with ^{3}H-thymidine, virtually all of the taste cells that differentiate in the fetus become labeled, and some of them remain labeled for over 6 weeks. Nevertheless, the renewal of cells is so important that the structural and functional integrity of the taste bud is lost if the proliferation of stem cells is inhibited. For example, the arrest of mitotic activity with colchicine interferes with the renewal of cells in the taste buds of a rat (5). Such organs lose their responsiveness to chemical

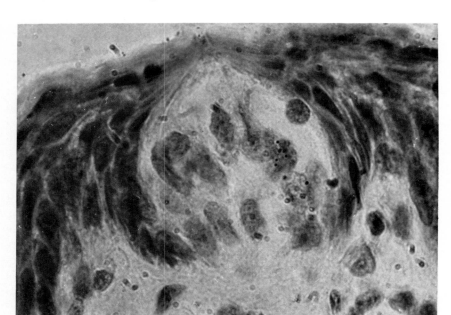

Fig. 24.   Autoradiograph of a taste bud from a rat fungiform papilla showing labeled nuclei 200 hours after injection of ^{3}H-thymidine. (Courtesy of Dr. Lloyd M. Beidler, Florida State University.)

stimuli, an effect that is reversible on discontinuation of treatment. Permanent absence of taste buds is brought about by exposure to X-rays. This results in the loss of the sense of taste and the disappearance of taste buds in about 10 days, presumably because they run out of cells.

The loss of taste buds following denervation is a well-known phenomenon (Fig. 25) (28). It occurs in the taste barbels of the catfish, as well as in the tongues of mammals. The rate of degeneration is temperature-dependent. In the catfish, denervated taste buds degenerate in about 25 days at 6°C, but are lost in only 6 days at 30°. In the rat they disappear in a week or so, but may take more than 2 weeks in the rabbit. The process of degeneration is an active one involving phagocytosis as well as the actions of lysosomes and autophagic vacuoles. Acid phosphatase, which is involved in the breakdown process, increases during the first week of degeneration. Alkaline phosphatase and ATPase decrease (32). Fragments not disposed of by phagocytosis are lost by desquamation. There is a concomitant thinning of the surrounding epidermis in the vicinity of the disappearing taste buds. It is clear that loss of taste buds following denervation is not to be explained entirely in terms of reduced mitotic activity or by dedifferentiation of the taste cells.

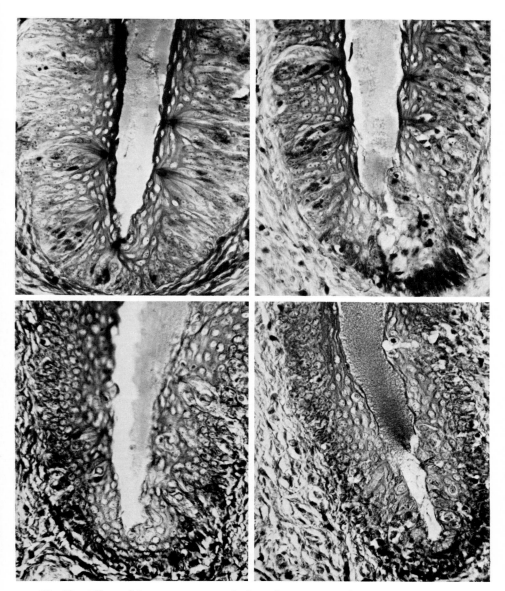

Fig. 25. Effects of denervation on taste buds on the rat circumvallate papilla (28). Normally innervated taste buds are illustrated in the upper left photograph. Stages of degeneration 2, 4, and 7 days after section of the glossopharyngeal nerve are shown in the upper right, lower left, and lower right illustrations, respectively.

Instead, it is attributable to the premature demise of cells in the taste bud, suggesting that their very lives are normally dependent upon the nerves supplying them.

Despite the dramatic disintegration of denervated taste buds, new ones reappear in several weeks after reinnervation. Thus, the epidermis retains its responsiveness to the nerves long after they are gone. The competence of the epidermis to differentiate in response to nerves is more widespread than the normal distribution of taste buds. If the entire vallate papilla, for example, is removed along with its taste buds, new ones will differentiate from the wound epidermis when nerves grow back (67). Even epidermis on the ventral side of the tongue where taste buds are normally not present is capable of being induced by nerves to differentiate (68). Nontongue epidermis from other parts of the body (e.g., ear skin) is incapable of developing taste buds in mammals.

The specificity of taste bud innervation can be studied in two ways. The most direct is by cross-reinnervation whereby the proximal stump of some other nerve is anastomosed to the distal stump of the glossopharyngeal nerve which normally supplies the taste buds. In this manner, foreign nerve fibers can be induced to regenerate along the former pathways of taste nerves to determine if, when they reach the site formerly occupied by taste buds, they will induce the regeneration of new ones. This technique can be used to test the efficacy of such foreign nerves as may be close enough to be diverted into glossopharyngeal pathways. More distant nerves are studied by transplanting their ganglia together with tongue epidermis to the anterior chamber of the eye where, with the right combinations, taste buds can be induced (Fig. 26). Comparable results are possible in culture, although the incidence of success is considerably lower than *in vivo*.

By these various techniques a number of different nerves have been tested for their abilities to induce taste bud differentiation (68). The hypoglossal nerve, which is made up almost exclusively of motor fibers, fails to promote taste bud regeneration when made to grow along the distal segment of the sectioned glossopharyngeal nerve. Sensory nerves, however, are more successful. If the chorda tympani or the vagus nerve, which are primarily sensory, are made to reinnervate the tongue, taste buds regenerate. Spinal nerves and nodose ganglia are also effective when combined with tongue epidermis in the anterior chamber of the eye. An exception to the rule that sensory fibers can induce taste bud differentiation is the auriculotemporal nerve, which, for unexplained reasons, has not been found to induce taste bud regeneration. In general, however, there would appear to be a correlation between sensory versus motor nerves and the presence or absence of taste bud differentiation. What the active agent might be is suggested by the effects of hemicholinium-3, an inhibitor of acetylcholine synthesis. Chronic

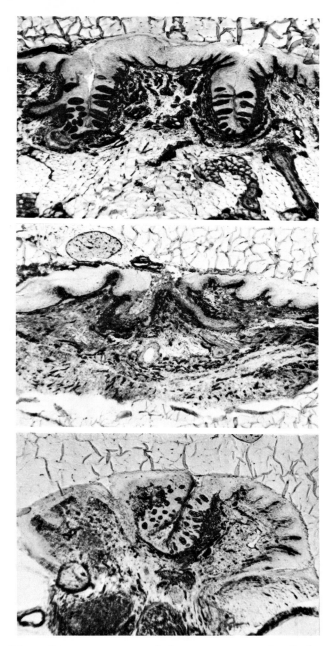

Fig. 26. Effects of innervation on taste bud regeneration in grafts of infant tongue to the anterior chamber of adult rat eyes (69). Top, normal taste buds lining the trench of a vallate papilla. Middle, tongue graft in which taste buds have disappeared in the absence of innervation. Bottom, regenerated taste buds in tongue graft supplied with nerves from a nodose ganglion transplanted next to it.

exposure of larval salamanders to this drug results in degeneration of their taste buds and lateral line organs (31).

Notwithstanding the apparent lack of discrimination between sensory nerves and taste buds in experimental situations, the fact remains that in the normal animal the specificities of both are matched to one another. Each taste bud is specified in its embryogenesis to respond to a particular class of compounds. Furthermore, its nerve must connect with the part of the brain where the afferent impulses can be interpreted appropriately. How this arrangement comes about is not easily explained (48). If the nerve were to impose its specificity on whatever taste bud it happened to innervate, one would expect that cross-reinnervation of taste buds with nerves normally supplying those of a different kind would result in a modification of the chemical sensitivies of the taste buds. Experiments have shown this not to be the case, for the electrophysiological properties of the chorda tympani and glossopharyngeal nerves change in accordance with the new taste buds they supply after cross-reinnervation. Therefore, the epidermal cells in the taste bud would appear to dictate the specificity of their nerves. Yet various branches of the same nerve fiber have been shown to innervate taste cells of the same type, indicating that the outgrowth of axons to taste buds may be a nonrandom process. If neither the receptor cell nor the neuron were to impose its specificity on the other, the establishment of correct connections between the two would presumably involve a selective process whereby the nerve fibers might seek out complementary taste cells. As in other parts of the nervous system where similar specificities seem to exist, the mechanism by which cells are able to recognize compatible connections is a problem more speculated about than understood.

REFERENCES

1. Argyris, T.S. Hair growth induced by damage. *Advan. Biol. Skin* **9**, 339–356 (1969).
2. Argyris, T.S. Unbalanced RNA accumulation in regenerating mouse epidermis following abrasion. *J. Invest. Dermatol.* **67**, 718–722 (1976).
3. Argyris, T.S., and Argyris B.F. Hair growth in skin grafts placed on hairless mice. *Anat. Rec.* **168**, 457–462 (1970).
4. Beidler, L.M. Physiological properties of mammalian taste receptors. *In* "Taste and Smell in Vertebrates" (J. Knight and G. Wolstenholme, eds.), pp. 51–70. Churchill, London, 1969.
5. Beidler, L.M., and Smallman, R.L. Renewal of cells within taste buds. *J. Cell Biol.* **27**, 263–272 (1965).
6. Berkovitz, B.K.B. Theories of tooth eruption. *In* "The Eruption and Occlusion of Teeth" (D.F.G. Poole and M.V. Stack, eds.), pp. 193–204. Butterworths, London, 1976.
7. Berkovitz, B.K.B. The order of tooth development and eruption in the rainbow trout (*Salmo gairdneri*). *J. Exp. Zool.* **201**, 221–225 (1977).

8. Berkovitz, B.K.B., and Moore, M.H. Tooth replacement in the upper jaw of the rainbow trout (*Salmo gairdneri*). *J. Exp. Zool.* **193**, 221–234 (1976).

9. Berkovitz, B.K.B., and Thomas, N.R. Unimpeded eruption in the root-resected lower incisor of the rat with a preliminary note on root transection. *Arch. Oral Biol.* **14**, 771–780 (1969).

10. Bhaskar, S.N., and Orban, B. Experimental occlusal trauma. *J. Periodont.* **26**, 270–284 (1955).

11. Breedis, C. Regeneration of hair follicles and sebaceous glands from the epithelium of scars in the rabbit. *Cancer Res.* **14**, 575–579 (1954).

12. Bryer, L.W. An experimental evaluation of the physiology of tooth eruption. *Int. Dent. J.* **7**, 432–478 (1957).

13. Bullough, W.S. The control of epidermal thickness. *Br. J. Dermatol.* **87**, 187–354 (1972).

14. Chiba, M., Tashiro, T., Tsuruta, M., and Eto, K. Acceleration and circadian rhythm of eruption rates in the rat incisor. *Arch. Oral Biol.* **21**, 269–271 (1976).

15. Christophers, E., Wolff, H.H., and Laurence, E.B. The formation of epidermal cell columns. *J. Invest. Dermatol.* **62**, 555–559 (1974).

16. Cohen, S., and Elliott, G.A. The stimulation of epidermal keratinization by a protein isolated from the submaxillary gland of the mouse. *J. Invest. Dermatol.* **40**, 1–6 (1963).

17. Cooper, J.S. Tooth replacement in amphibians and reptiles. *Br. J. Herpet.* **3**, 214–218 (1966).

18. Ebling, F.J., and Johnson, E. Systemic influence on activity of hair follicles in skin homographs. *J. Embryol. Exp. Morphol.* **9**, 285–293 (1961).

19. Eccles, J.D. The effects of reducing function and stopping eruption on the periodontium of the rat incisor. *J. Dent. Res.* **44**, 860–868 (1965).

20. Edmund, A.G. Tooth replacement phenomena in the lower vertebrates. *Contrib. Roy. Ont. Mus., Life Sci. Div.* **52**, 1–190 (1960).

21. Fisher, L.B., and Maibach, H.I. Physical occlusion controlling epidermal mitosis. *J. Invest. Dermatol.* **59**, 106–107 (1972).

22. Ghadially, F.N. The effect of transposing skin flaps on the hair growth cycle in the rabbit. *J. Pathol. Bacteriol.* **74**, 321–325 (1957).

23. Giacometti, L. The skin of the whale *Balaenoptera physalus*. *Anat. Rec.* **159**, 69–76 (1967).

24. Glasstone, S. The development of halved tooth germs. A study in experimental embryology. *J. Anat.* **86**, 12–15 (1956).

25. Goodwin, P., Hamilton, S., and Fry, L. A comparison between DNA synthesis and mitosis in uninvolved and involved psoriatic epidermis and normal epidermis. *Br. J. Dermatol.* **89**, 613–618 (1973).

26. Goss, R.J. The role of skin in antler regeneration. *Advan. Biol. Skin* **5**, 194–207 (1964).

27. Graver, H.T. Origin of the dental lamina in the regenerating salamander jaw. *J. Exp. Zool.* **189**, 73–84 (1974).

28. Guth, L. The effects of glossopharyngeal nerve transection on the circumvallate papilla of the rat. *Anat. Rec.* **128**, 715–731 (1957).

29. Hardy, M.H. The differentiation of hair follicles and hairs in organ culture. *Advan. Biol. Skin* **9**, 35–60 (1969).

30. Horton, D.L. The effect of age on hair growth in the CBA mouse: Observations on transplanted skin. *J. Gerontol.* **22**, 43–45 (1967).

31. Hui, F.W., and Smith, A.A. Degeneration of taste buds and lateral line organs in the salamander treated with cholinolytic drugs. *Exp. Neurol.* **34**, 331–341 (1972).

32. Iwayama, T., and Nada, O. Histochemical observation on phosphatase activities of degenerating and regenerating taste buds. *Anat. Rec.* **163**, 31–38 (1969).

33. Kollar, E.J., and Baird, G.R. Tissue interactions in embryonic mouse tooth germs. I.

Reorganization of the dental epithelium during tooth-germ reconstruction. *J. Embryol. Exp. Morphol.* **24**, 159–171 (1970).

34. Kollar, E.J., and Baird, G.R. Tissue interactions in embryonic mouse tooth germs. II. The inductive role of the dental papilla. *J. Embryol. Exp. Morphol.* **24**, 173–186 (1970).

35. Kusner, W., Michaeli, Y., and Weinreb, M.M. Role of attrition and occlusal contact in the physiology of the rat incisor: VI. Impeded and unimpeded eruption in hypophysectomized and magnesium-deficient rats. *J. Dent. Res.* **52**, 65–73 (1973).

36. Mackenzie, I.C. Spatial distribution of mitosis in mouse epidermis. *Anat. Rec.* **181**, 705–710 (1975).

37. Main, J.H.P., and Adams, D. Experiments on the rat incisor into the cellular proliferation and blood-pressure theories of tooth eruption. *Arch. Oral Biol.* **11**, 163–178 (1966).

38. Marks, R. The role of chalones in epidermal homeostasis. *Br. J. Dermatol.* **86**, 543–547 (1972).

39. Massler, M., and Schour, I. Studies on tooth development. Theories of eruption. *Am. J. Orthodont.* **27**, 552–576 (1941).

40. Menton, D.N. The effects of essential fatty acid deficiency on the skin of the mouse. *Am. J. Anat.* **122**, 337–356 (1968).

41. Menton, D. A minimum-surface mechanism to account for the organization of cells into columns in the mammalian epidermis. *Am. J. Anat.* **145**, 1–21 (1976).

42. Michaeli, Y., Pitaru, S., Zajicek, G., and Weinreb, M.M. Role of attrition and occlusal contact in the physiology of the rat incisor: IX. Impeded and unimpeded eruption in lathyritic rats. *J. Dent. Res.* **54**, 891–896 (1975).

43. Miles, A.E.W., and Poole, D.F.G. The history and general organization of dentitions. *In* "Structural and Chemical Organization of Teeth" (A.E.W. Miles, ed.), Vol. I, pp. 3–44. Academic Press, New York, 1967.

44. Ness, A.R. The response of the rabbit mandibular incisor to experimental shortening and to the prevention of its eruption. *Proc. Roy. Soc. London, Ser. B* **146**, 129 (1956).

45. Ness, A.R. Eruption rates of impeded and unimpeded mandibular incisors of the adult laboratory mouse. *Arch. Oral Biol.* **10**, 439–451 (1965).

46. Ness, A.R. Eruption—a review. *In* "The Mechanism of Tooth Support," pp. 84–88. Wright, Bristol, 1967.

47. Ness, A.R., and Smale, D.E. The distribution of mitoses and cells in the tissues bounded by the socket wall of the rabbit mandibular incisor. *Proc. Roy Soc. London, Ser. B* **151**, 106–128 (1960).

48. Oakley, B., and Cheal, M. Regenerative phenomena and the problem of taste ontogenesis. *In* "Olfaction and Taste" (D.A. Denton and J.P. Coghlan, eds.), Vol. 5, pp. 99–105. Academic Press, New York, 1975.

49. Oliver, R.F. Regeneration of the dermal papilla and its influence on whisker growth. *Advan. Biol. Skin.* **9**, 19–33 (1967).

50. Oliver, R.F. The induction of hair follicle formation in the adult hooded rat by vibrissa dermal papillae. *J. Embryol. Exp. Morphol.* **23**, 219–236 (1970).

51. Orsós, S., and Bartha, É. The effect of prolonged sleep on the eruption rate of the rat's incisor. *Acta Physiol. Hung.* **9**, 237–241 (1956).

52. Osborn, J.W. The evolution of dentitions. *Am. Sci.* **61**, 548–559 (1973).

53. Pinkus, H. Examination of the epidermis by the strip method of removing horny layers. I. Observations on the thickness of the horny layer and on mitotic activity after stripping. *J. Invest. Dermatol.* **16**, 383–386 (1951).

54. Potten, C.S. The epidermal proliferative unit: The possible role of the central basal cell. *Cell Tissue Kinet.* **7**, 77–88 (1974).

55. Potten, C.S. Epidermal cell production rates. *J. Invest. Dermatol.* **65**, 488–500 (1975).

56. Potten, C.S., and Allen T.D. The fine structure and cell kinetics of mouse epidermis after wound healing. *J. Cell Sci.* **17**, 413–447 (1975).
57. Rovee, D.T., Kurowsky, C.A., and Labun, J. Local wound environment and epidermal healing. Mitotic response. *Arch. Dermatol.* **106**, 330–334 (1972).
58. Sarnat, H., and Sciaky, I. Experimental lathyrism in rats: Effects of removing incisal stress. *Periodontics* **3**, 128–134 (1965).
59. Schour, I., and van Dyke, H.B. Changes in the teeth following hypophysectomy. I. Changes in the incisor of the white rat. *Am. J. Anat.* **50**, 397–433 (1932).
60. Smith, C.E., and Warshawsky, H. Cellular renewal in the enamel organ and the odontoblast layer of the rat incisor as followed by radioautography using ³H-thymidine. *Anat. Rec.* **183**, 523–562 (1975).
61. Smith, C.E., and Warshawsky, H. Movement of entire cell populations during renewal of the rat incisor as shown by radioautography after labeling with ³H-thymidine. The concept of a continuously differentiating cross-sectional segment. *Am. J. Anat.* **145**, 225–260 (1976).
62. Van Scott, E.J., and Ekel, T.M. Kinetics of hyperplasia in psoriasis. *Arch. Dermatol.* **88**, 373–381 (1963).
63. Wang, H. The morphogenetic functions of the epidermal and dermal components of the papilla in feather regeneration. *Physiol. Zool.* **16**, 325–350 (1943).
64. Weinreb, M.M., Michaeli, Y., and Berman, G. Role of attrition and occlusal contact in the physiology of the rat incisor: IV. Prevention of attrition in the articulating incisor. *J. Dent. Res.* **48**, 120–130 (1969).
65. Weiss, P., and Matoltsy, A.G. Wound healing in chick embryos *in vivo* and *in vitro*. *Dev. Biol.* **1**, 302–326 (1959).
66. Zajicek, G. The rodent incisor tooth proliferon. *Cell Tissue Kinet.* **98**, 207–214 (1976).
67. Zalewski, A.A. Regeneration of taste buds in the lingual epithelium after excision of the vallate papilla. *Exp. Neurol.* **26**, 621–629 (1970).
68. Zalewski, A.A. Neuronal and tissue specifications involved in taste bud formation. *Ann. N.Y. Acad. Sci.* **228**, 344–349 (1974).
69. Zalewski, A.A. Trophic function of neurons in transplanted neonatal ganglia. *Exp. Neurol.* **45**, 189–193 (1974).

3

Connective Tissues

Keloid: Gr., *chele*, crab claw

The coherence of the body is dependent on an assortment of tissues, the forms of which are as varied as the organs they hold together. The strength of connective tissue is derived from its collagen fibers, whose configurations range from the crisscross laminae in the dermis to the parallel array of tendon fibers. Resilience is provided by elastic fibers that lace the ground substance. Connective tissues can be as tough as leather, as sinewy as gristle, as soft as adipose tissue, as transparent as the cornea, or as liquid as the fluids filling the body cavities. Wherever such cavities occur, the cells of the connective tissue form a simple squamous epithelium at the interface. The ground substance of the connective tissue is populated by a rich variety of cells, some of them leukocytic tourists from the bloodstream, others permanent residents such as phagocytes, mast cells, and the ubiquitous fibroblasts.

COLLAGEN AND KELOIDS

The chief structural element of connective tissue is the collagen fiber, the basic unit of which is the tropocollagen molecule, itself a triple helix of polypeptides 15 Å wide and stretching 3000 Å in length. When such macromolecules are polymerized and aligned side by side, they give rise to a collagen fibril, the arrangement and dimensions of which vary with the locations and conditions under which it is formed. In the basement lamella beneath epithelia, the fibrils are ordered like plywood in an orthogonal grid (Fig. 27). The forces responsible for determining the regularity with which these fibrils are arranged in such precise register remain unknown. In tendons, collagen fibrils are packed in parallel bundles of considerable density.

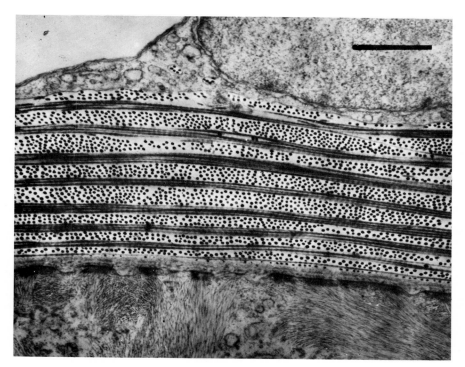

Fig. 27. Electron micrograph of basement lamella in the dermis of the tadpole integument. The orthogonal plies of collagen fibrils are seen in alternating cross and longitudinal sections beneath the epidermis (above). (Courtesy of Dr. Mac V. Edds, Jr.)

The diameters of collagen fibers in tendons increase with the growth of a maturing animal (*14*) and reflect the tensions to which they are subjected. In general, thicker muscles tend to have thicker tendons (*10*). Denervation or excision of its muscle reduces subsequent growth in the tendon (*9*). During development, tendons increase in length in part by internal expansion, a process presumably made possible by molecular turnover and remodeling of fibrils. Most of its elongation, however, is achieved at its junction with the muscle (*8*). At the opposite end, where the tendon inserts on a bone, reorganization allows for shifts in the site of attachment during the process of maturation (*43*).

Tendons are capable of repairing themselves in a fashion reminiscent of the mechanism by which bone fractures are healed. When one has been severed, DNA synthesis increases during the first day in the peritendinous sheath (*16*). By two or three days after operation, the fibroblasts within the severed stumps of the tendon are seen to multiply. Between the cut ends there then develops a callus within which new collagen fibrils are subsequently laid down. Owing to the proliferation of fibroblasts in the injured tendon, the regenerated portion of a tendon may possess a denser population of cells than was originally present, a condition that may later subside in growing animals (39). The persistence with which the severed stumps of tendons heal either to each other or, if separated by too great a distance, to adjacent tissues is as impressive as it is inconvenient. For this reason, experimental studies of the effects of tenotomy on skeletal muscle are difficult to carry on for much longer than the week or so it takes for tendinous adhesions to develop.

The repair of lesions in the dermis is as efficient as in tendons. Wound healing is faster in young animals than in old ones, and is said to be enhanced by treatment with histamines and vitamins B_{12} and C. It is not affected by prior denervation, but is interfered with by cortisone, which antagonizes protein synthesis, and by β-aminopropionitril, a lathyrogenic agent that interferes with synthesis of new collagen. It is the production of collagen in the granulation tissue of a dermal lesion that contributes to the tensile strength of the healed wound. The arrangement of collagen fibers in such scars lacks the orderly pattern characteristic of uninjured dermis. Nevertheless, a healed wound may often be stronger than adjacent regions of the skin.

Sometimes the process of integumental wound healing gives rise to hypertrophic scars or keloids (26). These unsightly growths occur more commonly in races with dark skins, and develop more readily in younger than older persons (Fig. 28). They are commonly encountered in those parts of the body where the tensions on the skin are greatest. The rate at which collagen is synthesized in such scars is greatly accelerated in comparison with that in normally healing wounds and the fibers which are formed are considerably smaller in diameter than normal (27). Of unexplained etiology, these dermal overgrowths are the bane of the plastic surgeon.

Keloids cannot be successfully treated by removing them surgically. They simply grow back in the new lesion (6). If shaved off down to the level of the surrounding skin, however, they can be held in abeyance by a thin graft of skin over the surface (22). Alternatively, the chronic application of pressure to a keloid results in the gradual degradation of its collagen (28), a process that may take months to yield cosmetic results. If keloids are surgically removed and grafted to a part of the body where they do not normally form, such as the abdominal wall, the transplanted scar may soften and become thinner in the course of the next couple of years (7). Perhaps the best treatment for keloids is to prevent their formation in the first place, which can be achieved by the local injection of hydrocortisone. Once formed, they have been successfully treated by direct injections of triamcinolone (34).

The stubborn persistence of keloids testifies to the relative inertness of collagen fibers. Next to the DNA molecule, collagen is probably one of the

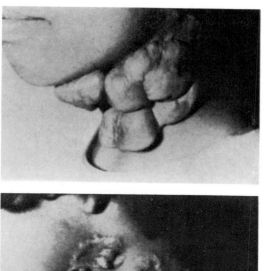

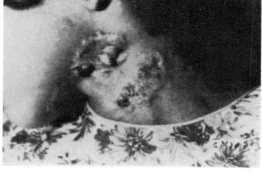

Fig. 28. Keloid formation in a 9-year-old girl as a result of a burn suffered 7 years earlier. When the keloids grew back (above) after excision of the original ones, triamcinolone acetonide was injected repeatedly into the growths causing their atrophy after 22 months of treatment (below) (*34*). From Maguire, H.C., *J. Am. Med. Assoc.* **192**, 325–326 (1965). Copyright 1965, American Medical Association.

most stable molecules in the body. Nevertheless, under certain circumstances it is known to undergo degradation, a process catalyzed by collagenase. This enzyme is normally absent from tissues, but is present wherever collagen is being resorbed or remodeled, as in the regression of tails in metamorphosing tadpoles, the amputation stumps of regenerating limbs in the newt, and in healing wounds in general. One wonders if it is present in normal amounts in wounds destined to develop keloids.

WOUND CONTRACTION

Integumental wounds heal by two mechanisms. One is the development of granulation tissue and the migration of epidermal cells over the lesion; the other is the contraction of the full thickness skin to reduce the dimensions of the wound. It is this contraction that, in the case of extensive injuries, can cause serious deformities to the surrounding tissues. For many years, the elusive nature of this contraction frustrated all attempts to identify its cause. Studies showed that the only way to prevent it was to graft normal skin into a wound, although if such grafts are not trimmed to fit exactly, the forces of contraction pull on the graft and the surrounding skin together (3).

Contraction in wounds cannot be attributed to the collagen fibers. Not only are these not contractile, but it has been shown that in the case of lathyrism, which reduces the tensile strength of wounds, contraction is not prevented (23). Indeed, when granulation tissue is explanted *in vitro*, it can still contract. On the other hand, preoperative exposure to X-rays retards contraction (43).

The quest for the contractile element in healing connective tissue wounds eventually yielded to the pharmacological approach (35) Granulation tissue maintained in culture was shown to contract under the influence of such compounds as 5-hydroxytryptamine, angiotensin, and epinephrine. These studies raised the possibility that cells in the granulation tissue might be endowed with contractile properties. Subsequent studies *in vivo* led to the discovery that topical application of β-diethylaminoethyldiphenolthioacetate (Trocinate), an antagonist of smooth muscle contraction, could inhibit the contraction of wounds (33). Such an effect depends upon the continued presence of the drug in the wound, and is reversible upon discontinuation of treatment. The interpretation is that the contractile cell in granulation tissue is the fibroblast which, in addition to being the source of collagen molecules, is capable of differentiating intracellular contractile molecules resembling those normally found in smooth muscle cells. In view of this important breakthrough in solving the problem of wound contraction, it is altogether fitting that such cells in the granulation tissue should be referred to as myofibroblasts (15).

THE UBIQUITOUS FIBROBLAST

One of the seemingly least differentiated cells in the body is the fibroblast. Usually recognized by its lack of distinguishing characteristics, this nondescript cell has much in common with embryonic mesenchyme, most notably its capacities for proliferation. When tissue fragments are explanted *in vitro*, it is the constituent fibroblasts that tend to be the first cells to grow out into the medium. Their proliferation in culture has been stimulated by a wide range of treatments, including exposure to serum, embryo juice, extracts of healing connective tissue (especially if inflamed), a fibroblast growth factor (FGF) derived from bovine pituitary glands (25), and epidermal growth factor (EGF) extracted from human urine (7). Proliferation is inhibited *in vitro* by cell-to-cell contacts established when confluency is attained.

The relatively unspecialized morphology of the fibroblast is not to be taken to indicate that this cell is not differentiated. First and foremost, fibroblasts are specialized for the synthesis of collagen molecules (37). As indicated above, they also develop contractile properties in healing wounds. Although all fibroblasts look alike, the connective tissues they are responsible for forming come in countless configurations. The question may be asked, Whence cometh such diversities if all fibroblasts are indeed identical? Can it be that the keratocytes in the cornea are fundamentally not different from the fibroblasts in an arterial wall, or from mesothelial cells lining the peritoneum? Connective tissue architecture derives either from its constituent fibroblasts or from supracellular morphogenetic influences. In the former alternative, it follows that there would be as many different kinds of fibroblasts as there are connective tissues, even to the extent that no two cells might be exactly alike. Otherwise, all fibroblasts could be equally pluripotent, the modulations of their differentiated states depending on where they happen to be located. A clue to the solution of this important problem may be found in the studies of wound healing in the peritoneum.

There is reason to suspect that fibroblasts may become organized into the simple squamous epithelia which line the surfaces of body cavities. The mesothelium, or peritoneal epithelium, with which the abdominal cavity is lined, is an especially interesting and important tissue, the integrity of which is largely responsible for preventing adhesions between visceral organs. Wound healing in the mesothelium has been studied in order to identify the source of the new cells which repair such lesions. Wounds can be inflicted by surgical excision or by application of a dry gelatin film which peels off the attached mesothelium when it is removed. Following such procedures, an exudate forms on the denuded surface and a fibrin clot develops. The healing process involves mitotic activity both in the wound itself and in surrounding tissues, reaching a peak two days after injury. Healing is usually complete in about a week (Fig. 29) (38), but it is perhaps significant that large wounds heal as quickly as small ones, an observation that argues against the centripetal migration of mesothelial cells as an important healing mechanism. Instead, there is evidence that subjacent connective tissue cells, presumably fibroblasts, transform into the newly forming mesothelium *in situ* (11, 38). Alternatively, there is evidence that cells from the peritoneal fluid may settle onto the fibrin network to differentiate into mesothelial cells (Fig. 30) (44). Indeed, if peritoneal fluid is introduced into Millipore filter diffusion chambers, or maintained *in vitro*, its cells do in fact differentiate into fibroblasts and mesothelium (12). It has also been suggested that mesothelial cells from opposing peritoneal surfaces may stick to an adjacent lesion and give rise to

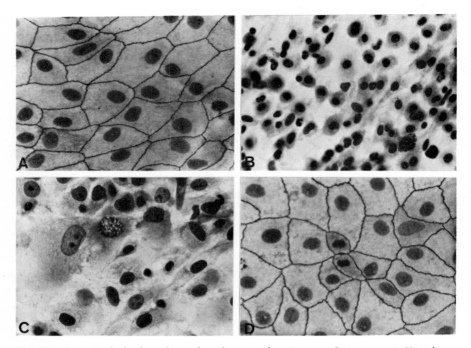

Fig. 29.   Stages in the healing of wounds in the parietal peritoneum of rats as seen in Häutchen preparations (38). In normal peritoneum (A), the cell boundaries are made visible by AgNO$_3$ staining. One day after injury (B), the wound surface is populated mostly by macrophages and monocytes. By the fifth day (C), there is a mixture of hard-to-classify cell types in areas not yet covered by migrating peritoneal cells. Healing may be complete after 8 days (D), although mitotic divisions persist in the reconstituted mesothelium.

islands of cells which seed the healing wound *(4)*. Whatever may turn out to be the correct explanation, the fact remains that the deceptive simplicity of the fibroblast is surpassed only by the versatility of forms into which it is capable of differentiating.

FAT CELLS: SIZE VERSUS NUMBER

There is a double standard in man's preoccupation with his corpulence. A plump child is considered healthy and well fed. An obese adult, discriminated against socially, is said to be a medical risk. While there is evidence of a causal relationship between the two, it is not altogether certain that at least moderate amounts of fat in adult mammals is necessarily undesirable except in commercial meat production. The tendency for older mammals to ac-

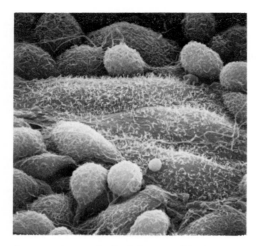

Fig. 30.   Scanning electron micrograph of a 3-day wound in the rat perietal peritoneum (44). The spherical cells, more numerous at earlier stages of healing, are being displaced by flattened cells with numerous microvilli on their surfaces. Reproduced with permission. © 1972 U.S.– Canadian Division of the International Academy of Pathology.

cumulate adipose tissue, especially in their abdominal fat depots, is more the rule than the exception. One suspects it may be a physiological adaptation to certain metabolic changes of aging.

Although the function of adipose tissue is that of fat storage, its lipids are in a constant state of physiological turnover. The conditions under which this turnover is regulated are as varied as the anatomical sites where adipose tissue is found. Interscapular or brown fat, otherwise known as the hibernating gland, may be an adaptation in some mammals to survive through the winter. Fat is a source of metabolic water in the camel's hump and the sheep's tail (45). For insulation against the frigid waters of the oceans, blubber serves to streamline the bodies of marine mammals. One can only guess what might be the functions of perirenal, epididymal, and parametrial fat bodies, but their association with specific organs is probably not without significance.

Studies of the growth of fat tissue are plagued by technical problems. One is the difficulty of determining the number of fat cells in the body in view of the widespread distribution of adipose tissue. Consequently, estimates must be made in biopsy samples which, when dissociated and fixed in osmium tetroxide, can be counted electronically. Measurements of cell size can also be made in such preparations. Attention is usually focused on a particular fat body of known size, the number of cells in which, together with their dimensions, can be determined with some accuracy.

A more serious technical problem relates to the early differentiation of fat cells. An adipocyte is recognizable as such only when lipid is stored in the cytoplasm. Therefore, the earliest stages of their differentiation are unidentifiable, however much they may already be programmed to differentiate into adipocytes. It is during these earliest stages that cellular proliferation is most prevalent (1). Once differentiation of an adipocyte becomes recognizable, little or no cell division occurs (21). Although estimates can be made of the rates of hyperplasia by monitoring the increasing numbers of fat cells, the actual proliferation of their precursors lies beyond the realm of technical feasibility.

The life history of an adipocyte is divisible into two stages, the hyperplastic and the hypertrophic. In the rat, hyperplasia continues for 3–4 months after birth (20, 24). In man, the number of fat cells increases up to about one year of age (5) (with a possible second phase of hyperplasia during the adolescent growth spurt). In the calf, proliferation of fat cells goes on for about 14 months after birth. As the rate of hyperplasia gradually declines, the sizes of the differentiated cells increase, a process that enables continued expansion of adipose tissue during maturation, if not beyond. In this sense, therefore, fat tissue may be classified with muscle and nerves as nonmitotic, although the possible restoration of proliferative potential in cells depleted of their lipid, though not as yet investigated *in vivo*, cannot be excluded. *In vitro* studies, however, have demonstrated that adipocytes from infants still in the hyperplastic phase are capable of division once their contained fat has been lost (1). In cultured adult adipose tissue, however, the number of cells does not increase. Thus, available evidence suggests that the normal adult complement of adipocytes is determined early in development when proliferation is still possible.

THE CAUSES OF OBESITY

The implications of this are important to our understanding of the problem of excessive obesity. In adults, a fat individual may have more cells of normal dimensions, a normal number of cells of large size, or a combination of the two. If the number of fat cells is increased, however, the cause must be sought early in development prior to the switch from hyperplasia to hypertrophy. This raises the question of whether such an increase is genetically determined or simply a physiological adaptation to overeating. In either case, the solution cannot be found without exploring how the normal complement of fat in the body is prescribed.

Assuming that the amount of fat to develop is at least partly under genetic control, it might be expected that the experimental reduction in the amount of adipose tissue should be subject to compensation. Excision of a fat body, however, is not followed by its regeneration *in situ*, nor does fat elsewhere in the body enlarge permanently to make up the difference (*13*). It must be admitted that if such compensatory growth were to occur throughout the remaining adipose tissue it would be exceedingly difficult to detect since even the most drastic lipectomies can only deplete the total mass of fat and the body to modest extents. There is evidence, however, that if the gonadal fat depots are removed and obesity is induced by experimental lesions in the hypothalamus to promote hyperphagia, compensatory hypertrophy of the remaining fat bodies is sufficiently exaggerated to become detectable (*31*).

The opposite approach is also instructive. Fat pads from infant animals can be grafted subcutaneously or to the peritoneal surface of the abdominal wall (*17*). Here they become vascularized and resume their growth, reaching nearly normal adult dimensions. If the host is made obese, the grafts enlarge accordingly (*30*). That such grafts participate in the overall regulation of adipose tissue is indicated by experiments in which pieces of adult fat tissue were grafted onto the surface of the liver (*32*). Such grafts were more success-ful in lipectomized than in intact hosts.

Genetic influences in fat cell regulation have been explored by exchanging pieces of epididymal fat pads between obese and lean strains of mice (*2*). When a fragment of adipose tissue from an obese strain is grafted beneath the kidney capsule of a lean mouse, the fat cells come to resemble those of the host, not the donor. Reciprocal transplants yield comparable results, indicating that the physiological environment in which adipocytes live can override a cell's hereditary predisposition to accumulate lipid.

In view of the central role played by fat in metabolic processes it is no surprise that the growth of adipose tissue should be subject to a wide variety of physiological influences. Hence, its growth is promoted by growth hor-mone (*21*) as well as insulin. In hamsters, melatonin or exposure to short photoperiods enhances the development of brown fat presumably in antici-pation of hibernation. Exercise tends to decrease the lipid content of fat cells and interferes with the normal rise in number of adipocytes in young rats still in the hyperplastic phase (*36*). Rats raised in the cold develop more numer-ous adipocytes but the dimensions of these cells remain smaller than in animals at room temperature (*41*). Brown fat in mice, on the other hand, accumulates more lipid and glycogen at lower temperatures. This increase in brown fat is partly prevented when mice are allowed to live in groups where body heat can be conserved, but it is more pronounced in hairless strains than in furred ones (*18*).

The state of nutrition, of course, exerts an overriding influence on adipose tissue development. Infant rats and mice can be raised in small versus large litters to compare the effects of increased and decreased nourishment, re-spectively. The well-fed animals in small litters not only develop larger adipocytes but the numbers of such cells increase (29). This suggests that dietary factors play an important role in regulating the proliferation of fat cell precursors in immature animals. The converse, however, is not necessarily true, for animals raised in large litters have smaller fat cells in normal numbers.

In adults, overeating can be experimentally induced by hypothalamic lesions made by the injection of gold thioglucose which selectively localizes in the hypothalamus where it damages neurons. This upsets appetite control as a result of which the animal overeats and becomes obese (Fig. 31) (30, 31). High fat diets have similar effects, and in either case the sizes of the adipocytes increase while their numbers remain unaffected (24, 40). Reduced food intake depletes the lipid content per cell reversibly without altering the total population of adipocytes (36). Thus, in genetically normal individuals the number of fat cells is affected by diet only during that part of the life cycle when fat cell precursors are normally capable of proliferation. Cell size is adaptable in young and old alike. Hence, overfeeding in infancy may augment the normal population of adipocytes predisposing the individual in later life to a greater tendency to become obese for dietary reasons (19).

The development of genetically obese strains of rats and mice has provided a useful model system in which to study human obesity, much of which may be assumed to be hereditary. Not only do these animals eat excessively and gain more weight than normal controls, but they acquire larger numbers of adipocytes, sometimes owing to protraction of the hyperplastic phase of adipose tissue development into adulthood (24). The

Fig. 31.   Experimentally induced obesity in mice. The two animals on the outside were given gold thioglucose to damage the hypothalamus and cause hyperphagia. Control mouse in the center. (Courtesy of Dr. Robert A. Liebelt, Northeastern Ohio University College of Medicine.)

abnormality in such mutants thus affects the regulation of cell proliferation itself, an effect of obvious clinical importance in understanding obesity but of even greater significance in explaining what turns proliferation on and off in other tissues of the body as well.

REFERENCES

1. Adebonojo, F.O. Studies on human adipose cells in culture: Relation of cell size and cell multiplication to donor age. *Yale J. Biol. Med.* **48**, 9–16 (1975).
2. Ashwell, M., Meade, C.J., Medawar, P., and Sowter, C. Adipose tissue: Contributions of nature and nurture to the obesity of an obese mutant mouse (ob/ob). *Proc. Roy. Soc. London* **195**, 343–353 (1977).
3. Billingham, R.E., and Medawar, P.B. Contracture and intussusceptive growth in healing of extensive wounds in mammalian skin. *J. Anat.* **89**, 114–123 (1955).
4. Bridges, J.G., and Whitting, H.W. Parietal peritoneal healing in the rat. *J. Pathol. Bacteriol.* **87**, 123–130 (1964).
5. Brook, C.G.D. Evidence for a sensitive period in adipose-cell replication in man. *Lancet* **2**, 624–627 (1972).
6. Calnan, J.S., and Copenhagen, H.J. Autotransplantation of keloid in man. *Br. J. Surg.* **54**, 330–335 (1967).
7. Carpenter, G., and Cohen, S. Human epidermal growth factor and the proliferation of human fibroblasts. *J. Cell Physiol.* **88**, 227–238 (1976).
8. Crawford, G.N.C. An experimental study of tendon growth in the rabbit. *J. Bone Jt. Surg., Br. Vol.* **32**, 234–243 (1950).
9. Elliott, D.H. The growth of tendon after denervation or excision of its muscle. *Proc. Roy. Soc. London, Ser. B* **162**, 203–209 (1965).
10. Elliott, D.H., and Crawford, G.N.C. The thickness and collagen content of tendon relative to the strength and cross-sectional area of muscle. *Proc. Roy. Soc., London, Ser. B* **162**, 137–146 (1965).
11. Eskeland, G., and Kjaerheim, A. Regeneration of parietal peritoneum. 2. An electron microscopical study. *Acta Pathol. Microbiol. Scand.* **68**, 379–395 (1966).
12. Eskeland, G., and Kjaerheim, A. Growth of autologous peritoneal fluid cells in intraperitoneal diffusion chambers in rats. 2. An electron microscopical study. *Acta Pathol. Microbiol. Scand.* **68**, 501–516 (1966).
13. Faust, I.M., Johnson, P.R., and Hirsch, J. Noncompensation of adipose mass in partially lipectomized mice and rats. *Am. J. Physiol.* **231**, 538–544 (1976).
14. Fitton Jackson, S. Structural problems associated with the formation of collagen fibrils *in vivo*. In "Connective Tissue" (R.E. Tunbridge, ed.), pp. 77–83. Blackwell, Oxford, 1957.
15. Gabbiani, G., and Montandon, D. Reparative processes in mammalian wound healing: The role of contractile phenomena. *Int. Rev. Cytol.* **48**, 187–219 (1977).
16. Greenlee, T.K., Jr., and Pike, D. Studies of tendon healing in the rat. Remodeling of the distal stump after severance. *Plast. Reconstr. Surg.* **48**, 260–270 (1971).
17. Hausberger, F.X. Quantitative studies in the development of autotransplants of immature adipose tissue of rats. *Anat. Rec.* **122**, 507–515 (1955).
18. Heldmaier, G. Temperature adaptation and brown adipose tissue in hairless and albino mice. *J. Comp. Physiol.* **92**, 281–292 (1974).
19. Hirsch, J. Cell number and size as a determinant of subsequent obesity. *In* "Childhood

Obesity. Current Concepts in Nutrition" (M. Winick, ed.), Vol. 3, pp. 15–21. Wiley, New York, 1975.

20. Hirsch, J., and Han, P.W. Cellularity of rat adipose tissue: Effects of growth, starvation and obesity. *J. Lipid Res.* **10**, 77–82 (1969).

21. Hollenberg, C.H., and Vost, A. Regulation of DNA synthesis in fat cells and stromal elements from rat adipose tissue. *J. Clin. Invest.* **47**, 2485–2498 (1968).

22. Hynes, W. The treatment of scars by shaving and skin graft. *Br. J. Plast. Surg.* **10**, 1–10 (1957).

23. Jacques, J. Wound contraction in experimental lathyrism. *Br. J. Exp. Pathol.* **50**, 486–489 (1969).

24. Johnson, P.R., Zucker, L.M., Cruce, J.A., and Hirsch, J. Cellularity of adipose depots in the genetically obese Zucker rat. *J. Lipid Res.* **12**, 706–714 (1971).

25. Kamely, D., and Rudland, P.S. Induction of DNA synthesis and cell division in human diploid skin fibroblasts by fibroblast growth factor. *Exp. Cell Res.* **97**, 120–126 (1976).

26. Ketchum, L.D., Cohen, I.K. and Masters, F.W. Hypertrophic scars and keloids. *Plast. Reconstr. Surg.* **53**, 140–154 (1974).

27. Kischer, C.W. Collagen dermal patterns in the hypertrophic scar. *Anat. Rec.* **179**, 137–146 (1974).

28. Kischer, C.W., Shetlar, M.R., and Shetlar, C.L. Alteration of hypertrophic scars induced by mechanical pressure. *Arch Dermatol.* **111**, 60–64 (1975).

29. Knittle, J.L., and Hirsch, J. Effect of early nutrition on the development of rat epididymal fat pads: Cellularity and metabolism. *J. Clin. Invest.* **47**, 2091–2098 (1968).

30. Liebelt, R.A. Response of adipose tissue in experimental obesity as influenced by genetic, hormonal, and neurogenic factors. *Ann. N.Y. Acad. Sci.* **110**, 723–748 (1963).

31. Liebelt, R.A., Ichinoe, S., and Nicholson, N. Regulatory influences of adipose tissue on food intake and body weight. *Ann. N.Y. Acad. Sci.* **131**, 559–582 (1965).

32. Liebelt, R.A., Vismara, L., and Liebelt, A.G. Autoregulation of adipose tissue mass in the mouse. *Proc. Soc. Exp. Biol. Med.* **127**, 458–462 (1968).

33. Madden, J.W., Morton, D., Jr., and Peacock, E.E., Jr. Contraction of experimental wounds. I. Inhibiting wound contraction by using a topical smooth muscle antagonist. *Surgery* **76**, 8–15 (1974).

34. Maguire, H.C. Treatment of keloids with triamcinolone acetonide injected intralesionally. *J. Am. Med. Assoc.* **192**, 325–326 (1965).

35. Manjo, G., Gabbiani, G., Hirschel, B.J., Ryan, G.B., and Statkov, P.R. Contraction of granulation tissue *in vitro*: Similarity to smooth muscle. *Science* **173**, 548–550 (1971).

36. Oscai, L.B., Babirak, S.P., Dubach, F.B., McGarr, J.A., and Spirakis, C.N. Exercise or food restriction: Effect on adipose tissue cellularity. *Am. J. Physiol.* **227**, 901–904 (1974).

37. Priest, R.E., and Davies, L.M. Cellular proliferation and synthesis of collagen. *Lab. Invest.* **21**, 138–142 (1969).

38. Raftery, A.T. Regeneration of parietal and visceral peritoneum. *Br. J. Surg.* **60**, 293–299 (1973).

39. Schmitt, W., Ervig, K., and Beneke, G. Die Sehnenregeneration im wachsenden und im ausgewachsenen Organismus. *Beitr. Pathol.* **141**, 261–275 (1970).

40. Stern, J.S., and Greenwood, M.R.C. A review of development of adipose cellularity in man and animals. *Fed. Proc.* **33**, 1952–1955 (1974).

41. Therriault, D.G., and Mellin, D.B. Cellularity of adipose tissue in cold-exposed rats and the calorigenic effects of norepinephrine. *Lipids* **6**, 486–491 (1971).

42. van den Brenk, H.A.S., Orton, C., Stone, M., and Kelly, H. Effects of X-radiation on growth and function of the repair blastema (granulation tissue). I. Wound contraction. *Int. J. Radiat. Biol.* **25**, 1–19 (1974).

43. Videman, T. An experimental study of the effects of growth on the relationship of tendons and ligaments to bone at the site of diaphyseal insertion. *Acta Orthopaed. Scand., Suppl.* **131**, 1–22 (1970).
44. Watters, W.B., and Buck, R.C. Scanning electron microscopy of mesothelial regeneration in the rat. *Lab. Invest.* **26**, 604–609 (1972).
45. Wells, H.G. Apidose tissue, neglected subject. *J. Am. Med. Assoc.* **114**, 2177–2183, 2284–2289 (1940).

4

Bone and Cartilage

Bone: G., *Bein*, leg

Vertebrates invented bone. Skeletal phylogeny has led to the evolution of a tissue that is sufficiently rigid to support the body but that is developmentally flexible enough to allow for growth. Cartilage is largely responsible for these versatile characteristics of skeletal tissues, but it has two important drawbacks. It is not vascularized and it has no way of turning over its population of cells once they are trapped in their matrix. With its growth zone limited to the perichondrium, cartilage is poorly equipped to remodel itself, and its capacity for repair and regeneration is correspondingly restricted. Although these shortcomings have been solved by the evolution of bone, the retention of the chondrogenic step in skeletal ontogeny even in the highest of vertebrates would seem to be more an accident of history than an indispensable stage in osteogenesis.

Nevertheless, it cannot be denied that vertebrates have not made good use of cartilage. It is the tough but resilient tissue needed in the construction of articulating surfaces. Furthermore, the cartilaginous plate in the long bones of higher vertebrates provides for their growth in length during the process of maturation. Even more important, the eventual disappearance of the cartilaginous plate from the bones of adult birds and mammals is an effective mechanism by which bodily stature can be regulated.

In cartilaginous fishes, the appositional nature of chondrogenesis allows for little intrinsic restraint in its potential for enlargement. In teleosts, although chondrogenesis is not limited to embryonic stages, more mature fishes possess bony skeletal parts. Here, a similar mode of enlargement of the bones occurs primarily by periosteal ossification at the surface. Such bones lack marrow cavities, and in some cases their matrix is completely acellular. They can continue to grow in length as well as width, however, expressing this potential at a decelerating pace for as long as the fish may live.

Amphibians evolved marrow cavities. They also acquired cartilaginous epiphyses, structures designed for articulation while providing the where-withal for skeletal elongation. Thus, in endochondral bones the remnants of a cartilaginous model are retained throughout life as sites where new bone can be formed. It is important to note, however, that a single marrow cavity extends from one end of the amphibian bone to the other, a situation that also prevails in most reptiles. This provision for growth may have accounted for the prodigious sizes attained by dinosaurs, if not the ripe old ages to which some extant reptiles are purported to live.

It was among the reptiles that the cartilaginous plate began to evolve. By segregating the part of the cartilaginous epiphysis responsible for growth from that concerned with making up the articular surface, it was possible to evolve a skeletal structure capable of elongating for a finite period of life, but which could be disposed of when the time came to stop growing. This adaptation was essential if terrestrial vertebrates were to avoid growing so large as to be encumbered by their own bulk. Able to turn off growth at sexual maturity, higher vertebrates became animals with determinate body sizes. The obverse of this is that they may also have initiated the aging process when they gave up the potential for growth (13).

There are two kinds of bone in the body, dermal and endochondral. Dermal bones, best represented by those in the skull, are formed by the direct ossification of connective tissues during ontogeny. They have little capacity for chondrogenesis. Endochondral bones, comprising most of the long bones in the body, develop first as cartilaginous models in the embryo and only secondarily do they become ossified. Both types of bone may consist of parts that are spongy or compact. The former consists of trabeculae most commonly associated with the marrow spaces. Compact bone may be made up of Haversian systems, or osteons. These are concentric layers of bone which have been laid down around a blood vessel occupying the central canal. In the world's smallest mammals, such as the shrews, the limb bones are avascular owing to their toothpick dimensions. Even in larger rodents the shaft of the bone may be too thin to permit the organization of osteons. In still larger mammals, osteons are not present in the bones of very young individuals, the diaphyses of which are originally formed by the laying down of concentric circumferential lamellae. Hence, osteons form only as a result of the secondary remodeling and turnover which characterizes bone. Following erosion by osteoclastic action around blood vessels, osteoblasts mediate the deposition of new layers of bone in a centripetal direction to give rise to the typical histological organization of the osteons. The same process is repeated in other locations at other times, and may eventually result in the destruction of previously laid down osteons.

The average diameters of osteons are roughly correlated with body size, ranging from 72 μm in the rat to 250 μm in the cow (22). The trade-off between bone resorption and deposition in the osteons is responsible for maintaining the strength and structural integrity of the bone itself. Not surprisingly, this turnover is affected by a variety of physiological influences. Turnover is maximal in infants, and the rate gradually declines to adulthood. Mechanical forces impinging upon a bone promote the formation of osteons. They are poorly developed in the bone stump of an amputated limb. During maturation, the rate of deposition exceeds that of resorption, but in old age the reverse may be true, though less so in individuals who exercise regularly. Thus, in the normal bone a balance is struck between the tendencies toward bone deposition (osteopetrosis) and resorption (osteoporosis). Osteopetrosis has been studied in mutant rats and mice (Fig. 32). When animals with congenital osteopetrosis are joined in parabiosis with normal littermates, or are injected with normal spleen or marrow cells, the density of their bones decreases owing to the acquisition of healthy cells (54). This restores the capacity of osteopetrotic animals to remodel their bones in normal ways, a remission of the disease that persists even after separation

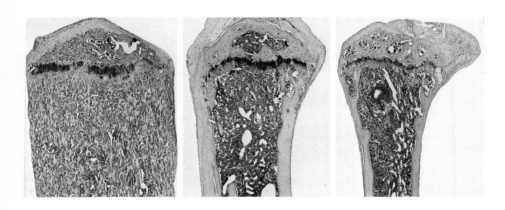

Fig. 32. Induction and cure of osteopetrosis in mice (54). Left, proximal tibia of normal mouse exposed to lethal radiation at 20 days of age and injected with unirradiated marrow cells from its osteopetrotic littermate. Excessive trabeculae of bone and cartilage fill the marrow cavity. Middle, control tibia from irradiated mouse injected with normal marrow. Right, tibia of an irradiated osteopetrotic mouse given normal marrow cells to prevent development of osteopetrosis.

from the parabiont. Osteoporosis, on the other hand, is encountered as a physiological consequence of skeletal disuse (4, 20), or the excessive resorption of mineral in hyperparathyroidism. Postmenopausal women may suffer varying degrees of osteoporosis, sometimes leading to "dowager's hump," due to the natural decline in estrogen secretion. Their osteons become excessively eroded (Fig. 33).

The role of sex hormones in regulating bone growth is nowhere more strikingly illustrated than in the case of the bone situated in the penis of the rat. Castration of the male results in resorption and decreased density of the penile bone, delaying the healing of its fractures (2). If testosterone is injected into infant females, it induces bone formation in the clitoris.

The balance between bone deposition and resorption is primarily under the control of calcitonin and the parathyroid hormone. The parathyroid hormone promotes the resorption of bone and the elevation of serum calcium levels. Under its influence, osteoclasts initiate the process of remodeling. Its effects can even be demonstrated *in vitro*, indicating its direct action on the target tissue (3). Conversely, parathyroidectomy shifts the calcium balance from the serum to the bone by promoting ossification. The deposition of

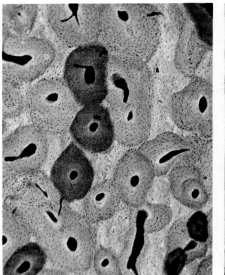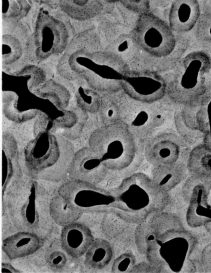

Fig. 33.   Effects of age on osteons. On the left is a section of femur from a 23-year-old woman showing the relatively solid condition of the bone (22). On the right is a comparable specimen from an 85-year-old woman whose osteons are irregularly shaped and Haversian canals more eroded.

bone is more directly mediated by calcitonin, however, a hormone secreted by cells that are interspersed between the thyroid follicles in higher vertebrates, but are consolidated into the ultimobranchial bodies of lower vertebrates. Calcitonin affects skeletal remodeling by promoting bone deposition. Not surprisingly, extirpation of the ultimobranchial bodies of frogs brings about an increase in the number of osteoclasts and promotes osteoporosis (42).

The pituitary, of course, exerts profound influences on growing bones, not so much by regulating the rate of turnover as by promoting growth and regeneration. Hypophysectomy retards the elongation of long bones in the body, a condition simulating that in pituitary dwarfs. The deficiency can be corrected by administration of growth hormone. Regeneration of the calvarium in infant rats, which can occur if the dura remains intact, is inhibited by hypophysectomy, but restored by growth hormone (47). The pituitary is equally important in promoting the appositional growth of bones, a role responsible for the skeletal symptoms of acromegaly in cases of pituitary tumors.

PHYSICAL REGULATION OF BONE GROWTH

Inasmuch as the principal function of bones is that of mechanical support, it is appropriate that they should be subject to local factors, not the least of which are the stresses and strains exerted on bones by the pull of gravity, the forces associated with other tissues attached to bones (e.g., tendons, muscles, teeth), and even the intermittent pulses of blood pressure in their vessels. Gravitational influences can be studied by exploring the effects of centrifugation or weightlessness on bone growth and maintenance. While centrifugation of chick embryos (39), or of their long bones in culture (29), promotes greater than normal elongation, the bones of adults chronically held under heightened g forces become heavier, but not longer (57). Weightlessness, on the other hand, promotes the loss of bone mass by demineralization (48).

Similar studies have been carried out by contriving to increase the load to be supported by a bone. One such method is to incapacitate the leg of an animal by denervation, tenotomy, or simply by mechanical immobilization. The bones of the opposite leg, forced to carry a load normally divided between two limbs, grow considerably thicker (but not longer). In still another experiment, lambs have been outfitted with saddlebags loaded with weights periodically adjusted to equal 40% of the body mass. As the animals matured, their leg bones grew to normal lengths but became markedly thicker (Fig. 34) (50). Thus, although longitudinal forces may be able to stimulate

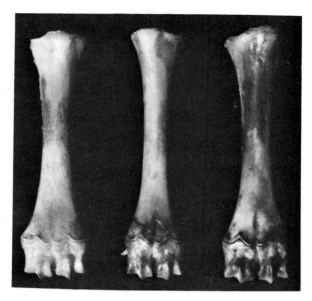

Fig. 34.   Influence of compression on metacarpal growth in artificially weighted-down lambs
(*50*). Although bone elongation is not affected by gravitational pull, the metacarpal from the
experimental animal (left) is considerably thicker than that from an unweighted control (right).

excessive elongation of embryonic bones, their influence on the growth of
bones in infants and adults is expressed only in lateral growth.

It is worth noting that the elongation of a bone can be arrested by excessive longitudinal compression. Thus, if the epiphysis and diaphysis are
stapled across the growth zone of the cartilaginous plate, the tissues of the
latter are unable to promote further elongation (*16*). It is not uncommon,
however, to find that in experiments of this kind the growth pressure of the
cartilaginous plate is sufficiently strong to open up some rather formidable
mental staples. Studies on the rabbit tibia have shown that elongation ceases
when the forces of compression equal 3 kg, or 15 g/mm^2 (*46*). Nevertheless,
compression of the cartilage plate seriously disrupts the normally precise
histological architecture of the columns of chondrocytes (*46*).

While static mechanical forces have profound effects on the growth and
maintenance of bone, the intermittent application of pressure has been
shown to affect the pathways of cellular differentiation in skeletal tissues (*15*).
Dermal bones, for example, are noted for the infrequency with which they
develop cartilage. When grown in culture, however, and subjected to frequent mechanical bending, the periosteal cells which normally would have
undergone ossification are now deflected into the pathways of chon-

drogenesis. One wonders, therefore, if the lack of cartilage in association with dermal bones might be attributed more to the normal immobility of such skeletal parts rather than to some innate incapacity for chondrogenesis. The corollary is that the chondrogenesis so frequently accompanying the growth of long bones may be explained at least in part by their greater mobility. The healing of fractures is a case in point. The callus first undergoes chondrogenesis and only secondarily does osteogenesis supervene. If the femur of a rat is fractured and subsequently bent on itself every day for a month or more, the cartilaginous callus is prevented from undergoing ossification (27). Instead, it continues to enlarge until its bulk precludes further refracturing. Only after cessation of the daily mobilization does such a callus finally ossify. Whether such mechanical influences are mediated via decreased blood supply or reduced oxygen levels, both of which are associated with cartilage formation, remains to be proved.

The experimental reduction of mechanical influences on skeletal structures might be expected to have equally interesting results on the growth and maintenance of bone. It is not technically possible to effect the complete disuse of bone *in vivo*, but the selective elimination of specific mechanical influences has proved feasible. As already mentioned, in the absence of g forces there is a net loss of mineral from bone (48). Similar effects are seen in other kinds of disuse. In human patients or volunteers committed to prolonged bed rest there is a similar decrease in the calcium content of bones. Osteoporosis typically occurs in the bones of amputated limbs (20). Immobilization of legs also promotes skeletal atrophy (characterized by osteoporosis) and reductions in thickness of cortical bone in the diaphysis (4). Denervation produces similar effects, with little or no influence on the elongation of bones (24).

Particularly meaningful findings have been obtained in studies designed to explore the role of muscle attachments on bone growth. These experiments have focused on the scapula, mandible, and skull. A variety of muscles originates on the various surfaces and margins of the scapula. The supraspinatus, for example, has its origin on the supraspinous fossa and inserts on the head of the humerus. If it is resected early in development, the dimensions of the supraspinous fossa are much reduced. Similarly, removal of the spinotrapezius and infraspinatus muscles results in the reduction of the scapular spine and infraspinous fossa, respectively, from which they originate. Resection of the rhomboideus muscle originating along the vertebral margin of the scapula is responsible for the shortening of the scapula in that direction (Fig. 35) (56).

Experiments on the mandible have yielded equally impressive results (40). Extraction of the lower incisors from infant rats leads to the development of considerably shortened mandibles owing to reduction of the anterior

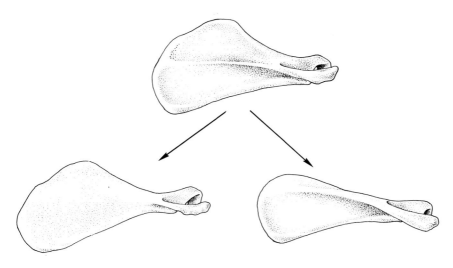

Fig. 35.   The morphology of the developing rat scapula is affected by the muscles which are attached to it (56). Above, normal scapula. Excision of the trapezius, infraspinatus and supraspinatus muscles from a neonatal rat yields a scapula lacking a spine (lower left). Removal of the suprapinatus reduces the subsequent area of the supraspinous fossa (lower right) from which it originates.

parts (Fig. 36), a condition simulating that which occurs in edentulous humans (Fig. 37). If the temporalis and/or masseter muscles are removed from newborn rats the coronoid and condyloid processes are greatly diminished or absent in subsequent mandibular development (Fig. 36). Resection of the internal pterygoid or masseter muscles reduces the angular process of the mandible, while the absence of the lateral pterygoid interferes with mandibular growth in the anterior–posterior direction. These and other muscles are also attached to various parts of the skull. In their absence the cranial protuberances which mark their margins of insertion are greatly reduced or absent. Indeed, the mass of the cranium, as well as the mandible, is significantly reduced in animals fed on a soft diet which obviates the need for mastication thus lessening the muscular pull on the bones of the skull and mandible (33).

The growth of the orbit is normally adjusted to the size of the eyeball. Enucleation in infant rabbits results in continued enlargement of the orbit but at a below normal rate. Attempts to enhance orbital growth by replacing the missing eye with an acrylic sphere have been unsuccessful probably owing to the fact that the implant, unlike the normal eyeball, is unable to grow (43). However, it was shown long ago that when the embryonic eyes of a large species of salamander were grafted in place of those in a smaller

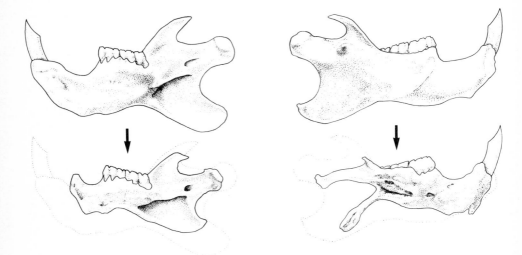

Fig. 36.   Influence of extracting the incisors (left) or of removing the temporalis and masseter muscles (right) from 10-day-old rats on subsequent growth of the mandible (*40*). In the absence of the incisors the anterior portion of the mandible fails to elongate normally. Without muscular attachments, the mandible is shortened at the expense of its posterior parts, the coronoid process nearly being eliminated.

species, the host orbit grew to extra large dimensions to accommodate the oversized eyeball (Fig. 38) (*55*).

A similar situation prevails with respect to the growth of the cranium. In hydrocephalic or microcephalic individuals, the size of the cranium is above or below normal, respectively, remaining precisely matched to the mass of its contents. Experimentally, it has been shown that if the brain is removed from infant rats by aspiration *in utero* the subsequent growth of the skull is reduced. Conversely, expansion of the cranial contents by the injection of a kaolin suspension stimulates extra growth of the skull bones (*58*).

The foregoing examples bear witness to the importance of forces acting upon bone in determining the growth, shape, and density of skeletal tissues which form in the developing organism, and to a lesser extent in the non-growing adult. These influences have little more than a modulating effect on the shapes of bones, however. While causing increases or decreases in the dimensions of the bumps and processes adorning the surfaces of bones, they have seldom been shown to result in the total absence of such structures. Their influences, therefore, are more quantitative than qualitative. One gains the impression that factors of a more genetic nature, not necessarily bound up with local mechanical tensions, might be responsible for the basic morphogenesis of skeletal structures. It must be admitted, however, that in

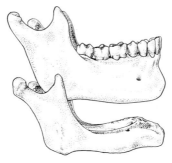

Fig. 37. Effect of functional dentition on maintenance of the mandible. Compared with a normal human jaw (above), the edentulous one (below) has suffered considerable bone resorption.

the absence of experimental methods for eliminating all forces impinging on skeletal structures, categorical conclusions as to the basic nature of bone growth are premature.

However important mechanical influences may be, their existence demands a unifying explanation. Wolff's law has been cited to account for the capacity of bones to remodel themselves so as to resist the effects of distorting pressures. A bone bent by the application of pressure at right angles to its longitudinal axis will tend to undergo resorption on the convex surface while

Fig. 38. Effects of transplanting the optic vesicle and associated ectoderm from the tail-bud embryo of *Ambystoma tigrinum* (a large species) in place of the embryonic eye of *A. punctatum* (a small species). The grafted eye develops autonomously but causes the host orbit to become larger than normal (55).

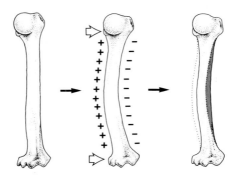

Fig. 39. Effects of mechanical forces (open arrows) on bone remodeling (*1*). Owing to the piezoelectric properties of hydroxyapatite crystals, electric charges are established as shown. Bone tends to be resorbed where positive charges predominate on the convex surface and deposited (shaded area) in the vicinity of negative charges on the concave side.

depositing extra layers of bone on the concave side (Fig. 39). This is correlated with the piezoelectric properties of collagen and hydroxyapatite crystals in bone (*1*). Piezoelectricity is the current generated in certain crystalline materials when subjected to mechanical distortions. In the case of the deformed bone, the convex surfaces becomes positively charged and the concave side negative. In accordance with the hypothesis that the resorption and deposition of bone might be directly affected by the electrical charges into which the mechanical forces acting on bone are translated, numerous experimental attempts have been made to promote the remodeling of bone by the direct application of electrical charges in the absence of mechanical stimuli. Mindful of the effects which the implantation of electrodes into the substance of a bone will have in terms of repair reactions, it is important to interpret the results of such experiments with caution since new bone is often formed at both the anode and cathode. Nevertheless, in a significant number of studies more bone is usually produced at the cathode than the anode, although resorption at the anode has not been observed (*37*). Moreover, the application of pulsed electric current has been found to be more effective in stimulating osteogenesis than constant electric fields (*26*). Both clinical and experimental applications of these techniques to the healing of fractured bones have also yielded positive results (*10*). In view of the many variables inherent in such research, not to mention the difficulty of conducting controlled experiments under clinical conditions, the promising future of this field can be realized only if it is investigated with the utmost care and accuracy, even to the extent of balancing positive results with the publication of negative ones as well.

Bones react in interesting, yet unexplained, ways to alterations in their blood supply. This has been explored by such methods as creating ar-

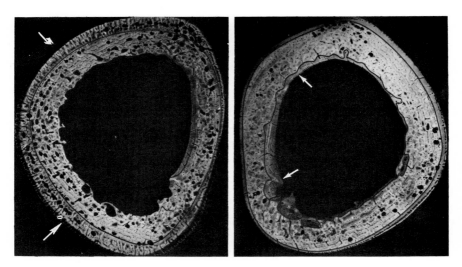

Fig. 40.   Periosteal ossification induced by a venous tourniquet on a dog's leg as seen in cross sections of experimental (left) and contralateral control (right) tibiae 12 weeks after operation (23). On the operated side the extent of periosteal deposition (larger arrows) exceeds that on the untreated leg. Venous ligation has promoted endosteal resorption as shown by the near absence of new endosteal bone in the experimental tibia compared with the unoperated one (smaller arrows).

teriovenous fistulas, ligating veins, or even placing tourniquets around limbs. The venous stasis created by these interventions stimulates bone growth distal to the occlusion. Long bones in growing animals attain greater lengths than their controls. They also tend to grow thicker. The latter effect is due to the increased deposition of periosteal bone on the surface, a response accompanied by endosteal resorption (Fig. 40) (23). Fracture healing in such bones is accelerated. How one explains these phenomena, unless it is by the effects of elevated temperature or reduced oxygen, is not known.

THE UNSOLVED PROBLEM OF BONE ELONGATION

Throughout the years of research on bone growth, no facet of the problem has resisted explanation more successfully than has the problem of how the elongation of bones is regulated. Although numerous physiological influences are known to affect elongation, the fact remains that the growth of bones in length is a remarkably autonomous process.

Elongation can be inhibited by a variety of methods which are known to interfere with the growth of the cartilaginous plate. As noted earlier, its direct constraint by stapling or by passing a wire loop through the diaphysis and epiphysis is very effective in halting elongation (16, 46). If removed and transplanted back into the bone upside down, the growth of the cartilaginous plate is likewise arrested (41). X-Irradiation interferes with the proliferative potential of its cells. It also decreases the uptake of ^{32}P and the activity of phosphatase while increasing the number of osteoclasts in the cartilaginous plate, the net result of which is the histological disorganization of the cartilage thus retarding further growth of the bone (7). The injection of crude papain also interferes with the elongation of bones (17). This proteolytic enzyme depletes the chondromucoprotein in cartilage, as well as its sulfur content, causing a loss in metachromasia and basophilia. In young rabbits, for example, it has the interesting effect of causing the ears to become floppy owing to the lack of cartilaginous rigidity (49). In long bones, papain brings about the premature closure of the cartilaginous plates, thus inhibiting the lengthening of bones. Somatotrophin is necessary to promote growth in the cartilaginous plate, an effect demonstrated by hypophysectomy and reversed by the administration of the missing hormone. Exposure to low oxygen tensions also retards the elongation of bones. Clearly, skeletal growth is a sensitive process easily interfered with.

More instructive are the mechanisms by which it can be accelerated. As already noted, the elongation of bones may be enhanced by such experimental techniques as centrifugation, growth hormone, hyperoxia, and venous stasis. Increased exercise will also promote excess growth of bones in length, a fact that may explain the extra elongation of the femurs in bipedal rats forced to walk on their hind legs following the amputation of their forelimbs in infancy (5). Especially interesting results have come from experiments involving the stripping away of the periosteum from the shaft of a long bone. Under these circumstances the elongation of such bones is accelerated, suggesting that tensions normally exerted by the fibrous tissue of the periosteum might normally be communicated to the cartilaginous plate thus slowing its growth (19). If tendons to the bone are cut, or the limb amputated through the bone, increased elongation also ensues perhaps owing to the severance of the periosteum. Trauma to the shaft of the bone by fracture or the insertion of ivory screws likewise accelerates elongation. This effect could be mediated either by disruption of the periosteum or by interrupting the blood flow.

Most long bones possess a cartilaginous plate at either end, the growth rates of which are not usually equivalent. Sometimes it is the proximal end of the bone that grows faster and sometimes the distal one. There is evidence, however, that the growth rates at the two ends of the bone may be to some extent dependent upon each other. If one cartilaginous plate is incapacitated by resection, stapling, fracturing, or X-irradiation, the other end compen-

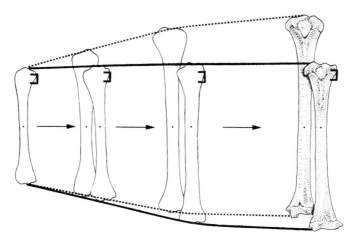

Fig. 41. Compensatory elongation of the rabbit tibia (*16*). Stapling the proximal epiphyseal plate arrests its growth (in comparison with controls shown to the left and behind the experimentals). The distal end of the tibia grows extra long in bones prevented from elongating proximally. Growth curves of control (broken lines) and experimental bones (solid lines) show the differences in elongation after 20, 60, and 140 days.

sates (*7, 16*). The increased growth at the opposite end of the bone does not completely make up for the deficiency, but its partial compensation suggests the existence of physiological communication from one end to the other (Fig. 41).

Despite the variety of ways in which the elongation of bones can be accelerated or retarded, the overriding fact remains that even under the most artificial conditions immature bones will continue to fulfill their growth potentials insofar as conditions permit. Even in tissue culture, embryonic bones continue to elongate until interrupted by the constraints of their avascular conditions. It may be asked to what extent the tissues with which the bone is normally attached play a role in regulating growth. To test this possibility, the bones of fetal or infant animals have been transplanted to other sites in the body, such as the brain, spleen, kidney capsule, or sucutaneous tissues. In these ectopic locations, such bones continue their development usually enlarging to dimensions which seldom equal, but compare favorably with, the normal sizes they were destined to attain (Fig. 42) (*9, 30, 36*). Although the growth of these bones is possible in both infant and adult hosts, there is evidence for faster growth in younger animals suggesting the importance of systemic, perhaps endocrine, influences (*36*). Nevertheless, the persistence of longitudinal growth, and even the preservation of the fracture healing abilities of transplanted bones (*6*), testifies to the essentially autonomous nature of bone growth.

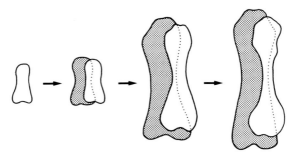

Fig. 42. Continued elongation of subcutaneously transplanted rat phalanges (unshaded) (9). Despite disuse, such bones continue to grow in length almost as much as controls (shaded). Grafted bones remain thinner than normal. Postimplantation intervals of 2, 34, and 200 days.

BROKEN BONES

Few phenomena of tissue repair are so impressive as the healing of fractures. Though this has much in common with processes of wound healing in other parts of the body, in terms of the considerable amounts of new material laid down it is almost an example of regeneration rather than tissue repair.

When an endochondral bone is broken, numerous blood vessels are ruptured and the ensuing hemorrhage results in the formation of a sometimes massive blood clot at the site of the fracture. If a callus is to fill the gap between the broken ends, then the clot must be removed by phagocytosis concomitant with the proliferation of callous cells. The histogenesis of the callus itself has been a bone of contention for many years. The demonstration that bones exposed to heavy doses of X-rays and transplanted to unirradiated tissues are not able to heal their fractures argues against the surrounding soft tissues as a source of callous cells. Conversely, when unirradiated bones are transplanted to irradiated sites they are capable of healing their fractures normally, suggesting that the callus is derived from tissues in the bone itself (6). Histological evidence emphasizes that the endosteum and periosteum are primarily responsible for giving rise to the callus. Soon after injury, the periosteum is normally lifted off the underlying bone by the hemorrhage. It is in such sites of separation that chondrification of the callus begins. Farther away from the site of injury where the periosteum remains in contact with the bone there is no healing response. It would appear not to be the separation of the periosteum from the bone *per se* which triggers the callus forming response, however, for when a flap of periosteum is reflected from the parietal bone and imbedded in the nearby trapezius muscle it does not give rise to new cartilage or bone (31). Therefore, the proximity of the periosteum to preformed bony matrix seems to be an important prerequisite for callus formation.

Once a mass of undifferentiated cells accumulates in the fracture site and the clot is resorbed, chondrogenesis of the callus occurs. The cartilaginous callus is an outsized mass of tissue which acts as a provisional union between the two broken ends. If the gap between the ends is too great for fusion of the callous cells from both sides to take place, then nonunion results. The calluses thus produced from each fragment heal only to themselves, rounding off the broken ends to create a pseudarthrosis.

The cartilaginous callus undergoes endochondral ossification not unlike that which occurs during the original ontogeny of the bone. The innermost chondrocytes become hypertrophic, their matrix calcifies, and they are gradually eaten out by invading blood vessels. Osteoblasts of uncertain origin then replace the cartilaginous callus with bony tissue to form a more solid junction between the broken ends. Remodeling of the healed fracture is responsible for the resorption of excess peripheral bone.

The rate at which fracture healing occurs depends on the size of the bone and the species of animal. The small bones of laboratory rodents may complete the entire process of fracture healing within a few weeks. In larger animals such as humans, months may be required. Younger individuals, of course, may heal their broken bones more quickly than the elderly.

Fracture healing may be enhanced by a variety of conditions some of which, such as decreased blood flow, electrical stimulation, and hyperoxia, have already been alluded to. Parathyroidectomy, or calcitonin administration, may also promote fracture healing, especially in the osteogenic phase (59). The interposition of decalcified bone or the transplantation of bladder mucosa to a fracture site, both of which are capable of inducing ectopic ossification, have been shown to assist in the healing of fractures (1). The repair of bone can be retarded or inhibited by X-irradiation, the administration of cortisone, or ultimobranchialectomy. Healing may also be interfered with by experimental methods favoring condrification while inhibiting calcification. These include parathyroidectomy, hypoxia, and the repeated mobilization of a fracture. Neither denervation nor the transplantation of bones prevents the repair reaction.

Dermal bones are remarkedly deficient in their capacities for healing defects. Archeological evidence of trephining in humans testifies to the inability of the skull to fill in full thickness holes sawed through it, even in those individuals fortunate enough to have survived the immediate aftereffects of such primitive surgery (Fig. 43). Experiments in rats have confirmed that the healing of fractures or holes in the adult skull is very limited indeed. Cartilage is rarely produced; usually a fibrous scar is all that unites the broken surfaces, the formation of which is enhanced if the edges are in direct contact with each other (8). The skulls of infant rats are not without some capacity for regeneration, however. Resection of large portions of the cal-

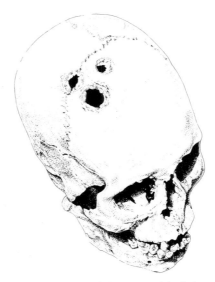

Fig. 43.   Trephined Inca skull. The remodeled margins of the holes testify to the postoperative survival of the individual. (Courtesy of Dr. George E. Erikson, Brown University.)

varium is followed by almost complete regeneration, but only provided the dura remains intact (47). Even this process, however, is inhibited in the absence of growth hormone.

The skull is composed of a number of flattened bones which originate independently at different locations but which expand toward each other during maturation. Suture lines as uneven as jigsaw puzzles mark the margins along which the occipital, parietal, and frontal bones unite. The eventual fusion along the suture lines evidently depends upon the growth pressure of the bones on either side, for when parts of cranial bones including their sutures are transplanted to intracerebral locations where compression is alleviated, no fusion occurs (35).

As the vault of the skull enlarges during maturation to accommodate the growing brain, considerable remodeling of the cranium is required, a remodeling that involves the sutural margins as well as the main substance of the bones. There is reason to believe that the abutment of one bone with another may establish the boundary conditions limiting the lateral expansion of each element. This interpretation is suggested by experiments in which one of the bones is removed in a young animal (34) or its margin cauterized to inhibit further growth (12). Under these circumstances the intact adjacent bone overgrows the territory it would normally have covered on the surface of the skull (Fig. 44). This evidence, coupled with the extra growth encountered in hydrocephalics, suggests that the so-called normal dimensions of the

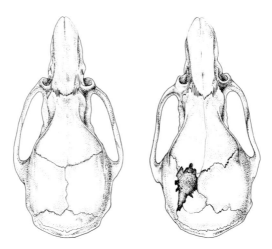

Fig. 44. Overgrowth of one parietal bone in the absence of the other (*12*). Left, control rat skull. Right, the skull of a rat 59 days after cauterization of the left parietal bone at 3 days of age. Its right parietal has expanded across the midline.

bones in the skull are not entirely predetermined by genetic constitution. Their areas are established and their boundaries fixed by a mutual interplay of forces not unlike the military confrontations responsible for determining boundaries between adjacent countries.

There is reason to suspect that the failure of holes in the skull to heal may in part be caused by the lack of chondrification in dermal bones. In broken long bones the early chondrification of the callus is an important step in fracture healing. Evidence that repeated mechanical stimulation can promote the differentiation of cartilage even in dermal bones is not inconsistent with the interpretation that the normal immobility of cranial defects might be correlated with their lack of chondrogenesis (*15*). Conversely, the greater mobility of fractured endochondral bones might in fact be causally related to the development of a cartilaginous callus in such sites. Indeed, the chondrogenic effects of repeated mobilization of fractured bones suggest that ossification may itself depend on the immobilization of the fracture provided by the establishment of a cartilaginous callus (*27*). To test these possibilities, it is necessary to inflict injuries on endochondral bones in such a way that the cut edges are immobile. This can be achieved by means of saw cuts or drill holes through only part of the diaphysis of a long bone, the remaining intact circumference of the shaft acting as an effective immobilizing splint. Analysis of healing in such defects confirms that little or no chondrogenesis takes place (*32, 38*). Instead, such callous cells as may be produced undergo direct ossification, not unlike the intramembranous type characterizing dermal bone formation. This does not explain why similar holes in the skull are not filled by new bone formation, but the proximity of underlying marrow in long bones may be a factor in promoting the induction of ossification in such defects.

THE REGROWTH OF CARTILAGE

The regenerative ability of cartilage suffers by comparison with that of bone. With few exceptions, defects in the cartilage of adults are repaired only by the reestablishment of periochondrium, rarely by the production of new cartilage to fill the gap. Embryonic cartilage, however, repairs itself more completely. Transected cartilaginous elements of embryos reunite by chondrogenesis. The cartilaginous nasal septum of infant rats, even if almost completely removed at birth, may fill in the excised area with new cartilage (25). Partial regeneration may occur after operation at 7 days of age, but little or no chondrogenesis ensues in 2-week-old rats. Only slight cartilage regeneration occurs in nasal septum defects in neonatal guinea pigs, presumably owing to the maturity of these animals at birth.

Cartilage regeneration in adults is a rare but interesting phenomenon, one which can best be observed in the regrowth of holes punched through the ears of rabbits (14, 21). In the vast majority of other mammals, full thickness holes through the pinna of the ear simply heal around the edges but remain patent. In the rabbit, the tissues form a regeneration blastema instead of a scar at the margins, a blastema whose cells proliferate as the wound margins converge eventually to obliterate the aperture (Fig. 45). Meanwhile, the internal tissues undergo chondrogenesis in continuity with the margins of the cartilaginous sheet sandwiched between the inner and outer layers of ear skin. Chondrogenesis does not occur in internal defects of ear cartilage not contiguous with healing epidermis. This may be taken to indicate that some interaction between the healing of the overlying epidermis and injury to the internal cartilage may be responsible for initiating the process of regeneration.

This exception to the general rule that mammals are not supposed to regenerate histologically complex appendages is not limited to the ears of the rabbit. Other lagomorphs, namely, hares and pikas, fill in ear holes in similar manner. Pikas are small animals which live in rock slides in mountainous terrains. Since they have short ears, one cannot attribute the unique regenerative ability of rabbits and hares to the fact that they have such relatively large ears. Indeed, it is now known that the domestic cat can heal holes punched in its external ear in much the same way. Bats, on the other hand, respond somewhat differently. Fruit bats, which fly by night vision,

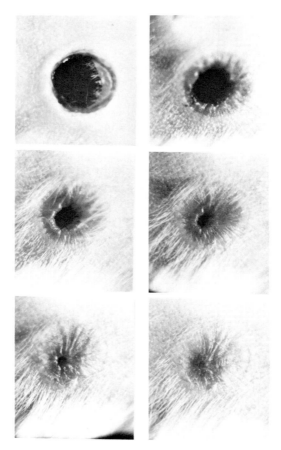

Fig. 45.   Successive stages at weekly intervals of the centripetal regeneration of tissue from the margins of a 1-cm hole through a rabbit ear. One week after injury the margins of the hole are swollen. By 2 weeks new tissue is seen growing in from the edges, and the aperture is closed by 6 weeks. The ingrowth includes restoration of the cartilaginous sheet.

are not capable of regenerating ear tissue, while species which fly by echolocation are. However, they do so by the centripetal growth of skin from the margins of the hole in the complete absence of new cartilage formation.

ECTOPIC OSSIFICATION

There are some tissues and organs of the body which are by nature predisposed to form bone. Ossification sometimes occurs in association with tendons, perhaps reflecting their intimate association with skeletal structures. It is also induced inside glass cylinders implanted subcutaneously in rats (Fig. 46) (44). Kidneys may form bone, especially under conditions of ischemia following arterial ligation (28). This may relate to their association with urothelium, a tissue well known to promote ectopic bone formation. The injection of zinc chloride into the testis also induces ossification. In arteries, various kinds of irritants are known to promote cartilage or bone formation in the tunica media. This sometimes occurs in association with arteriosclerosis or gangrene, and has been induced experimentally by treatment with $AgNO_3$, $CuSO_4$, or carrageenin. Mechanical factors may have similar effects, for if the arterial wall is pierced by a steel wire, cartilage may develop in nearby tissues, perhaps caused by the repetitious movements of the wire with each pulse beat. The occasional production of cartilage and bone in the heart itself may have much in common with that occurring in arteries. Skeletal differentiation occurs normally in the hearts of certain large ungulates. It can be induced experimentally in rats by tying off the apex of the ventricles where the ischemic tissues form a layer of bone during the next month or two (Fig. 47) (45). This may relate to the occasional discovery of bone at autopsy in association with prior coronary infarcts in humans.

The other way to investigate ectopic bone formation is to explore the ability of certain kinds of tissue grafts to induce bone when transplanted elsewhere in the body. It has long been known that grafts of the bladder wall can induce bone to form following transplantation to a variety of sites, includ-

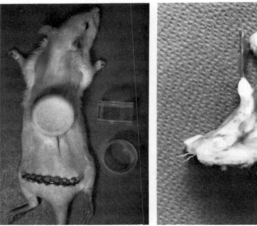

Fig. 46. Induction of bone, cartilage, and marrow within a subcutaneously implanted glass cylinder (44). Left, operated rat. Right, skeletal structure formed within the cylinder 60 days later.

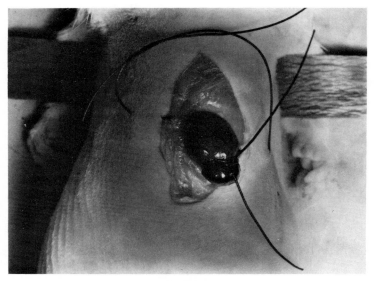

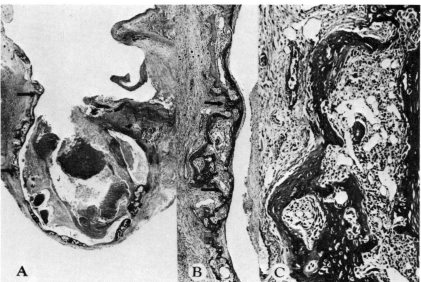

Fig. 47. Induction of bone in the rat heart (45). Above, heart is exterioized and the apex of its ventricles ligated. Below, two months later bone and cartilage are formed in association with cardiac connective tissue adjacent to the chambers. A, B, and C are progressively enlarged views of such areas as indicated by the arrows.

ing the granulation tissue of wounds, subcutaneous connective tissues, peritoneum, muscle, regenerating tendon, knee joint, skull defects, and fractured long bones. It is the urothelium in such grafts that is responsible for inducing the adjacent tissues at the transplantation site to undergo osteogenesis. Cartilage is not induced. The effect operates across a Millipore filter, but depends on the continued survival of the transplanted bladder. When bladder allografts are immunologically rejected, the bone they had induced eventually disappears (18).

Grafts of bone may likewise induce skeletal differentiation in surrounding tissues, but again the induced bone does not persist after the rejection of allografts. Nevertheless, studies have indicated that devitalized bone and cartilage grafts may be more effective in inducing the differentiation of cartilage and bone than are living tissues (Fig. 48) (51). Even demineralized dentine has been proved effective. In general, demineralization of the bony matrix of the graft promotes its capacity to induce the differentiation of cartilage and bone on its surface from adjacent tissue at the host site. Evidence suggests that there are substances emanating from such grafts that are

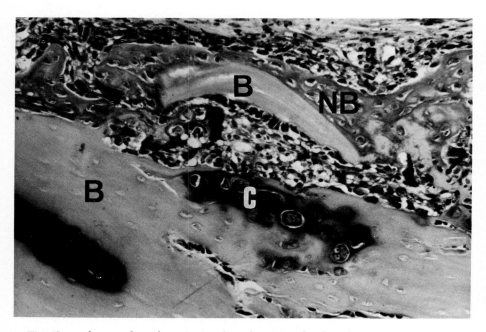

Fig. 48.   Induction of new bone (NB) and cartilage (C) under the influence of demineralized allogeneic rat bone matrix (B) three weeks after implantation into muscle. (Courtesy of Marshall R. Urist, M.D., University of California at Los Angeles School of Medicine.)

endowed with the remarkable capacity for effecting the metaplasia of surrounding tissue cells, deflecting their differentiation in the direction of cartilage, then bone (52). This reaction can even occur across cellulose acetate membranes (53). The nature of such substances, and their mode of action, is a promising field of investigation with far-reaching clinical implications.

REFERENCES

1. Bassett, C.A.L. Electrical effects in bone. *Sci. Am.* **213**(4), 18–25 (1965).
2. Beresford, W.A. The influence of castration on fracture repair in the penile bone of the rat. *J. Anat.* **112**, 19–26 (1972).
3. Brown, R.M., and Cameron, D.A. Acid hydrolases and bone resorption in the remodeling phase of the development of bony fracture callus. *Pathology* **6**, 53–61 (1974).
4. Burkhart, J.M., and Jowsey, J. Parathyroid and thyroid hormones in the development of immobilization osteoporosis. *Endocrinology* **81**, 1053–1062 (1967).
5. Colton, H.S. How bipedal habit affects the bones of the hind legs of the albino rat. *J. Exp. Zool.* **53**, 1–11 (1929).
6. Cooley, L.M., and Goss, R.J. The effects of transplantation and X-irradiation on the repair of fractured bones. *Am. J. Anat.* **102**, 167–182 (1958).
7. Dawson, A., and Kember, N.F. Compensatory growth in the rat tibia. *Cell Tissue Kinet.* **7**, 285–291 (1974).
8. Eggers, G.W.N., Shindler, T.O., and Pomerat, C.M. The influence of the contact-compression factor on osteogenesis in surgical fractures. *J. Bone Jt. Surg., Am. Vol.* **31**, 693–716 (1949).
9. Felts, W.J.L. Transplantation studies of factors in skeletal organogenesis. 1. The subcutaneously implanted immature long-bone of the rat and mouse. *Am. J. Phys. Anthropol.* **17**, 201–213 (1959).
10. Friedenberg, A.B., and Brighton, C.T. Electrical fracture healing. *Ann. N.Y. Acad. Sci.* **238**, 564–574 (1974).
11. Gilbert, G.H., and Gorman, H.A. Transplantation of urinary bladder mucosa for osteogenic effect. *J. Am. Vet. Med. Assoc.* **158**, 77–81 (1971).
12. Girgis, F.G., and Pritchard, J.J. Effects of skull damage on the development of sutural patterns in the rat. *J. Anat.* **92**, 39–51 (1958).
13. Goss, R.J. Aging versus growth. *Perspect. Biol. Med.* **17**, 485–494 (1974).
14. Goss, R.J., and Grimes, L.N. Tissue interactions in the regeneration of rabbit ear holes. *Am. Zool.* **12**, 151–157 (1972).
15. Hall, B.K. Cellular differentiation in skeletal tissues. *Biol. Rev.* **45**, 455–484 (1970).
16. Hall-Craggs, E.C.B., and Lawrence, C.A. The effect of epiphysial stapling on growth in length of the rabbit's tibia and femur. *J. Bone Jt. Surg. Br. Vol.* **51**, 359–365 (1969).
17. Hulth, A., and Westerborn, O. The effect of crude papain on the epiphysial cartilage of laboratory animals. *J. Bone Jt. Surg., Br. Vol.* **41**, 836–847 (1959).
18. Ioseliani, D.G. The use of tritiated thymidine in the study of bone induction by transitional epithelium. *Clin. Orthop. Relat. Res.* **88**, 183–196 (1972).
19. Jenkins, D.H.R., Cheng, D.H. F., and Hodgson, A.R. Stimulation of bone growth by periosteal stripping. *J. Bone Jt. Surg., Br. Vol.* **57**, 482–484 (1975).
20. Jenkins, D.P., and Cochran, T.H. Osteoporosis: The dramatic effect of disuse of an extremity. *Clin. Orthop. Relat. Res.* **64**, 128–134 (1969).

21. Joseph, J., and Dyson, M. Tissue replacement in the rabbit's ear. *Br. J. Surg.* **53**, 372–380 (1966).

22. Jowsey, J. Studies of Haversian systems in man and some animals. *J. Anat.* **100**, 857–864 (1966).

23. Kelly, P.J. The effects of thyroid and parathyroid deficiency on bone remodelling distal to a venous tourniquet. *J. Anat.* **110**, 349–361 (1971).

24. Kharmosh, O., and Savile, P.D. The effect of motor denervation on muscle and bone in the rabbit's hind limb. *Acta Orthop. Scand.* **36**, 361–370 (1965).

25. Kvinnsland, S., and Breistein, L. Regeneration of the cartilaginous nasal septum in the rat, after resection. Its influence on facial growth. *Plast. Reconstr. Surg.* **51**, 190–195 (1973).

26. Levy, D.D. A pulsed electrical stimulation technique for inducing bone growth. *Ann. N.Y. Acad. Sci.* **238**, 478–490 (1974).

27. Lindholm, R.V., Lindholm, T.S., Toikkanen, S., and Leino, R. Effect of forced interfragmental movements on the healing of tibial fractures in rats. *Acta Orthop. Scand.* **40**, 721–728 (1970).

28. Lindholm, T.S., Hackman, R., and Lindholm, R.V. Histodynamics of experimental heterotopic osteogenesis by transitional epithelium. *Acta Chir. Scand.* **139**, 617–623 (1973).

29. McMaster, J.H., and Weinert, C.R., Jr. Effects of mechanical forces on growing cartilage. *Clin. Orthop. Relat. Res.* **72**, 308–314 (1970).

30. Meikle, M.C. The influence of function on chondrogenesis at the epiphyseal cartilage of a growing long bone. *Anat. Rec.* **182**, 387–400 (1975).

31. Melcher, A.H., and Accursi, G.E. Osteogenic capacity of periosteal and osteoperiosteal flaps elevated from the parietal bone of the rat. *Arch. Oral Biol.* **16**, 573–580 (1971).

32. Mindell, E.R., Rodbard, S., and Kwasman, B.G. Chondrogenesis in bone repair. A study of the healing fracture callus in the rat. *Clin. Orthop. Relat. Res.* **79**, 187–196 (1971).

33. Moore, W.J. Masticatory function and skull growth. *J. Zool.* **146**, 123–131 (1965).

34. Moss, M.L. Growth of the calvaria in the rat. The determination of osseous morphology. *Am. J. Anat.* **94**, 333–362 (1954).

35. Moss, M.L. Fusion of the frontal suture in the rat. *Am. J. Anat.* **102**, 141–166 (1958).

36. Noel, J.F., and Wright, E.A. The growth of transplanted mouse vertebrae. Effects of transplantation under the renal capsule, and the relationship between the rate of growth of the transplant and the age of the host. *J. Embryol. Exp. Morphol.* **28**, 633–645 (1972).

37. O'Connor, B.T., Charlton, H.M., Currey, J.D., Kirby, D.R.S., and Woods, C. Effects of electric current on bone *in vivo*. *Nature (London)* **222**, 162–163 (1969).

38. Radden, B.G., and Fullmer, H.M. Morphological and histochemical studies of bone repair in the rat. *Arch. Oral Biol.* **14**, 1243–1252 (1969).

39. Redden, D.R. Chronic acceleration effects on bone development in the chick embryo. *Am. J. Physiol.* **218**, 310–313 (1970).

40. Reisenfeld, A. The adaptive mandible. An experimental study. *Acta Anat.* **72**, 246–262 (1969).

41. Ring, P.A. Excision and reimplantation of the epiphyseal cartilage of the rabbit. *J. Anat.* **89**, 231–237 (1955).

42. Robertson, D.R. The ultimobranchial body of *Rana pipiens*. VIII. Effects of extirpation upon calcium distribution and bone cell types. *Gen. Comp. Endocrinol.* **12**, 479–490 (1969).

43. Sarnat, B.G., and Shanedling, P.D. Orbital growth after evisceration or enucleation without and with implants. *Acta Anat.* **82**, 497–511 (1972).

44. Selye, H., Lemire, Y., and Bajusz, E. Induction of bone, cartilage and hemopoietic tissue by subcutaneously implanted tissue diaphragms. *Wilhelm Roux' Arch. Entwicklungsmech. Org.* **151**, 572–585 (1960).

45. Selye, H., Mahajan, S., and Mahajan, R.S. Histogenesis of experimentally induced myositis ossificans in the heart. *Am. Heart J.* **73**, 195–201 (1967).

46. Sijbrandij, S. Inhibition of tibial growth by means of compression of its proximal epiphysial disc in the rabbit. *Acta Anat.* **55**, 278–285 (1963).

47. Simpson, M.E., Van Dyke, D.C., Asling, C.W., and Evans, H.M. Regeneration of the calvarium in young normal and growth hormone-treated hypophysectomized rats. *Anat. Rec.* **115**, 615–625 (1953).

48. Stubbs, D. Skeletal function and weightlessness. A mechanism for hypogravic skeletal atrophy. *Aerospace Med.* **41**, 1126–1128 (1970).

49. Thomas, L., McCluskey, R.T., Potter, J.L., and Weissmann, G. Comparison of the effects of papain and vitamin A on cartilage. I. The effects on rabbits. *J. Exp. Med.* **111**, 705–718 (1960).

50. Tulloh, N.M., and Romberg, B. An effect of gravity on bone development in lambs. *Nature (London)* **200**, 438–439 (1963).

51. Urist, M.R. The substratum for bone morphogenesis. *Dev. Biol. Suppl.* **4**, 125–163 (1970).

52. Urist, M.R., Iwata, H., and Strates, B.S. Bone morphogenetic protein and proteinase in the guinea pig. *Clin. Orthop. Relat. Res.* **85**, 275–290 (1972).

53. Urist, M.R., Granstein, R., Nogami, H., Svenson, L., and Murphy, R. Transmembrane bone morphogenesis across multiple-walled diffusion chambers. New evidence for a diffusible bone morphogenetic property. *Arch. Surg.* **112**, 612–691 (1977).

54. Walker, D.G. Control of bone resorption by hematopoietic tissue. The induction and reversal of congenital osteopetrosis in mice through use of bone marrow and spleen transplants. *J. Exp. Med.* **142**, 651–663 (1975).

55. Washburn, S.L., and Detwiler, S.R. An experiment bearing on the problems of physical anthropology. *Am. J. Phys. Anthropol.* [NS]**1**, 171–190 (1943).

56. Wolffson, D.M. Scapula shape and muscle function, with special reference to the vertebral border. *Am. J. Phys. Anthropol.* [NS]**8**, 331–341 (1950).

57. Wunder, C.C., Briney, S.R., Kral, M., and Skaugstad, C. Growth of mouse femurs during continual centrifugation. *Nature (London)* **188**, 151–152 (1960).

58. Young, R.W. The influence of cranial contents on postnatal growth of the skull in the rat. *Am. J. Anat.* **105**, 383–415 (1959).

59. Zeigler, R., and Delling, G. Effect of calcitonin on the regeneration of a circumscribed bone defect (bored hole in the rat tibia). *Acta Endocrinol.* **69**, 497–506 (1972).

5

Turnover of Blood Cells

Poietin: Gr., *poiein*, to make

ERYTHROPOIESIS

The ideal tissue in which to study growth and proliferation would be one that could be sampled from time to time without triggering a significant compensatory reaction, one bearing a specific molecule capable of being quantitated colorimetrically and preferably containing an element of diagnostic nature that could be isotopically labeled. Erythropoietic tissues fulfill these criteria.

Few cells in the body are so specialized for a single function as are erythrocytes. With over 90% of their protein in the form of respiratory pigment, and lacking all cytoplasmic organelles, these cells are stripped-down models equipped for little more than oxygen transport. Were it not for the presence of enzymes and their metabolic activity, there would be little reason to believe that they were alive at all, especially in the absence of a nucleus. Even in the nucleated erythrocytes of nonmammalian vertebrates, the lack of ribosomes precludes the possibility of protein synthesis in fully differentiated cells. It is interesting to note that the capacity for protein turnover, so much an integral part of other cells, has been dispensed with in erythrocytes in favor of renewal at the cell level.

Red Cell Life Span

Unable to synthesize hemoglobin or even to repair damage to itself, the days of a red blood cell are numbered (65, 81). Whether its eventual demise is the result of wear and tear in the relentless rush hour traffic of the circulatory system, or to the programmed depreciation of the aging process, cannot be categorically determined. The fact remains, however, that the average red cell life span tends to be species-specific. For example, labeling studies have shown that red cells survive approximately four months in man, three months in sheep, about two months in the rat and rabbit, and around one month in the mouse. In pigeons and chickens, red cells may survive for five weeks or more. Marmot erythrocytes have a comparable life span during the summer months, but they live longer during hibernation. In cold-blooded vertebrates, which are at the mercy of the seasons, there is reason to suspect that red cells may survive the better part of a year. The hematocrit of frogs is maximal in the fall and minimal in the spring when there is a burst of erythropoiesis in the spleen, suggesting the possibility of a nearly wholesale turnover in the red cell population each year (37). In larvae the turnover may be more rapid, for red cells have been shown to survive for only about 100 days in bullfrog tadpoles (27). At metamorphosis, amphibian red cells are completely replaced as hemoglobin synthesis switches from the larval to the adult type. So completely different are these two varieties of hemoglobin that they exhibit no immunological cross-reactions, nor do they coexist in the same cell (52, 84). This change in hemoglobin reflects a shift in the sites of red cell production. Erythrocytes are produced primarily in the liver of tadpoles but in the bone marrow of postmetamorphic frogs. In the urodele, *Triturus*, the conversion from larval to adult hemoglobin synthesis takes place during the transformation of the larva into a red eft, when gills are resorbed and lungs develop, (not at the prolactin-induced eft-to-newt metamorphosis characterized by sexual maturation and a return to water) (13).

A somewhat comparable situation prevails during mammalian development (8). Here there are at least three different kinds of hemoglobins characteristic of the embryo, fetus, and adult, presumably reflecting shifts in the sites of erythropoiesis as well as adaptations to changing physiological conditions during prenatal and postnatal development. These changes in hemoglobin types are not abrupt ones. Although about three-fourths of the red cells in the newborn human contain fetal hemoglobin, adult hemoglobin may be detected much earlier in pregnancy. Similarly, fetal hemoglobin may linger several months after birth as one population of cells gradually replaces the other. Unlike our amphibian ancestors, the changeover in hemoglobin types may result in the presence of fetal and adult hemoglobin in one and the same cell (84).

Erythropoietin Production

The phenomenal rate at which red cells are renewed, approaching 2–3 million cells per second in humans, requires a control mechanism precise enough to hold the number of circulating cells constant. Clearly, there must be some way to monitor the number of peripheral cells as well as transmit the signal to the stem cells in the marrow or spleen. In 1906, Carnot and Deflandre reasoned that whatever regulates erythropoiesis might logically be carried in the plasma (17). To test their hypothesis, they bled some rabbits and later recovered the plasma and injected it into normal assay animals where it was found to accelerate erythropoiesis. The factor responsible for this reaction they called "hémopoïétine." Further research into the nature of factors regulating red cell production was not to be resumed until after World War II when hematologists became acutely aware of how little was known about the kinetics of red cell replacement in blood donors. Boycott had shown in 1934 that young rats and rabbits made good the loss of blood more rapidly than did adults, and that the larger of these two animals regenerated its blood cells more slowly than the smaller (11). Studies on humans in 1942 confirmed these findings (28). In general, the rat may replace lost blood in a week or so, while a rabbit requires several weeks and humans may take up to two months, depending on how much blood was removed in the first place. With the availability of isotopic tracers, it later became possible to quantitate the rate of hemoglobin synthesis by means of ^{59}Fe incorporation into differentiating erythroblasts. Hemoglobin synthesis commences in the marrow in nucleated precursors of red cells and even continues for a time after extrusion of the nucleus. The resulting reticulocyte continues to take up ^{59}Fe into the hemoglobin it synthesizes for about one day after its release into the peripheral circulation, a process it can carry on owing to the long-lived mRNA which persists after the nucleus is gone.

Techniques such as this have made it possible to prove that erythropoiesis is stimulated by a blood-borne factor, erythropoietin. Thus, the loss of blood is followed after about 18 hours by a marked increase in the incorporation of ^{59}Fe into the marrow. It takes a few days, however, for the descendants of stem cells, when stimulated by erythropoietin, to proliferate and differentiate into reticulocytes in the circulation. Hence, the increase in reticulocyte count, together with the elevated hematocrit, lags a couple of days behind the rise in ^{59}Fe incorporation. The accuracy of such assays, however, is reduced by the fact that animals are normally engaged in erythropoiesis at all times. Accordingly, techniques were developed to turn off erythropoiesis, thus reducing the background level of hemoglobin synthesis to nil, permitting assessment of a response measured from a base line of zero. Prior starvation of an assay animal reduces, but does not abolish, erythropoiesis. Exposure to hypoxic conditions for a few weeks, however, stimulates the production of excess red cells. When returned to atmospheric conditions, the plethora of red cells persists for the duration of their life spans and the hemopoietic tissues of such polycythemic animals cease to produce red cells until their peripheral counts return to normal. Similar results can be produced by direct transfusion of excess erythrocytes by injection into the peritoneal cavity where they are subsequently taken up into the circulation. Hypertransfusion polycythemia also abolishes erythropoiesis in animals used to assay erythropoietin.

The rate of erythropoiesis and the production of erythropoietin upon which it depends are closely bound up with oxygen supply and demand (41). When the number of circulating red cells is experimentally reduced (either by bleeding or by hemolysis following the injection of phenylhydrazine) oxygen supply is decreased although its demand by the tissues of the body may remain normal. This imbalance results in an increase in the production of erythropoietin. Comparable results can be achieved by exposure to hypoxia. Under these circumstances, the number of circulating red cells is "normal" but the decrease in oxygen supply elicits an erythropoietic response leading eventually to a degree of polycythemia proportional to the magnitude and duration of hypoxic exposure. Hypoxia is just as effective when the oxygen tension is reduced at atmospheric conditions as it is at high altitude. In sea level animals exposure to high altitude may increase the red cell count as much as 50% above normal (15), although the oxygen capacity per cell remains unchanged. Animals endemic to high altitudes, such as the llama, not only have extra-high red cell counts, but also have hemoglobin molecules with a greater affinity for oxygen. The latter adaptation is genetic, the former physiological (79).

Another way to increase erythropoiesis is to hold the supply of oxygen constant while increasing the demands for it by the tissues of the body. Triiodothyronine, for example, stimulates metabolic activity, thus creating a greater respiratory demand for oxygen. Not surprisingly, this triggers an increase in erythropoietin formation and red cell production, even in the absence of a decrease in the oxygen supply or a reduction in the number of circulating red cells.

The reverse conditions produce opposite effects. When the demand for oxygen is lowered but the supply remains normal, the rate of erythropoiesis declines. This can be achieved by acute starvation or by hypophysectomy, thus lowering the metabolic rate and creating a physiological situation in which the oxygen-carrying capacity of the blood exceeds the need. Erythropoietin production drops, erythropoiesis is reduced, and the population of circulating red cells gradually declines to otherwise anemic levels.

The rate of erythropoiesis also falls when the supply of oxygen is increased while the demand remains normal. Hyperoxia tips the scales in this direction by creating a physiological superabundance of red cells. The same effect is produced in transfusion-induced polycythemia. In both of these situations, more oxygen is delivered to the tissues than they need as a result of which the production of erythropoietin is turned off and the rate of erythropoiesis drops.

The foregoing experimental conditions emphasize the fact that it is neither the availability of oxygen nor the red cell count *per se* which is responsible for regulating the rate of red cell production. It is the relative need by the body for oxygen. So long as the demand exceeds the supply, erythropoiesis will be stimulated. It follows that there is no such thing as a "normal" hematocrit or red cell count except with reference to prevailing environmental conditions. Thus, the so-called normal human red cell count of approximately $5 \times 10^6/mm^3$ is not genetically determined, for it is no more constant than is the boiling point of water. In the course of evolution, efficiency has been enhanced not by a genetically fixed number of circulating red cells but by inheriting the ability to adjust this number to variations in oxygen supply and demand.

Erythropoiesis-regulating agents circulate in the plasma. Although much attention has been focused on the erythropoiesis stimulating factor (ESF), or erythropoietin, there is also evidence for the existence of inhibitors of red cell production. The latter have been found in the plasma of individuals normally acclimated to high altitude following their descent to sea level, as well as in the plasmas of newborn infants, patients with polycythemia vera and hypertransfused rabbits (62, 92). Neither the source nor the mode of action of such factors is known, but it would seem that erythropoietic inhibition may not be due solely to the lack of erythropoietin.

In contrast, more is known about erythropoietin than probably any other growth regulating factor with the possible exception of tropic hormones from the pituitary. It is present in a wide variety of species from fishes to mammals, and erythropoietin produced in one class of vertebrates is sometimes, but not always, found to be effective in another (102). Though undoubtedly present in normal serum at concentrations too low to be detectable by existing assay techniques, it is readily found in the plasma of animals following bleeding or phenylhydrazine-induced hemolysis. It is present in the plasma of animals exposed to high altitude, and in certain kinds of pathological anemias. In thalassemia, or Cooley's anemia, there is a high concentration of erythropoietin in the plasma as well as in the urine. It is found in the plasma of animals made polycythemic by treatment with cobalt, a respiratory poison that creates an artificial demand for oxygen (93). Polycythemia vera is also characterized by the production of abnormally high concentrations of erythropoietin. The plasma of fetal animals contains higher concentrations of erythropoietin than does that of normal adults, presumably owing to their more rapid rates of erythropoiesis.

It is not detectable by ordinary techniques in the plasma or urine of normal individuals nor in that of animals made polycythemic by hypertransfusion. Although anemic animals may have considerable erythropoietin in their plasma, it is absent in such cases following exposure to 100% oxygen (43). Erythropoietin fails to be detected in extracts of a wide variety of organs and tissues, including liver, spleen, lung, muscle, bone marrow, and kidney.

Analyses have shown that erythropoietin is stable except at the extremes of the pH range, suggesting that it is not a protein. This is confirmed by the fact that it is not precipitated by perchloric acid, and only in high concentrations of ethanol or ammonium sulfate. Nevertheless, it is digested by proteolytic enzymes, migrates electrophoretically as an α-globulin, and is nondialyzable. It can be hydrolyzed to a variety of monosaccharides, one of the most important of which is neuraminic acid. This evidence, combined with its metachromasia when stained with toluidine blue, suggests that it is a mucoprotein. Its molecular weight may be as high as 60,000 (47, 64).

The quest for the source of erythropoietin has required some intricate experimental detective work, most of which points to the kidney as the primary site of erythropoietin production. Bilateral nephrectomy not only retards the rate of erythropoiesis but reduces the normal rise of erythropoietin in response to hypoxia (87). Administration of $CoCl_2$, which normally promotes erythropoietin production leading to polycythemia, fails to cause either of these responses in nephrectomized animals (60). None of these effects is inhibited by ligating the ureters. The renal origin of erythropoietin is further substantiated by experiments in which the renal artery has been partially constricted or plastic microspheres injected into the kidney to clog the capillaries. The resulting ischemia and renal infarcts lead to the production of extra-high levels of erythropoietin in the plasma (2). This effect suggests that the kidney possesses cells sensitive to the availability of oxygen, cells that mediate the synthesis of erythropoietin under conditions of oxygen deficiency. The fact that homogenates of kidneys lack erythropoietic activity (22), however, indicates that the renal factor may only be a precursor molecule. This has been confirmed by studies showing that if serum is added to extracts of kidneys from hypoxic animals, the combination possesses erythropoietic activity (101). Moreover, perfusion of kidneys with serum likewise yields an erythropoietically active factor (22). Finally, Wilms' tumor, a renal adenocarcinoma, is commonly responsible for the production of extra-high levels of erythropoietin (63). This would suggest that the tubular epithelial cells of the kidney may be a source of erythropoietin. It has been shown, however, that isolated glomeruli from the kidneys of goats will also produce erythropoietin when maintained in culture (14), as will the renal medulla grown *in vitro*. More than one cell type would appear to be involved.

Although much evidence favors the renal origin for erythropoietin, the fact remains that erythropoiesis does not come to a complete halt in the absence of the kidneys. Even anephric patients maintained on hemodialysis do not remain anemic. This evidence supports the notion that the kidney may not be the only organ in the body capable of producing erythropoietin, although it is undoubtedly the most important. Attempts to pinpoint the extrarenal origins of erythropoietin have disclosed that even after bilateral nephrectomy an animal can still produce some erythropoietin in response to hypoxia (76). When such animals are partially hepatectomized, however, this reaction is abolished.

Once the kidney had been identified as the principal source of erythropoietin, some investigators realized the possibility that this might be related to, if not equivalent with, the production of renin by the juxtaglomerular apparatus (JGA). This was supported by studies showing that both urine and plasma with erythropoietic activity also possess the ability to raise the blood pressure. Injection of renin into rats increases the concentration of erythropoietin and angiotensinogen in the plasma (31), and JGA hypertrophy in humans has been associated with heightened levels of erythropoietin, as well as polycythemia. Moreover, stenosis of the renal artery in rabbits not only promotes renin secretion by the JGA, but tends to accelerate erythropoiesis in rats treated with plasma from such animals. On the negative side, however, increased erythropoietin production is not correlated with alterations in the JGA, nor does the administration of erythropoietin affect these cells (25). Injection of angiotensin II fails to stimulate hemoglobin synthesis nor does renin affect the levels of erythropoietin in the plasma. The contradictory nature of evidence relating to renin and erythropoietin does not inspire confidence in the possibility that the two may be physiologically related.

One can only speculate as to why the source of erythropoietin should be localized in the kidney, and perhaps in the liver too. The alternative would be to have the hemopoietic tissues directly responsive to the balance between oxygen supply and demand thus eliminating the need for a humoral factor such as erythropoietin. A disadvantage of such an arrangement would be that the very tissue responsible for correcting a deficiency in oxygen would have to undergo proliferation and differentiation at the same time it would be handicapped by lowered metabolism. It is more logical to locate the control mechanism elsewhere in the body. The cells of the kidney would be particularly suited to this task if their sensitivity to oxygen were more acute than the cells of other organs. There may also be some phylogenetic advantage to localize more than one vital function in the same organ, the strategy being that an animals's survival is just as seriously jeopardized whether it is deprived of erythropoietin, renin, or excretory capacities. Still another explanation for this arrangement might relate to the phylogenetic origin of hemopoietic tissues. Prior to the evolution of marrow cavities, blood cells were produced in some of the visceral organs, notably the spleen, kidney, and liver. It still goes on in these sites in fishes, as well as in the immature stages of higher vertebrates. When hemopoiesis moved out of the liver and kidney to more radio-opaque recesses of the skeletal system, the control centers may have remained in the original organs.

The Stimulation of Red Cell Production

Whether erythropoietin exerts its effects on erythropoiesis by stimulating differentiation or proliferation of hemocytoblasts is a difficult question to answer. The reason is that erythroid precursors normally divide repeatedly during the course of their differentiation. It is significant, however, that the target cell of erythropoietin appears to be the undifferentiated stem cell. Once differentiation is under way the cells are insensitive to erythropoietin, as the unresponsiveness of proerythroblasts and normoblasts testifies. Indeed, erythropoietin appears to be necessary only to initiate erythropoiesis, not to sustain it. Temporary exposure to hypoxia, or withdrawal and replacement of red cells, effects a pulse of erythropoietin production which stimulates some of the hemocytoblasts to embark upon the course of erythroid differentiation. Although the original stimulus may have been withdrawn, the affected cells complete their maturation into erythrocytes. Erythropoietin may interact with receptors on the stem cells (30) and cause the derepression of genes for hemoglobin synthesis. How the concomitant stimulation of DNA synthesis relates to the onset of differentiation remains uncertain.

Pursuant to the stimulation of erythropoiesis, it is worth noting that the role of erythropoietin can be bypassed under certain circumstances. Several viruses have been discovered that, when they infect mice, elicit an erythropoietic reaction. The polycythemic virus and the Friend virus both cause excessive production of red blood cells, but they do not do so by triggering the production of excess erythropoietin. Indeed, their stimulation of stem cell differentiation can occur in the absence of erythropoietin as, for example, in mice treated with antibodies to it (61). Since these viruses are equally effective on spleen or marrow cells in culture, they are believed to exert their effects directly on hemocytoblasts, perhaps by mimicking the action normally reserved for erythropoietin. In contrast, Rauscher virus stimulates erythropoiesis by increasing erythropoietin production (71).

Erythropoiesis in Infancy

The foregoing control mechanisms are those which prevail in adult animals. Immature ones exhibit some interesting differences. One is the postnatal anemia which characterizes the blood picture in a variety of animals from the rat to the human (90). The reason for this is that mammals experience a relative plethora of red cells when their systems are flooded with oxygen at birth. Their rates of erythropoiesis drop until such time as enough circulating erythrocytes disappear to reduce the peripheral count to a level commensurate with the respiratory needs of the infant.

Perhaps the most curious attribute of immature animals is the failure of their hemopoietic tissues to respond to hypoxia the way adults do. In chick embryos incubated under hypoxic conditions, neither the red cell count, the hematocrit, nor the hemoglobin content of the blood rises above normal during the first two weeks, although the area vasculosa may expand (42). Thereafter, the hemoglobin increases at low oxygen tensions and decreases at high ones. Moreover, treatment with cobalt exerts a lesser effect in embryos than in adult birds. It is not known if these embryos are responsive to erythropoietin, but it may be significant that toward the end of the second week of incubation the excretory functions of the embryo shift from the mesonephros to the metanephros.

A similar situation obtains in the case of infant rats. Exposure to hypoxic conditions fails to elicit erythropoiesis during the first month or so after birth (16). Another paradox is that bilateral nephrectomy of infant rats fails to alter the rate of erythropoiesis. Not until the animal is about 40 days old does its production of red cells become dependent upon the kidneys as in adults (51). These findings, however, should not be taken to indicate that infant erythropoiesis is an uncontrolled process. The rate of red cell production increases in response to the loss of blood, and even in the fetus a heightened erythroblastic reaction typically accompanies the destruction of Rh-positive fetal red cells by antibodies from an Rh-negative mother. Conversely, hypertransfusion of red cells into infant rats depresses their rates of erythropoiesis.

It is important to know if erythropoietin is normally produced by infants and whether or not their hemopoietic tissues are responsive to it. The answer to both questions is affirmative. Significant concentrations of erythropoietin can be recovered from fetal plasma. Moreover, the rate of erythropoiesis is increased by erythropoietin when injected into infant rats previously made polycythemic by hypertransfusion (100). The conclusion is inescapable, therefore, that there must be some extrarenal source of erythropoietin. By extrapolation from experiments on adults, the liver would seem to be the leading candidate for this role. What remains to be explained, however, is the unresponsiveness of infants to hypoxia. If it is not oxygen deficiency that triggers the production of erhythropoietin, it is a mystery what the nature of the feedback control might be.

Anemias

There is much to be learned by studying nature's mistakes, especially those cases of hereditary anemia, the causes of which are attributable to lesions somewhere in the sequence of events leading to the production of red blood cells. Anemia is a relative condition, for what is normal under one set of circumstances may be anemic or polycythemic under others. The most extreme case in point is the ice fish, an inhabitant of the supercooled waters of the Antarctic Ocean (85). This creature somehow survives in the total absence of erythrocytes. It does so presumably by virtue of its very low metabolic rate, its elevated cardiac output, its large gills, and the resistance of its tissues to hypoxia. The amount of oxygen it can carry in physical solution in the plasma is sufficient for its needs. Other species of Antarctic fishes contain red cells in their circulation, but even these may have fewer than those inhabiting less frigid waters. One wonders if the ice fish retains the potential for erythropoiesis and the capacity to produce erythropoietin.

The natural anemia of the ice fish is rivaled by the ability of salamanders and tadpoles to survive experimentally imposed total anemia (26). Treatment with appropriate doses of phenylhydrazine can hemolyze all of the animal's red cells, reducing its erythrocyte count to zero. Despite such a drastic hematological insult, the results are not fatal. Indeed, these amphibians can survive the month or more required to repopulate their blood streams with red cells, a process characterized by the proliferation and differentiation of precursors in the peripheral circulation itself (27, 32).

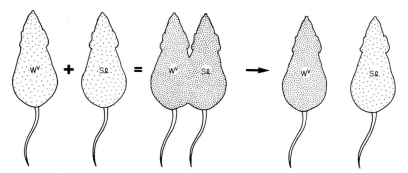

Fig. 49.   The effects of parabiosing mice with different kinds of anemia (Sl and W^v). Because each possesses what the other lacks, both are cured in parabiosis. When separated again, the W^v mouse remains cured because its defective hemocytoblasts have been replaced by normal Sl cells. The Sl mouse reverts to the original anemic condition because its deficiency is noncellular in nature and therefore not permanently reverse.

Of the many hereditary anemias which have been bred in mice, the W/W^v and the Sl/Sl^d strains have been particularly informative (7, 45). Both of these strains of mice are anemic, but for different reasons. Accordingly, when individuals of each kind are grafted together in parabiosis, both of their anemias are cured by virtue of the fact that each provides what the other lacks. When such mice are later separated the W^v individual remains cured while the Sl mouse reverts to its original anemic condition (Fig. 49). The recovery of the W^v mouse can be explained by the fact that normal hemopoietic cells are acquired from its parabiotic partner to repopulate its own hemopoietic tissues. In other words, the Sl stem cells compete with the host's own W^v cells in the marrow and spleen, eventually replacing them and giving rise to a chimeric W^v mouse with Sl blood cells in its circulation. The failure of the Sl anemic mouse to remain cured after separation from its parabiotic partner is less easily explained. The Sl anemia is not due to a deficiency in the hemopoietic stem cells, but in the conditions necessary for their normal development.

In most anemias the mechanisms which normally regulate erythropoiesis are not altogether absent. Exposure to hypoxia elicits an erythropoietic response in anemic animals as it does in normal ones. Since the red cell count is below par, however, the hypoxic response is correspondingly inadequate. Thus, the principal difference between the normal condition and hereditary anemias is that their "hemostats" are set at different levels.

In the case of the W^v anemic mouse treated with healthy stem cells by parabiosis or injection of marrow, the pattern of blood forming tissues is different from normal (54). Instead of repopulating the host's hemopoietic tissues in a random distribution, the new cells give rise to clones of descendants which develop into macroscopically visible colonies in the spleen. The number of spleen colonies formed is proportional to the number of marrow cells injected.

A similar phenomenon can be demonstrated in animals exposed to whole body irradiation to block the proliferation of their hemopoietic tissues. Injection of unirradiated replacement cells, derived from marrow, spleen or fetal liver, results in the development of spleen colonies (Fig. 50), most of which represent specific types of blood cells derived from single ancestral colony forming cells. As a result of the transplanted hemopoietic cells, the host is enabled to survive the otherwise lethal effects of the radiation. The transplanted cells are not rejected by the host, even if they are from a different species, because the immunological competence of the host is knocked out by the radiation. It is a perfect case of two wrongs making a right. Despite the extensive proliferation and differentiation of cells in the spleen colonies, enough of them retain the capacity to form new colonies when cells from the spleen of the first host are injected into a second irradiated animal (48).

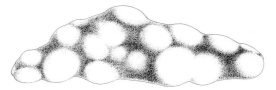

Fig. 50.   Spleen of irradiated mouse injected with healthy marrow cells. The latter give rise to clones which develop into hemopoietic spleen colonies, the externally visible ones being of the erythroid type.

Erythroid colonies can also be grown in cultures of marrow or spleen cells, although they do not reach the sizes attained by those *in vivo* (57). In either case, the growth of erythroid colonies is dependent upon the same regulatory factors that promote the growth of normal erythropoietic cells (*10, 70*). Their development is stimulated by exposure to hypoxia, injection of erythropoietin or infection with the polycythemic or Friend viruses. Conversely, erythroid colony formation *in vivo* is less impressive in healthy unirradiated hosts and is markedly suppressed in plethoric animals. Colony formation therefore flourishes under conditions conducive to erythropoiesis, but can take place *in vivo* only when competition with the resident population is eliminated by radiation or congenital anemia.

LEUKOPOIESIS

There are two categories of white blood cells, granular and nongranular. The former are produced primarily in the myeloid tissues of the bone marrow. The latter originate in the various lymphoid organs of the body. The functions of these different kinds of cells are as dissimilar as the tissues that produce them. The mechanisms by which their production is regulated may have equally little in common. It is only by histological accident that they mingle with one another in the peripheral circulation and are therefore collectively referred to as white blood cells.

Granulocyte Production

Leukocyte turnover is presumed to be controlled by mechanisms analogous to those known to regulate erythropoiesis, that is, by various leukopoietins (*96*). With respect to granulocytes, this may in fact be the case. Unfortunately, it is technically less feasible to analyze granulopoiesis by the methods applied so successfully to erythropoiesis, namely, by depletion of the peripheral population and treatment of assay animals with the resulting serum. Not only is it difficult to remove granulocytes selectively from the bloodstream (leukapheresis), but the lack of an easily monitored cell-specific molecule (such as hemoglobin) does not facilitate measuring the rate of granulocyte production. In spite of these handicaps, significant progress in the exploration of granulopoiesis has been made in recent years.

Much of what is known about granulocytes applies to their most common representative, the neutrophil (80, 88). This cell plays an important role in the body's defenses against local infections, leaving the bloodstream to take up positions in tissues adjacent to a lesion. The daily loss of countless neutrophils is balanced by their replenishment from the body's store of reserve cells in the marrow. This reservoir of granulocytes, many times larger than the number actually in circulation, is not easily depleted. Derived from stem cells in the marrow, the maturing neutrophil divides repeatedly during the early days of its differentiation. At the end of this proliferative phase its nucleus may undergo lobulation in acquiring the characteristic polymorphonuclear configuration of the mature neutrophil. Once it is released into the peripheral circulation its life span may be measured in hours or days. It is this rapid turnover of neutrophils that has apparently necessitated the establishment of such a sizable reserve of cells to be drawn upon in emergency when the peripheral needs are greater than can be met by the relatively slow process of granulopoiesis alone. Thus, there are two control mechanisms, one regulating the production of new cells, the other their actual release into the circulation.

The search for granulopoietins has depended upon the development of techniques for leukapheresis, preferably so as to deplete neutrophils selectively. The most direct method is to centrifuge repeatedly withdrawn samples of blood, discard the buffy coat containing white blood cells, and reinject the reconstituted erythrocytes. While large numbers of leukocytes can be removed in this way, such a technique is like sweeping back the tide because peripheral levels are promptly restored from the pool of noncirculating white blood cells (9, 40). X-Irradiation (94), pyrogens (97), and intravenous administration of starch (24) all induce a general leukopenia. More specific effects can be achieved by the intraperitoneal infusion of saline or glucose in .01 M HCl (23). Peritoneal exudates subsequently withdrawn from such animals contain high concentrations of neutrophils. When repeatedly carried out, this technique permits the marked depletion of these cells from the circulation. Still another approach is to administer antineutrophil serum to destroy the cells *in vivo* (50). With the exception of irradiation, these techniques elicit a granulopoietic response in the myeloid tissues. The effects of X-rays are more long-lasting due to the sensitivity of the stem cells to irradiation. Plasma from animals made neutropenic in a variety of ways has been shown to stimulate granulopoiesis in normal recipients into which it is injected, suggesting the presence of a granulopoietic factor (94).

The opposite approach to the problem is to promote granulocytosis, or the peripheral increase in the numbers of granular leukocytes. Histamine is known to cause neutrophilia and eosinophilia. The latter is also associated with parasitic infections. Although serum from such animals does not necessarily inhibit white cell production, the possibility of circulating factors capable of counteracting stimulators is an important part of the problem, especially in view of the granulopoietic effects of plasma from patients with leukemia. Of particular significance may be the findings that extracts of granular leukocytes are capable of inhibiting DNA synthesis in marrow cells (86).

Despite a number of interesting advances, it must be admitted that *in vivo* studies of granulopoiesis have not as yet yielded clear-cut answers to the question of how the differentiation and proliferation of granulocytes are controlled. The relative stability of the neutrophil count in the peripheral blood suggests that their numbers must somehow be monitored. It is not yet known whether it is the concentration of neutrophils, their total numbers in the circulation, or those in the circulation plus the multitudes sequestered in the marrow, to which the stem cells are responsive. There is evidence in favor of the production of stimulators under conditions of depletion, as well as inhibitors from mature neutrophils themselves. How such factors act on granulopoiesis in the stem cell compartment of the marrow is not known, but answers may be forthcoming from the investigation of granulocyte colonies.

If an animal's hemopoietic competence is knocked out by total body irradiation, the otherwise lethal effects can be avoided by injecting healthy marrow cells. As previously noted, such cells are not rejected as foreign tissue grafts because the host's immunological competence has been blocked by the irradiation, although graft-versus-host reactions may eventually supervene. Some of the healthy marrow cells injected into the irradiated recipient establish clonal colonies in the spleen. Although the most conspicuous colonies are erythroid, smaller ones are made up of neutrophils or eosinophils, while still others consist of megakaryocytes or macrophages (91). It is a curious thing that the development of various kinds of blood cells should be so compartmentalized under these conditions when during normal ontogeny they remain intermingled in the myeloid tissues. The histology of hemopoiesis would be greatly simplified if each cell type were to be produced in its own specific organ more or less equivalent to the spleen colonies. Nevertheless, granulocyte colonies in the spleen do not lend themselves to the close scrutiny and quantitative investigation upon which the detailed investigation of granulopoiesis depends.

It was a major breakthrough in this field when techniques were developed for growing such colonies *in vitro*. If hemopoietic cells are suspended in semisolid agar they give rise to colonies of descendants during the following week or two, each one of which may be made up of thousands of cells if the growth conditions are right (72, 103). Some colonies consist of macrophages, others of granulocytes, each one representing a clone of cells descended from a single colony-forming cell (CFC). Although CFC's are most abundant in the marrow, they can also be derived from the spleen and even from the circulating blood (103). CFC's are believed to be undifferentiated cells with narrower potentialities for differentiation than possessed by cells capable of giving rise to spleen colonies *in vivo*. Although two types of colonies are observed to develop in semisolid agar seeded with marrow cells, it is the granulocyte which appears to differentiate first and which in some cases transforms into macrophages. Accordingly, colonies of mixed cells are not uncommon.

Except for the earliest stages of proliferation, such colonies are incapable of growing independently and will die out in several days if certain growth factors are absent from the medium. Early investigations showed that colony growth is dependent upon a "feeder layer" of cells. Embryonic cells from a variety of organs are sufficient to promote colony growth by producing a colony stimulating factor (CSF) (59, 95) the continued presence of which is necessary for sustained colony growth. CSF is derived from a wide variety of sources. Various organs in the body, including kidney, lung, spleen, and especially the submaxillary salivary gland, can produce CSF. It is also present in urine, suggesting that it is cleared from the system via the kidneys. Normal serum also contains CSF, but serum from patients with leukemia or from animals rendered neutropenic by irradiation or treatment with antineutrophil serum is a particularly rich source. Chemically, it is a glycoprotein component of the α-globulin fraction of the serum. The impressive effects of CSF *in vitro* are not matched by *in vivo* responses, suggesting that it would be misleading to equate CSF with granulopoietin.

Antigenic Stimulation of Lymphopoiesis

The capacity for colony formation exhibted by other types of blood cells under experimental conditions is a natural attribute of the lymphocyte. Though lymphocytes may sometimes be produced in the myeloid tissues of the marrow, the majority originate in a variety of lymphatic organs throughout the body, the basic histological structure of which is the lymph nodule. The latter may occur singly or in unorganized clusters as found in association with the digestive tract (e.g., Peyer's patches, tonsils) or in lymph nodes where the nodules are organized around the lymphatic vasculature. In the spleen the lymph nodules, or white pulp, envelop arterioles. The thymus is made up of medullary and cortical tissue reminiscent of the histological organization of lymph nodules. Lymphoblasts are situated centrally in the lymph nodule. As they proliferate, their descendants decrease in size during the course of their differentiation while being displaced peripherally in the nodule, eventually to be released into the lymphatic circulation. The superficial resemblance between lymph nodules and spleen colonies of other blood cells is not easily ignored, but whether or not their cells are clonal is unknown. It is well established, however, that lymphocytes produced in the so-called secondary lymphatic organs were originally derived from the thymus during immature stages of development (98). In the rat and mouse, thymic lymphocytes colonize the other lymphatic organs during the first two weeks after birth. It is conceivable, though not proved, that each nodule may have originated from a single thymic lymphocyte.

The thymus is more important in the neonatal animal than it is in the adult. Thymectomy in adults is inconsequential. Neonatal thymectomy, however, has profound effects leading to immunological incompetence and the wasting disease that gives rise to runts. Runting is iteself a secondary symptom of the immunological deficiencies of thymectomized animals. If the thymus is removed from newborn mice maintained under germfree conditions, runting does not occur despite the poor development of their lymphatic organs. Not until such animals are exposed to conventional conditions do they display overt symptoms of their inabilities to combat infections, symptoms that also underlie their failure to reject foreign tissue grafts.

What is most unique about lymphocytes is their capacity to detect antigens and to mediate the production of specific antibodies. Whether this specificity is achieved by selection or instruction has been a matter for conjecture. The fact that each lymphocyte is responsible for the production of one kind of antibody favors the view that there might be a separate kind of lymphocyte for every conceivable antigen, the latter simply selecting those cells of a complementary type. In view of the numbers of cells that would have to be involved, however, this interpretation is less likely than one involving the modulation of lymphocytes by antigens to which they are exposed. In either case, antigenic stimulation results in the amplification of the population of antibody-producing cells by a series of mitotic episodes. Hence, subsequent exposure to the same antigen results in a more prompt and massive immunological reaction made possible by larger numbers of presensitized lymphocytes. Such cells are even capable of producing antibodies *in vitro* when exposed to their complementary antigens, but cells not previously sensitized can do so only *in vivo*. Other cells may therefore be required to effect the primary reaction to an antigenic stimulus. This conclusion is consistent with the fact that lymphocytes from genetically distinct sources will react to each other *in vitro* in mixed lymphocyte cultures (MLC). Such lymphocytes are transformed into lymphoblasts capable of proliferation. The intensity of this blastogenic transformation is a function of the genetic disparity between the donor sources (3), and is markedly enhanced if the animals have been previously sensitized to each other by exchange of tissue grafts (20). Blastogenesis depends upon the production into the medium of proteins responsible for provoking an immunological reaction.

Not all lymphocytes are involved in the MLC reaction. Cells from thymectomized individuals, for example, fail to undergo blastogenic transformation in MLC (46). Further, if marrow is exchanged between newborn animals of different strains, their lymphocytes do not react against each other in MLC after maturity, presumably owing to the tolerance between cells incapable of recognizing each other as foreign (89). The MLC technique is of clinical significance in assessing the immunological incompatibilities between potential organ donors and hosts by pretreating the donors' lymphocytes with mitomycin C. This arrests their capacity for DNA synthesis without interfering with their competence to stimulate blastogenesis in other lymphocytes in MLC. The intensity of the response can then be scored exclusively on the basis of how the lymphocytes from the prospective recipient react.

The response observed in MLC is a special case of a more general phenomenon. Whatever stimulates the antigenic sensitivity of lymphocytes in culture triggers a transformation characterized by cell enlargement, proliferation, and nucleic acid and protein synthesis. Lymphocytes exposed to antigens to which they have been previously sensitized become lymphoblasts which, in addition to proliferating, produce antibodies against the antigen in culture.

A number of substances extracted from plants have been found to elicit lymphocyte transformations similar to those seen following antigenic exposure (Fig. 51). These immunomimetic substances are phytomitogens inasmuch as they promote lymphocyte proliferation. The first of these to be discovered was phytohemagglutinin (PHA), a derivative of the kidney bean (77, 99). Comparable effects are produced by concanavalin A, isolated from the jack bean, and extracts of pokeweed and wisteria seeds (6). These and other substances stimulate lymphocytes in culture to revert to lymphoblasts capable of mitotic activity. The mitogenic effects of these stimulants is accompanied by the production of lysosomes and increased acid phosphatase activity. RNA synthesis may rise above normal as early as one hour after treatment, a reaction eventually followed by protein production. The rate of DNA synthesis is maximum after three days. Although such reactions resemble those following exposure to specific antigens, there is no evidence that lymphocytes produce antibodies against these nonspecific mitogens.

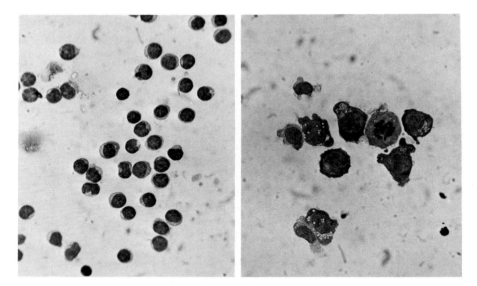

Fig. 51. Left, human peripheral lymphocytes grown *in vitro* without exposure to phytohemagglutinin (PHA). Right, the effect of PHA on similar lymphocytes after 72 hours in culture. The cells are enlarged, have acquired cytoplasmic basophilia, and are dividing. (Courtesy of Dr. Barbara Barker, Rhode Island Hospital.)

Nevertheless, their readiness to proliferate under such a wide variety of stimuli emphasizes the sensitivity of these cells to foreign substances.

The ease with which lymphocytes in culture can be stimulated to divide is reminiscent of the effects of CSF on granulocyte colony formation (38). The study of such phenomena has been useful in revealing the cellular mechanisms of proliferation. However impressive these studies may be, they do not necessarily explain how leukocyte proliferation is regulated in the body. Mitogenic substances that work *in vitro* do not necessarily exert the same effects *in vivo*. Therefore, we are still far from discovering the true nature of such leukopoietins as may exist, and until we do the regulation of leukocyte production will remain as challenging a problem as it always has been.

THROMBOPOIESIS

For reasons yet to be explained, mammals have taken the unprecedented step of dispensing with nuclear material not only from their red cells but also from their clotting cells. In all other vertebrates, including birds, thrombo-

cytes are nucleated, being produced in the conventional way by differentiation from hemocytoblasts. In amphibians, they are even capable of dividing
by mitosis in the peripheral circulation when the spleen, presumably their
normal site of origin, is removed (44).

Platelet Turnover

Platelets are produced from megakaryocytes, one of nature's most uncon-
ventional cells. Formed by fragmentation from the cytoplasm of the
megakaryocyte (Fig. 52), platelets represent virtually the only instance in
the animal kingdom in which such small, nonnucleated pieces of a cell are
capable of survival. As part of a renewing population, their turnover requires
a feedback communication from the differentiated compartment in the circulation to the germinative cells in the marrow. It might be anticipated that

Fig. 52.   Cultured mouse megakaryocyte in the process of fragmenting into platelets, one of
which (above) has separated from the megakaryocyte cytoplasm. Others are in various stages of
separating as demarcation channels develop (57).

platelet homeostasis would involve mechanisms not unlike those responsible for regulating the turnover of other blood elements.

To determine the life span of circulating platelets, it is necessary to label them. This can be done by incubating them with ^{51}Cr, ^{32}P (in diisopropylfluorophosphate, DFP), or ^{35}S (29). When such labeled platelets are infused back into the animal from which they came their rate of disappearance can be monitored in samples of blood taken at intervals thereafter. These techniques have shown that rat platelets may normally survive 4–5 days (39), while in humans they enjoy a life span twice as long (1). The removal of superannuated platelets from the blood occurs in the spleen where they are destroyed by phagocytosis. Hence, the enlargement or removal of the spleen has predictable effects on platelet counts. Splenomegaly, for example, can be produced experimentally by intraperitoneal injections of methylcellulose. Repeated treatments may triple the size of the spleen in just a few months. In such animals the platelet life span is shortened and their concentration in the circulating blood is decreased (39). Concomitantly, the rate at which megakaryocytes are produced in the marrow is increased (82). Splenectomy has the opposite effect of increasing platelet counts (74). It reverses the thrombocytopenia which occurs naturally in hibernating ground squirrels (78), or which is a pathological symptom of idiopathic thrombocytopenic purpura (ITP) (4). Remission of ITP by splenectomy (36) is characterized by lengthened platelet life spans, increased peripheral counts, and enhanced thrombopoiesis. Perhaps it is an oversimplification to explain these effects by the loss of the platelet graveyard. More subtle factors may be at play, for the administration of spleen extracts (both normal or from ITP patients) tends to decrease the peripheral platelet count (19).

Megakaryocyte Production

The control of the circulating platelet population must ultimately operate at the level of the megakaryocyte. There are two checkpoints. One is where platelets are released from mature megakaryocytes. The other controls the rate at which hemocytoblasts differentiate into platelet-producing cells. Once they embark on the pathway of differentiation, megakaryocytes do not amplify their numbers by proliferation. Instead, each cell replicates its DNA repeatedly. This leads to progressive increments of polyploidy, some cells reaching the 32N stage (67, 73). The nuclei become increasingly lobulated as the cytoplasmic mass grows to keep pace with the replicating nuclei. No other cell in the bone marrow is so large, nor so easily recognized, as the megakaryocyte. Its entire course of maturation may take place in only 24–42 hours, whereupon DNA synthesis ceases and the platelets are released. This self-immolation may occur as early as the 8N stage or as late as 32N (73).

Megakaryocyte production in the marrow is heightened when the population of circulating platelets is below normal (69). This can be brought about experimentally by depleting platelets from the blood (Fig. 53) or by inducing splenomegaly to lower the platelet count. It is also characteristic of ITP. The number of megakaryocytes may be decreased in the marrow if the platelet count is artificially elevated by hypertransfusion (68). There is convincing evidence, therefore, that the mechanism responsible for regulating the production of megakaryocytes is sensitive to the numbers of circulating platelets.

One way to explore the mechanisms by which thrombopoiesis is regulated is to deplete the numbers of circulating platelets artificially in order to stimulate their regeneration. The most direct way to achieve this is by thrombopheresis. Blood is withdrawn from an animal, centrifuged, and the plasma then allowed to coagulate. The clot, containing the platelets, is discarded, the blood cells are resuspended, and the defibrinated blood is then injected back into the animal. When performed repeatedly, the platelet counts can be reduced to about 10% of normal in this way. The megakaryocyte population in the marrow promptly increases (83), leading to the accelerated production of platelets. The rate at which this regeneration occurs is proportional to the degree of peripheral depletion (69). Within a few days there is an overshoot in the production of platelets which subsequently subsides to the normal range. A similar reaction ensues other methods of inducing experimental thrombocytopenia, as for example by treatment with antiplatelet serum or infection with Friend virus (18). Total body irradiation also brings about a drastic reduction in the platelet count (66), and although this is followed by an early increase in the numbers of megakaryocytes in the marrow, the thrombopoietic response is not sustained presumably due to the greater sensitivity of stem cells to X-rays compared with their descendants which had already begun to differentiate at the time of exposure.

Idiopathic Thrombocytopenic Purpura

Idiopathic thrombocytopenic purpura (ITP) is characterized by a marked reduction in the platelet count (4). This decrease is caused not by a lesion in the production facilities, but by the abbreviation of platelet life span. In patients with ITP the platelets may survive only a day or two instead of the normal 8–9 days (Fig. 54). It is of no use to transfuse extra platelets from healthy individuals because these too are promptly destroyed, suggesting that the problem may be systemic rather than a defect in the platelets themselves. The crucial breakthrough in discovering the etiology of ITP was achieved by injecting serum from ITP patients into healthy human volunteers (35). The results were dramatic, for their platelet counts promptly

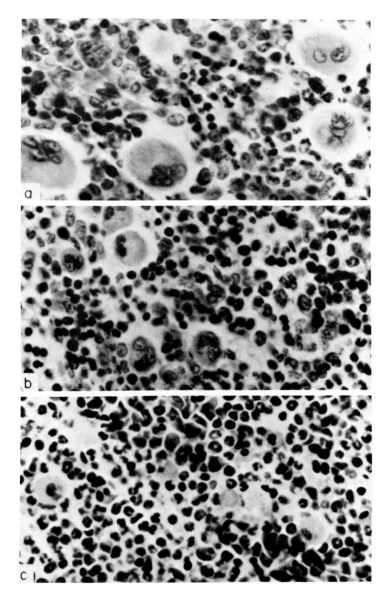

Fig. 53.   Sections of rat bone marrow showing (a) increased numbers of megakaryocytes following injection of anti-platelet serum to cause acute thrombocytopenia. (b) Normal marrow. (c) Reduced megakaryocyte population in a rat transfused with excess platelets (75).

the source of thrombopoietin, the explanation for this is as elusive as is its role in the production of erythropoietin.

One can only speculate about the mechanism by which thrombopoietin stimulates hemocytoblasts to differentiate into megakaryocytes in the marrow (21). Like other hemopoietins, it may derepress the genes for differentiation in that particular direction. Equally mysterious is the control of thrombopoietin production itself (Fig. 55). Whatever organ in the body may be responsible for its synthesis, there must be some way to monitor the population of platelets in the circulation. It would seem reasonable that thrombopoietin might be produced in response to physiological factors related to the functions of platelets. Of the many factors involved in the concatenation of events which leads to the formation of a blood clot, there are some that might be produced by the platelets themselves. If so, then a deficiency in one of these might trigger thrombopoietin production thereby restoring the platelet count and the production of clotting factors to normal. Clearly, the direction of future research on the problem of how thrombopoiesis is controlled must probe the relationship between platelet production and clotting factors, perhaps by testing the possible effects of anticoagulants.

REFERENCES

1. Aas, K.A., and Gardner, F.H. Survival of blood platelets labeled with chromium[51]. *J. Clin. Invest.* **37**, 1257–1268 (1958).
2. Abbrecht, P.H., Malvin, R.L., and Vander, A.J. Renal production of erythropoietin and renin after experimental kidney infarction. *Nature (London)* **211**, 1318–1319 (1966).
3. Bain, B., and Lowenstein, L. Genetic studies on the mixed leukocyte reaction. *Science* **145**, 1315–1316 (1964).
4. Baldini, M. Idiopathic thrombocytopenic purpura. *New Engl. J. Med.* **274**, 1245–1251, 1301–1306, 1360–1367 (1966).
5. Baldini, M.G., and Ebbe, S. (eds.) "Platelets: Production, Function, Transfusion and Storage." Grune & Stratton, New York, 1974.
6. Barker, B.E., and Farnes, P. Mitogenic property of *Wistaria floribunda* seeds. *Nature (London)* **215**, 659 (1967).
7. Bernstein, S. Modification of radiosensitivity of genetically anemic mice by implantation of blood-forming tissue. *Radiat. Res.* **20**, 695–702 (1963).
8. Betke, K. Ageing changes in blood cells. *In* "Structural Aspects of Ageing" (G.H. Bourne, ed.), pp. 227–245. Hafner, New York, 1961.
9. Bierman, H.R. Homeostasis of the blood cell elements. *In* "Functions of the Blood" (R.G. MacFarlane and A.H.T. Robb-Smith, eds.), pp. 349–418. Academic Press, New York, 1961.
10. Bleiberg, I., Liron, M., and Feldman, M. Reversion by erythropoietin of the suppression of erythroid clones caused by transfusion-induced polycythemia. *Transplantation* **3**, 706–710 (1965).

11. Boycott, A.E. Regeneration of red corpuscles. *Trans. Roy. Soc. Trop. Med. Hyg.* **27**, 529–532 (1934).

12. Branehög, I. Platelet kinetics in idiopathic thrombocytopenic purpura (ITP) before and at different times after splenectomy. *Br. J. Haematol.* **29**, 413–426 (1975).

13. Brown, E.R., and DeWitt, W. Hemoglobin changes during metamorphosis in *Triturus viridescens. Comp. Biochem. Physiol.* **35**, 495–497 (1970).

14. Burlington, H., Cronkite, E.P., Reincke, U., and Zanjani, E.D. Erythropoietin production in cultures of goat renal glomeruli (renal erythropoietic factor/Fe-uptake assay). *Proc. Nat. Acad. Sci. U.S.* **69**, 3547–3550 (1972).

15. Burton, R.R., and Smith, A.H. The effect of chronic erythrocytic polycythemia and high altitude upon plasma and blood volumes. *Proc. Soc. Exp. Biol. Med.* **140**, 920–923 (1972).

16. Carmena, A.O., Howard, D., and Stohlman, F., Jr. Regulation of erythropoiesis. XXII. Erythropoietin production in the newborn animal. *Blood* **32**, 376–382 (1968).

17. Carnot, P., and Deflandre, C. Sur l'activité hémopoiétique du serum au cours de la régénération du sang. *C.R. Acad. Sci.*, Paris **143**, 384–386 (1906).

18. Cooper, G.W., Cooper, B., Ossias, A.L., and Zanjani, E.D. A hypertransfused mouse assay for thrombopoietic factors. *Blood* **42**, 423–428 (1973).

19. Cronkite, E.P. Further studies of platelet reducing substances in splenic extracts. *Ann. Intern. Med.* **20**, 52–62 (1944).

20. Dutton, R.W., and Mishell, R.I. Lymphocytic proliferation in response to homologous tissue antigens. *Fed. Proc.* **25**, 1723–1726 (1966).

21. Ebbe, S. Editorial: Thrombopoietin. *Blood* **44**, 605–608 (1974).

22. Erslev, A.J. Renal biogenesis of erythropoietin. *Am. J. Med.* **58**, 25–30 (1975).

23. Estes, F.L., Smith, S., and Gast, J.H. A method for obtaining polymorphonuclear leukocytes from intraperitoneal exudates. *Blood* **13**, 1192–1197 (1958).

24. Fehér, I., and Gidáli, J. Quantitative changes in the level of the myelopoiesis-stimulating agent in rabbit sera. *J. Lab. Clin. Med.* **66**, 272–279 (1965).

25. Fisher, E.R., and Balcerzak, S.P. Effect of exogenous erythropoietin on juxtaglomerular cells. *Proc. Soc. Exp. Biol. Med.* **132**, 367–371 (1969).

26. Flores, G., and Frieden, E. Structural requirements for the hemolytic effect of phenylhydrazine derivatives on amphibian red cells. *J. Pharmacol. Exp. Therapy.* **174**, 463–472 (1970).

27. Forman, L.J., and Just, J.J. The life span of red blood cells in the amphibian larvae, *Rana catesbeiana. Dev. Biol.* **50**, 537–540 (1976).

28. Fowler, W.M., and Barer, A.P. Rate of hemoglobin regeneration in blood donors. *J. Am. Med. Assoc.* **118**, 421–427 (1942).

29. Gardner, R.H., and Cohen, P. Platlet life span. *Transfusion* **6**, 23–31 (1966).

30. Goldwasser, E. Erythropoietin and the differentiation of red blood cells. *Fed. Proc.* **34**, 2285–2292 (1975).

31. Gould, A.B., Goodman, S.A., and Green, D. An *in vivo* effect of renin on erythropoietin formation. *Lab. Invest.* **28**, 719–722 (1973).

32. Grasso, J.A. Erythropoiesis in the newt, *Triturus cristatus* Laur. I. Identification of the "erythroid precursor cell." *J. Cell Sci.* **12**, 463–489 (1973).

33. Harker, L.A. Thrombokinetics in idiopathic thrombocytopenic purpura. *Brit. J. Haematol.* **19**, 95–104 (1970).

34. Harker, L.A. Regulation of thrombopoiesis. *Am. J. Physiol.* **218**, 1376–1380 (1970).

35. Harrington, W.J., Minnich, V., Hollingsworth, J.W., and Moore C.V. Demonstration of a thrombocytopenic factor in the blood of patients with thrombocytopenic purpura. *J. Lab. Clin. Med.* **38**, 1–10 (1951).

36. Harrington, W.J., Sprague, C.C., Minnich, V., Moore, C.V., Ahlvin, R.C., and Duback,

R. Immunologic mechanisms in idiopathic and neonatal thrombocytopenic purpura. *Ann. Intern. Med.* **38**, 433–469 (1953).

37. Harris, J.A. Seasonal variation in some hematological characteristics of *Rana pipiens. Comp. Biochem. Physiol.* **A 43**, 975–989 (1972).

38. Havemann, K., Schmidt, M., and Rubin, A.D. Humoral regulation of lymphocyte growth *in vitro. In* "Humoral Control of Growth and Differentiation. Vertebrate Regulatory Factors" (J. LoBue and A.S. Gordon, eds.), Vol. 1, pp. 183–212, Academic Press, New York, 1973.

39. Hjort, P.F., and Paputchis, H. Platelet life span in normal, splenectomized and hypersplenic rats. *Blood* **15**, 45–51 (1960).

40. Hollingsworth, J.W., Berend, J.A., Silbert, D.R., and Finch, S.C. Leukocyte mobilization in normal, splenectomized, and leukemic rats after replacement transfusion. *J. Lab. Clin. Med.* **50**, 36–44 (1957).

41. Jacobson, L.O., Goldwasser, E., Fried, W., and Plzak, L. Role of the kidney in erythropoiesis. *Nature (London)* **179**, 633–634 (1957).

42. Jalavisto, E. The development of responsiveness to erythropoietic stimuli during ontogeny. *In* "Control of Cellular Growth in Adult Organisms" (H. Teir and T. Rytömaa, eds.), pp. 139–147. Academic Press, New York, 1967.

43. Jaskunas, S.R., Stork, E.J., and Richardson, B. Effects of a hyperoxic environment on erythropoietin production. *Aerospace Med.* **44**, 1112–1116 (1973).

44. Jordan, H.E., and Speidel, C. The hemocytopoietic effect of splenectomy in the salamander, Triturus viridescens. *Am. J. Anat.* **46**, 55–90 (1930).

45. Keighley, G.H., Lowy, P., Russell, E.S., and Thompson, M.W. Analysis of erythroid homeostatic mechanisms in normal and genetically anaemic mice. *Br. J. Haematol.* **12**, 461–477 (1966).

46. Kisken, W.A., and Swenson, N.A. Unresponsiveness of mixed leucocyte cultures from thymectomized adult dogs. *Nature (London)* **224**, 76–77 (1969).

47. Krantz, S.B., and Jacobson, L.O. Erythropoietin and the Regulation of Erythropoiesis. University of Chicago Press, Chicago, Illinois, 1970.

48. Kretchmar, A.L., and Conover, W.R. Early proliferation of transplanted spleen colony-forming cells. *Proc. Soc. Exp. Biol. Med.* **129**, 218–220 (1968).

49. Krizsa, F. Study on the development of posthaemorrhagic thrombocytosis in rats. *Acta Haematol.* **46**, 228–231 (1971).

50. Lawrence, J.S., and Craddock, G., Jr. Stem cell competition: The response to antineutrophilic serum as affected by hemorrhage. *J. Lab. Clin. Med.* **72**, 731–738 (1968).

51. Lucarelli, G., Howard, D., and Stohlman, F., Jr. Regulation of erythropoiesis. XV. Neonatal erythropoiesis and the effect of nephrectomy. *J. Clin. Invest.* **43**, 2195–2203 (1964).

52. Maniatis, G.M., and Ingram, V.M. Erythropoiesis during amphibian metamorphosis. III. Immunochemical detection of tadpole and frog hemoglobins *(Rana catesbeiana)* in single erythrocytes. *J. Cell Biol.* **49**, 390–404 (1971).

53. McClure, P.D., and Choi, S. Thrombopoietin and erythropoietin levels in idiopathic thrombocytopenic purpura and iron-deficiency anaemia. *Br. J. Haematol.* **15**, 351–354 (1968).

54. McCulloch, E.A., Siminovitch, L., and Till, J.E. Spleen-colony formation in anemic mice of genotype WW^v. *Science* **144**, 844–846 (1964).

55. McDonald, T.P. Role of the kidneys in thrombopoietin production. *Exp. Hematol.* **4**, 27–31 (1976).

56. McDonald, T.P., Clift, R., Lange, R.D., Nolan, C., Tribby, I.I.E., and Barlow, G.H.

Thrombopoietin production by human embryonic kidney cells in culture. *J. Lab. Clin. Med.* **85,** 59–66 (1975).

57. McLeod, D.L., Shreeve, M.M., and Axelrad, A.A. Improved plasma culture systems for production of erythrocytic colonies *in vitro:* Quantitative assay method for CFU-E. *Blood* **44,** 517–534 (1974).

58. McLeod, D.L., Shreeve, M.M., and Axelrad, A.A. Induction of megakaryocyte colonies with platelet formation *in vitro. Nature (London)* **261,** 492–494 (1976).

59. Metcalf, D. The colony stimulating factor (CSF). *In* "Humoral Control of Growth and Differentiation. Vertebrate Regulatory Factors" (J. LoBue and A.S. Gordon, eds.), Vol. 1, pp. 91–118. Academic Press, New York, 1973.

60. Miller, M.E., Howard, D., Stohlman, F., Jr., and Flanagan, P. Mechanism of erythropoietin production by cobaltous chloride. *Blood* **44,** 339–346 (1974).

61. Mirand, E.A. Nonerythropoietin-dependent erythropoiesis. *In* "Regulation of Hematopoiesis" (A.S. Gordon, ed.), Vol. 1, pp. 635–647. Appleton-Century-Crofts, New York, 1971.

62. Moriyama, Y., Lertora, J.J.L., and Fisher, J.W. Studies on an inhibitor of erythropoiesis. I. Effects of sera from normal and polycythemic rabbits on heme synthesis in rabbit bone marrow cultures. *Proc. Soc. Exp. Biol. Med.* **147,** 740–743 (1974).

63. Murphy, G.P., Mirand, E.A., Johnston, G.S., Gibbons, R.P., Jones, R.L., and Scott, W.W. Erythropoietin release associated with Wilms' tumor. *Johns Hopkins Med. J.* **120,** 26–32 (1967).

64. Nakao, K., Fisher, J.W., and Takaku, F. (eds.) "Erythropoiesis." University Park Press, Baltimore, Maryland, 1974.

65. Neuberger, A., and Richards, F.F. Protein biosynthesis in mamalian tissues. II. Studies on turnover in the whole animal. *In* "Mammalian Protein Metabolism" (H.N. Munro and J.B. Allison, eds.), Vol. 1, pp. 243–296. Academic Press, New York, 1964.

66. Odell, T.T., Jr., Jackson, C.W., and Friday, T.J. Effects of radiation on the thrombocytopoietic system of mice. *Radiat. Res.* **48,** 107–115 (1971).

67. Odell, T.T., Jr., Jackson, C.W., and Gosslee, D.G. Maturation of rat megakaryocytes studied by microspectrophotometric measurement of DNA. *Proc. Soc. Exp. Biol. Med.* **119,** 1194–1199 (1965).

68. Odell, T.T., Jr., Jackson, C.W., and Reiter, R.S. Depression of the megakaryocyte-platelet system in rats by transfusion of platelets. *Acta Haematol.* **38,** 34–42 (1967).

69. Odell, T.T., Jr., and Murphy, J.R. Effects of degree of thrombocytopenia on thrombocytopoietic response. *Blood* **44,** 147–156 (1974).

70. O'Grady, L.F., Lewis, J.P., and Trobaugh, F.E., Jr. The effect of erythropoietin on differentiated erythroid precursors. *J. Lab. Clin. Med.* **71,** 693–703 (1968).

71. Okunewick, J.P., and Erhard, P. Accelerated clearance of exogenously administered erythropoietin by mice with Rauscher viral leukemia. *Cancer Res.* **34,** 917–919 (1974).

72. Paran, M., Ichikawa, Y., and Sachs, L. Feedback inhibition of the development of macrophage and granulocyte colonies. II. Inhibition by granulocytes. *Proc. Nat. Acad. Sci. U.S.* **62,** 81–87 (1969).

73. Paulus, J.-M. DNA metabolism and development of organelles in guinea-pig megakaryocytes: A combined ultrastructural, autoradiographic and cytophotometric study. *Blood* **35,** 298–311 (1970).

74. Pedersen, N.T. The effect of splenectomy on the megakaryocyte and platelet count in the blood of rats. *Scand. J. Haematol.* **12,** 291–297 (1974).

75. Penington, D.G., and Olsen, T.E. Megarkaryocytes in states of altered platelet production: Cell numbers, size and DNA content. *Br. J. Haematol.* **18,** 447–463 (1970).

76. Peschle, C., Rappaport, I.A., Jori, G.P., Chiariello, M., Gordon, A.S., and Condorelli, M. Sustained erythropoietin production in nephrectomized rats subjected to severe hypoxia. *Br. J. Haematol.* **25**, 187–193 (1973).
77. Phillips, B., and Roitt, I.M. Evidence for transformation of human B lymphocytes by PHA. *Nature (London) New Biol.* **241**, 254–256 (1973).
78. Reddick, R.I., Poole, B.I., and Penick, G.D. Thrombocytopenia of hibernation: Mechanism of induction and recovery. *Lab. Invest.* **28**, 270–278 (1973).
79. Reynafarje, C., Faura, J., Paredes, A., and Villavicencio, D. Erythrokinetics in high-altitude-adapted animals (llama, alpaca, and vicuna). *J. Appl. Physiol.* **24**, 93–97 (1968).
80. Robinson, W.A., and Mangalik, A. Kinetics and regulation of granulopoiesis. *In* "Neutrophil Physiology and Pathology" (J.R. Humbert, P.A. Miescher, and E.R. Jaffe, eds.), pp. 5–23. Grune & Stratton, New York, 1975.
81. Rodnan, G.P., Edbaugh, F.G., Jr., and Fox, M.R.S. The life span of the red blood cell and the red blood cell volume in the chicken, pigeon and duck as estimated by the use of $Na_2Cr^{51}O_4$. With observations on red cell turnover rate in the mammal, bird and reptile. *Blood* **12**, 355–366 (1957).
82. Rolovic, Z., and Baldini, M. Megakaryocytopoiesis in splenectomized and "hypersplenic" rats. *Br. J. Haematol.* **18**, 257–268 (1970).
83. Rolovic, Z., Baldini, M., and Dameshek, W. Megakaryocytopoiesis in experimentally induced immune thrombocytopenia. *Blood* **35**, 173–188 (1970).
84. Rosenberg, M. Electrophoretic analysis of hemoglobin and isozymes in individual vertebrate cells. *Proc. Nat. Acad. Sci. U.S.* **67**, 32–36 (1970).
85. Ruud, J.T. Vertebrates without erythrocytes and blood pigments. *Nature (London)* **173**, 848–850 (1954).
86. Rytömaa, T. Role of chalone in granulopoiesis. *Br. J. Haematol.* **24**, 141–146 (1973).
87. Schooley, J.C., and Mahlmann, L.J. Evidence for the de novo synthesis of erythropoietin in hypoxic rats. *Blood* **40**, 662–670 (1972).
88. Schultz, E.F., Lapin, D.M., and LoBue, J. Humoral regulation of neutrophil production and release. *In* "Humoral Control of Growth and Differentiation. Vertebrate Regulatory Factors" (J. LoBue and A.S. Gordon, eds.), Vol. 1, pp. 51–68. Academic Press, New York, 1973.
89. Schwarz, M.R. The mixed lymphocyte reaction: An *in vitro* test for tolerance. *J. Exp. Med.* **127**, 879–890 (1968).
90. Seeley, V.S., Cantor, L., and Gordon, A.S. Response of the neonatal rat to erythropoietin (ESF). *Biol. Neonate* **19**, 108–117 (1971).
91. Silini, G., Pons, S., and Pozzi, L.V. Quantitative histology of spleen colonies in irradiated mice. *Br. J. Haematol.* **14**, 489–500 (1968).
92. Skjaelaaen, P., and Halvorsen, S. Inhibition of erythropoiesis by plasma from newborn infants. *Acta Paediat. Scand.* **60**, 301–308 (1971).
93. Smith, R.J., and Fisher, J.W. Effects of cobalt on the renal erythropoietic factor and kidney hydrolase activity in the rat. *Blood* **42**, 893–905 (1973).
94. Sodicoff, M., and Binhammer, R.T. Leukocytosis-inducing factor in the blood of X-irradiated rats. *Radiat. Res.* **33**, 82–93 (1968).
95. Stanley, E.R., Cifone, M., Heard, P.M., and Defendi, V. Factors regulating macrophage production and growth: Identity of colony-stimulating factor and macrophage growth factor. *J. Exp. Med.* **143**, 631–647 (1976).
96. Till, J.E., Price, G.B., Mak, T.W., and McCulloch, E.A. Regulation of blood cell differentiation. *Fed. Proc.* **34**, 2279–2284 (1975).
97. Trubowitz, S., Moschides, E., and Feldman, D. Alkaline phosphatase activity of the

polymorphonuclear leukocyte in rapidly induced leukopenia and leukocytosis. *J. Lab. Clin. Med.* **57**, 747–754 (1961).

98. Turpen, J.B., and Volpe, E.P. On the origin of thymic lymphocytes. *Am. Zool.* **15**, 51–61 (1975).

99. Younkin, L.H. *In vitro* response of lymphocytes to phytohemagglutinin (PHA) as studied with antiserum to PHA. I. Initiation period, daughter-cell proliferation, and restimulation. *Exp. Cell Res.* **75**, 1–10 (1972).

100. Zaizov, R., and Matoth, Y. Regulation of erythropoiesis in the newborn rat. *Is. J. Med. Sci.* **7**, 846–849 (1971).

101. Zanjani, E.D., Gordon, A.S., Wong, K.K., and McLaurin, W.D. The renal erythropoietic factor (REF). X. The question of species and class specificity. *Proc. Soc. Exp. Biol. Med.* **131**, 1095–1098 (1969).

102. Zanjani, E.D., Yu, M.-L., Perlmutter, A., and Gordon, A.S. Humoral factors influencing erythropoiesis in the fish (Blue gourami—*Trichogaster trichopterus*). *Blood* **33**, 573–581 (1969).

103. Zucker-Franklin, D., and Grusky, G. Ultrastructural analysis of hematopoietic colonies derived from human peripheral blood. A newly developed method. *J. Cell Biol.* **63**, 855–863 (1974).

6

Vascular Expansion

Capillary: L., *caput*, head + *pilus*, hair

As the body grows its circulatory system must expand to keep pace. This it does by sprouting new capillaries and by increasing the diameters of arteries and veins. There is also some provision for the lengthening of blood vessels. Despite their remarkable capacity for expansion, arteries and veins can originate *de novo* only from capillaries. Presumably it is the blood pressure kinetics that determines whether a capillary will develop into a vein or an artery. Hemodynamic forces may also prescribe the dimensions to which such blood vessels grow. The impression cannot be avoided, however, that the ultimate size of a blood vessel is indeterminate. The stovepipe dimensions of a whale's aorta, big enough for a baby to crawl through, originated as a capillary in the embryo.

ANGIOGENESIS

The differentiation of capillaries can be studied in a variety of situations, especially those that facilitate direct visualization of the neovascularization process. One of the earliest systems to be developed for this purpose was the rabbit ear chamber. This consists of two coverslips inserted across a hole punched through the ear of a rabbit and held in place by a frame attached to the ear around the edges of the hole. The two windows are separated from each other by a thin space into which connective tissue and blood vessels can grow. Such chambers can be viewed by transmitted illumination on a microscope stage to record the day-by-day progress of capillary development and associated blood flow (19). The rabbit ear is also useful in studying the regeneration of blood vessels across a gap by making a transverse cut across the entire width of the ear except for an isthmus at one margin to preserve

120

minimal blood flow, and resuturing the skin of the separated portions (Fig. 56). It has been learned that continuity of the original blood vessels is reestablished after 4 days (20). This is in close agreement with comparable studies in which arteries and veins on the backs of rabbits were isolated inside tubes of skin up to 9 cm long. When these pedicles were transected and the skin resutured, the severed blood vessels reunited in an average of 4.2 days (27). Following a second interruption, only 3 days were required for the vessels to grow together.

The normal absence of blood vessels in the cornea, coupled with its potential for vascularization under certain conditions, lends itself to the study of capillary regeneration. Traumatization of the cornea tends to elicit the ingrowth of new blood vessels. This can be seen following direct mechanical injury, inflammation, or the application of $AgNO_3$, whereupon new capillaries appear within a few days and grow to the center of the cornea by the end of a week (23, 33). In the rabbit, it has been possible to analyze a variety of substances for their capacity to induce corneal vascularization. If a curved plastic tube is introduced into the corneal stroma parallel to the surface so that its tip is some distance from the point of entry, certain amines can be

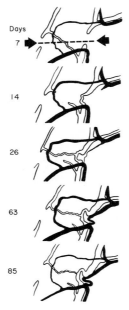

Fig. 56. Vascular patterns at various intervals as indicated after partially severing a rabbit ear (arrows) and allowing the blood vessels to regenerate across the gap (20). Arteries in black, veins open.

infused through the tube and into the surrounding connective tissue (41). Acetylcholine, serotonin, histamine, and bradykinin stimulate capillary growth, while saline controls are ineffectual. The new capillaries grow as far as the tip of the plastic tube during the first week, and may later extend into the lumen of the tube for distances of 6–7 mm in some cases (Fig. 57).

It has been hypothesized that chemical influences may be responsible for the growth of blood vessels into tumors, the vascularization of which is derived from the host. If a rapidly growing tumor is to remain viable its blood supply must keep pace with its increase in mass. The central necrosis characteristic of some tumors testifies to the fact that the growth of blood vessels is not always optimal. Nevertheless, the promotion of capillary production by cancerous tissues suggests the possibility of controlling neoplastic growth by arresting its vascularization. To test the capacity of tumors and other tissues to stimulate the differentiation of blood vessels, various kinds of cancerous tissues have been transplanted to sites where neovascularization can be monitored (10). In some cases they have been grafted into the cornea (13), in others to a subcutaneous air sac (3). The expectations were justified, for some kinds of cancerous tissues elicited a remarkable ingrowth of blood vessels. Normal tissues, except for placenta and certain rapidly growing embryonic tissues, fail to provoke capillary growth. Even when separated from surrounding tissues in a Millipore filter chamber, cancers elicit a response presumably by producing a "tumor angiogenesis factor" (TAF) which can be extracted from tumors and shown to promote vascular proliferation. If a way can now be found to antagonize the effects of TAF without doing

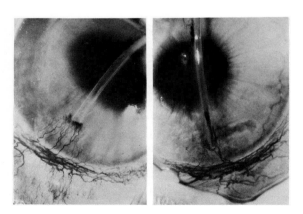

Fig. 57.   Neovascularization of the rabbit cornea in response to instillation of 1.5% histamine. After 8 days (left), new blood vessels have grown in from the limbus as far as the tube through which the histamine is administered. By 32 days (right), vessels have invaded the tube itself (41).

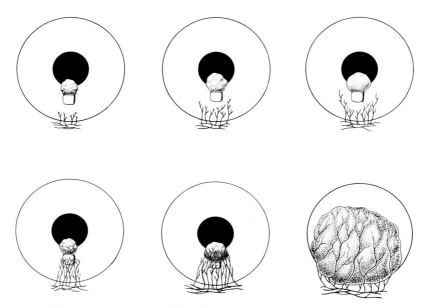

Fig. 58.   The growth of tumor transplants depends on vascularization (2). When grafted along with a piece of cartilage to the cornea of a rabbit's eye, the ingrowth of capillaries which would otherwise occur is held in abeyance by a factor from the cartilage (above). If the fragment of cartilage is previously devitalized, the unimpeded invasion of blood vessels from the limbus permits unchecked growth of the tumor (below).

violence to the blood supplies to normal tissues, it might be possible to limit the growth of tumors to diameters of a millimeter or so, which is as large as they grow in tissue culture without benefit of a vascular supply (9). Indeed, some such mechanism may explain why cartilage is not normally vascularized, for it has been shown that the stimulation of capillary growth in corneas to which tumors have been transplanted is inhibited by extracts of cartilage (Fig. 58) (2).

THE ATROPHY OF BLOOD VESSELS

While capillary formation would appear to be regulated by exogenous factors, the growth of blood vessels once they are formed seems to be subject to intraluminal influences. These influences are most acutely felt in the tunica intima and its endothelial lining, and in the tunica media, which is made up of smooth muscle laced with elastic and collagenous fibers. Although blood pressure is of primary importance in determining the size of a blood vessel and the thickness of its walls, it is not the only factor responsible for vascular growth. In the chick embryo, for example, angiogenesis commences before the blood begins to flow. Indeed, removal of the embryonic heart results in continued development of the vitelline veins, although the omphalomesenteric arteries and veins cease development (4). Mounting evidence suggests that the constriction of the ductus arteriosus at birth is triggered by increased oxygenation of the blood (14). Decreased oxygen will delay its closure in the rat fetus.

How the closure of the ductus venosus is to be explained is not known. This vessel, which functions in the fetus to shunt blood from the umbilical vessels past the liver to the inferior vena cava, usually closes soon after birth (25). Its surgical occlusion in the prenatal lamb, however, seems to have little or no effect on fetal circulatory physiology. Indeed, the ductus venosus spontaneously ceases to function early in the gestation of the horse and pig. Although it would seem not to be a very essential blood vessel, its eventual

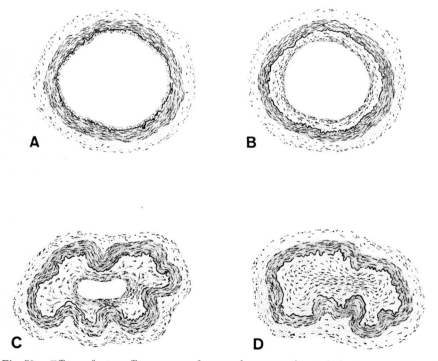

Fig. 59.   Effects of tying off a segment of a carotid artery in the rat (24). A, cross section of normal artery. B, after 14 days the tunica intima has increased in thickness. C, D, subsequent changes (28 and 90 days) result in collapse of the artery with eventual ingrowth of the intima to fill the lumen.

atrophy is correlated with the reduction in blood flow when the cord is cut. The uterine artery also illustrates the reaction of blood vessels to reductions in the flow of blood, something that happens after each pregnancy. The formerly enlarged uterine vessels undergo marked collapse following delivery, including disintegration and replacement of the elastic tissue (22).

Perhaps the most direct way in which to study the reaction of blood vessels to reduced blood flow is to isolate a segment of artery or vein between two ligatures (24). Deprived of intraluminal pressure, the tunica media of a doubly ligated artery atrophies. Concomitantly, the tunica intima becomes progressively thickened and may eventually fill the lumen completely (Fig. 59). In due course, capillaries may penetrate the wall of the original vessel to establish new ones in place of the original.

BLOOD PRESSURE AND ARTERIAL HYPERTROPHY

If reduced blood pressure results in atrophy and eventual disappearance of a blood vessel, high blood pressure should have the opposite effect. It is well known from clinical observations that hypertension is associated with thickening of the arterial wall, including the media and intima, sometimes to the extent of reducing the size of the lumen further aggravating the hypertensive condition. It is noteworthy that in experiments on rats made hypertensive by renal arterial stenosis, smooth muscle cells have been found in arterial intimas presumably derived from those in the tunica media via fenestrations in the internal elastic membrane (37). This situation is reminiscent of what happens in the formation of atherosclerotic plaques.

A particularly favorable system in which to study the effects of experimentally induced hypertension is in the pulmonary arteries, each of which carries about half the blood pumped into the main pulmonary artery by the right ventricle. If one of these vessels is tied off, as in the case of unilateral pneumonectomy, the remaining pulmonary artery must suddenly accommodate twice its normal blood flow (18). This has profound effects on the arterial tree in the remaining lung. The immediate result is a greatly increased vascular resistance. The long-term effect is a doubling of the cross-sectional area of the remaining pulmonary artery. In addition, there is considerable medial hypertrophy in even the smallest arteries of the lung. A comparable situation prevails in association with other conditions that increase pulmonary blood flow. For example, in newborn infants with congenital left ventricular atrophy, most of the blood pumped by the heart passes through the pulmonary circulation. Such infants exhibit hypertrophy of the tunica media in the arterial trees of their lungs (28). A similar effect is produced in dogs following anastomosis of one pulmonary artery to the aorta to cause a marked increase in blood flow through that lung. The resulting pulmonary hypertension causes a narrowing of the smaller pulmonary arteries due to medial hypertrophy (26). Evidence exists that the noncellular components of the media are also susceptible to the effects of increased blood pressure. In cultured segments of rabbit aorta subjected to periodic stretching, there is a two- to fourfold increase in the rates at which collagen, hyaluronate, and chondroitin sulfate are synthesized (21).

Inasmuch as the histological differences between veins and arteries are more quantitative than qualitative, it is tempting to speculate that the thinner and more flexible walls of veins are attributable to the relatively lower venous blood pressures, compared with those in arteries. It would be difficult to refute the contention that all capillaries have the potential for becoming either arteries or veins, the direction of their development depending upon the hemodynamic conditions under which they develop. Perhaps it would be too much to expect a vein to turn into an artery if it were subjected to the pulsations of arterial blood pressures. Experiments have shown, however, that under such conditions the walls of veins tend to become thickened. This happens when a segment of jugular vein in the dog is grafted to the aorta, the increased thickness one year later occurring primarily in the media (5). If the subclavian artery is anastomosed to the pulmonary vein, the venous walls thicken while the lumen decreases in diameter (26). Thus, although veins, like arteries, have the capacity to adapt their histology to physiological needs, they have not as yet exhibited the capacity actually to become arteries.

In the case of the uterine blood vessels the hypertrophy that occurs is transient with each episode of pregnancy. What the stimulus for this enlargement might be is not known, although it has been established that when pregnancy occurs in only one uterine horn the blood vessels leading to the nonpregnant side do not enlarge. Studies on the sow have shown that with each pregnancy the tunica intima of the uterine artery thickens, but does not completely regress between successive pregnancies. Therefore, there builds up a series of concentric layers of intimal connective tissue which may eventually threaten to occlude the lumen (12).

COLLATERAL CIRCULATION

One of the most interesting, and useful, examples of vascular enlargement is the establishment of a collateral circulation past a site of arterial occlusion. Credit for the discovery of this important phenomenon goes to John Hunter who demonstrated it in a stag at London's Richmond Park in the summer of 1785. The growing antler is supplied by a branch of the carotid artery. So copious is the arterial blood flow to an antler in velvet that it feels almost hot to the touch. Overcoming what one can only imagine must have been a herculean task of restraining even a tame animal, Hunter succeeded in tying off the carotid artery supplying blood to one antler. Immediately thereafter the temperature of that antler dropped. A week later, however, it was back to normal and growth had resumed. Subsequent dissection revealed that other arteries, normally of lesser dimensions, had enlarged to restore the normal flow of blood to the growing antler. What happens here is typical of reactions elsewhere in the body.

Studies of collateral circulation have focused on a variety of anatomical sites, including the uterus, external ear, hind leg, heart, and lungs. Wherever an artery is occluded, there is rapid dilation of collaterals to restore blood flow to the distal parts. This eventually leads to permanent enlargement of the collateral vessels by growth of the tissue layers in their walls. Two principal theories have been advanced to account for this response. One is that the temporarily ischemic tissues downstream from the occlusion produce a substance that stimulates the growth of the vessels upon which their survival depends. A more mechanistic explanation suggests that the growth of collateral vessels past an arterial obstruction is the result of hemodynamic reactions to a pressure gradient between the hypertensive proximal region and the decreased resistance distal to the site of operation.

It is conceivable that the lack of oxygen *per se* or the production of some metabolite in response to hypoxia might provide a chemical stimulus to vascular growth. Experiments have shown, however, that perfusion of limbs distal to an arterial ligation either with hypoxic or hyperoxic blood does not affect the development of collateral circulation past the ligated femoral artery of the dog (39). Moreover, if the leg is amputated immediately distal to the site of occlusion of the femoral artery, removing much of the presumed source of substances purported to stimulate blood flow and vascular development, collateral circulation in the vicinity of the obstruction still develops.

The role of mechanical factors in promoting collateral circulation is more consistent with available experimental evidence (Figs. 60 and 61). Since blood flows in the direction of least resistance, the creation of a pressure gradient along collateral vessels past an occluded artery inevitably leads to a rise in blood flow through these channels (17). Not unexpectedly, the creation of an artificial bypass reverses such collateral circulation as may have been established (39). Finally, the effects of an arteriovenous (AV) fistula are particularly relevant (15). In this situation much of the arterial blood is shunted into the adjacent vein and returned directly to the heart without passing through the distal capillary beds. As in the case of arterial ligation, tissues distal to the fistula are deprived of much of their blood supply. If they

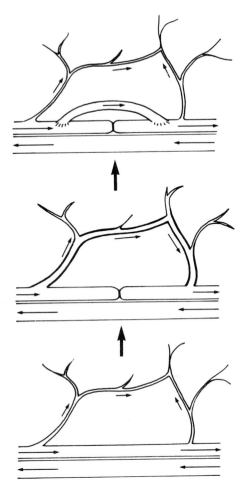

Fig. 60. Collateral circulation. Ligation of an artery forces blood to flow through side branches to bypass the occlusion. The collateral vessels enlarge and develop thickened walls to accommodate the increased pressure. If the occlusion is artificially bypassed the formerly enlarged collaterals revert to their previous dimensions.

were to produce a substance in response to oxygen starvation to elicit increased blood flow, one would expect to observe the development of collateral circulation past an AV fistula. This is exactly what happens, although for different reasons. In bypassing the fistula, the collateral blood flow returns to the segment of artery downstream from the site of operation, then in retrograde fashion back to the fistula to be returned to the heart via the venous route. This pattern of flow is in keeping with the tendency of blood to follow

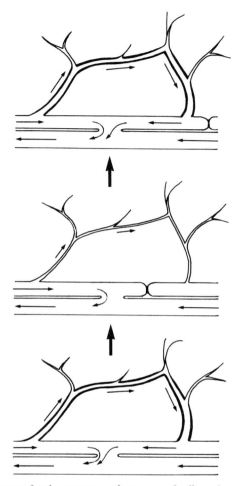

Fig. 61. An arteriovenous fistula promotes enlargement of collateral vessels because of pressure differentials which force blood to flow through side branches and return to the artery downstream from the fistula. Ligating the artery immediately distal to the fistula interferes with this retrograde arterial flow thereby preventing collateral circulation. If the artery is ligated distal to the side branch, the collateral circulation persists.

the path of least resistance. Since the arterial pressure distal to a fistula is lower than that on the proximal side, blood will naturally reverse its direction in the side branches originating from the part of the artery downstream from the fistula. Prevention of this retrograde flow by ligating the artery immediately distal to the fistula so as to exclude access via side branches precludes the establishment of collateral circulation (16). If the artery distal to the fistula is ligated at a level distal to one or more side branches, an extensive collateral circulation still develops. These findings offer compelling arguments in favor of the mechanistic explanation of collateral circulation. If the vascular resistance of a ligated artery is relieved by an adjacent AV fistula the pressure gradient which would otherwise be responsible for the establishment of collateral blood flow is exceeded by the AV shunt. However much the distal tissues deprived of oxygen might require a collateral circulation, their needs cannot be filled in the absence of appropriate hemodynamic pressures.

The lung is the only visceral organ naturally supplied with a collateral arterial circulation. It thus lends itself to the elimination of one of these sources of blood in order to observe the response of the other. Each lung receives unoxygenated blood from the right ventricle via the pulmonary artery. It also receives oxygenated blood from the bronchial arteries which branch off the aorta. Confluency between these two circuits is limited, if it exists at all. Whether or not the lung can survive in the absence of its bronchial circulation has not been tested, but ligation of the pulmonary artery is well tolerated. Numerous studies have confirmed that interruption of the pulmonary arterial blood flow to a lung causes dilation of the bronchial circulation (38), a reaction not logically attributable to oxygen starvation in an organ that is normally well ventilated and merely deprived of venous blood. Although such conditions elicit compensatory growth in the opposite lung, its bronchial circulation is not known to react like that on the operated side. Concomitant with the increased blood flow through, and enlargement of, the bronchial arteries, anastomoses develop with the pulmonary arteries beyond the level of constriction. Arterial blood can thus return to the heart via the pulmonary veins. In addition, such lungs tend to form adhesions to the pleural walls where new vessels develop. The latter reaction, involving the *de novo* formation of capillaries, may be an expression of the apparently natural tendency of adjacent capillary networks to form cross connections. The expansion of preexisting vessels of the bronchial circulation in response to occlusion of the pulmonary artery may represent a hyperemic response to increased functional demands coupled with a tendency for bronchial blood under high pressure to be channeled into low pressure pathways through the pulmonary arterioles and veins which have been cut off from their normal source of blood flow.

The physiological factors thus far considered are ones of relatively wide distribution affecting blood vessels collectively. Each segment of vessel, however, is ultimately responsive to locally prevailing conditions. For a closer look at such responses, it is necessary to study local interventions, one of which is to constrict an artery to approximately two-thirds of its normal cross-sectional area (30). Proximal to the stenosis where the blood pressure is

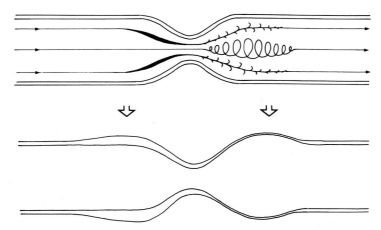

Fig. 62.   Aortic stenosis creates turbulence in the otherwise laminar flow of blood. Drag forces on the arterial wall are responsible for poststenotic dilatation with thinning of the vessel wall. Proximal to the stenosis the heightened blood pressure causes hypertrophy of the tunica media (30).

elevated, the wall of the artery becomes thickened due to hypertrophy of its tunica media. Downstream, there is a typical poststenotic dilatation where the arterial wall becomes thinner than normal (Fig. 62). Here the blood pressure is reduced, accounting in part for the reduction in thickness of the arterial wall (6). The explanation for the dilatation, however, is not so obvious. One possibility is that the local turbulence in the blood flow distal to the construction may create drag forces which somehow cause expansion of the arterial lumen (30). Alternatively, it is possible that reduction in lateral pressure and consequent thinning of the wall may so weaken its tissues as to permit dilatation of the vessel wall.

THE SENSITIVITY OF THE ENDOTHELIUM

A common kind of injury to blood vessels is the puncture wound produced either for purposes of injection or as vascular sutures. The tunica adventitia heals by scar formation while the intima usually thickens and becomes invaded by smooth muscle cells from the tunica media. Bone or cartilage formation is sometimes observed in the media in association with various kinds of injuries, including treatment with $AgNO_3$ or $CuSO_4$, injection of carrageenin, or insertion of a steel wire through the arterial wall, the pulsed movements of which stimulate chondrogenesis in the adjacent media. Few insults fail to elicit a proliferative response in the endothelium.

Situated as it is at the interface between the blood and the tissues of the vessel wall, the endothelial lining is subjected to the incessant abrasions of flowing blood yet must preserve its structural integrity as a barrier between the plasma in the lumen and the interstitial fluids on the intimal side. The endothelium is a renewing tissue, the rate of turnover varying with age and location (34). It is estimated that 10–20% of the endothelial cells in a new-born rat aorta divide each day, the rate dropping to about 1% in adults. Not uncommonly, the distribution of endothelial cells synthesizing DNA is non-random. The tendency for ^{3}H-thymidine labeled endothelial nuclei to occur in "clutches," coupled with the higher incidence of labeling in the vicinity of arterial bifurcations, suggests that the proliferation of these cells may not be entirely spontaneous. Exposure to Evans Blue, a dye selectively taken up by dead cells, reveals that the site of maximum cell death is in the aortic arch, a location where one might expect the greatest turbulence and where the incidence of endothelial proliferation is highest (40). It would seem, there-fore, that endothelial cells are sensitive to disturbances in blood flow. They are also sensitive to chemically induced injuries.

Intravenous injection of endotoxin has been shown to damage the en-dothelial cells lining the aorta (11). Within a matter of minutes, such cells are to be found circulating in the blood. Presumably it is this loss of cells from the endothelium which is responsible for the increased proliferation ob-served within several days after exposure to endotoxin. In general, larger arteries exhibit a higher incidence of ^{3}H-thymidine labeled nuclei than do capillaries. A high cholesterol diet, known for its association with atherosclerotic plaques, stimulates DNA synthesis in the endothelial cells of arteries (8).

THE ATHEROSCLEROTIC REACTION

An important hypothesis to explain the mystery of what initiates the forma-tion of atherosclerotic plaques is the possibility that the local loss or destruc-tion of endothelial cells may be involved. Accordingly, a number of recent investigations have focused on the capacity of the endothelium to heal wounds. Such wounds can be inflicted in a variety of ways. One is to use a balloon catheter, which involves injecting a needle into an artery and inflat-ing a balloon on its end intraluminally. By withdrawing the balloon to the point of entry the endothelium can be denuded by abrasion, after which the balloon is deflated and withdrawn (31, 35, 36). Another method depends upon desiccation. This is achieved by clamping off a length of artery, with-drawing the trapped blood and blowing a stream of air in one end and out the other, effectively destroying the dried endothelium (7). In clinical situations,

Fig. 63.   Advancing edge of migrating endothelium healing a rat carotid artery 4 days after denudation by desiccation. The as yet unhealed intima (lower part of picture) is covered with platelets (7). Reproduced with permission. © 1975 U.S.-Canadian Division of the International Academy of Pathology.

segments of an artery are sometimes replaced by a fabric prosthesis, the inner surface of which eventually becomes covered with endothelium.

The length of time required for a denuded artery to become resurfaced with endothelium varies with the area to be covered, but usually takes weeks or months to go to completion. Endothelial healing is achieved by a combination of cellular migration and proliferation (Fig. 63). What happens in the underlying layers, however, is particularly significant (Fig. 64).

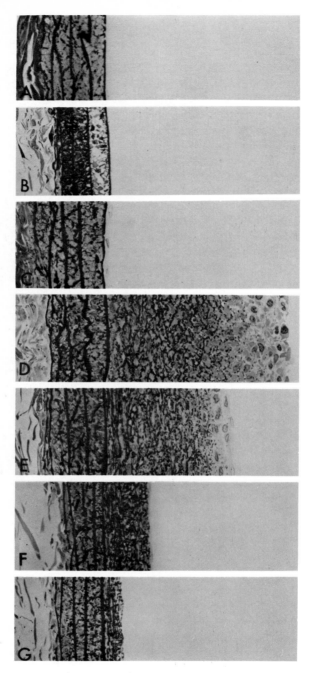

Fig. 64. Cross sections of rat carotid arteries (intima and inner media) in control (A) and experimental animals 4, 7, 14, 17, 30, and 90 days (B–G) after loss of the endothelium. The denuded intima becomes covered with new endothelium by 7 days (C), followed by maximal thickening of the intima after 14 days (D). This myointimal thickening subsides gradually thereafter (E–G) (7). Reproduced with permission. © 1975 U.S.–Canadian Division of the International Academy of Pathology.

In the absence of an endothelium, the remnants of the tunica intima and media are exposed to plasma constituents from which they are normally separated by the endothelium. It is this abnormal situation which may play a role in the reactions of these layers to injury. In the normal tunica intima there are no smooth muscle cells. Following injury, however, they make their appearance in the intima, presumably by migration from the media through the fenestrations in the inner elastic membrane (7, 29, 31, 35, 36). By a combination of migration and proliferation, these smooth muscle cells may serve to create a "false" endothelium on the intimal surface until true endothelial cells can reestablish their continuity. The thickening of the intima following injury is a transient phenomenon, but one which may take months to revert to normal. At its maximal thickness it may become more than twice as wide as the media itself (7, 36). This reaction would be of little more than academic interest were it not for the fact that intimal thickenings at sites of endothelial wounds bear a close resemblance to atherosclerotic plaques (Fig. 65) (1). Here, too, there is a buildup of smooth muscle cells in the thickened intima, often in association with an accumulation of lipids. Indeed, studies have shown that hyperlipidemia potentiates plaque formation at sites of endothelial injury. Whether it is the trauma of wounding, the presence of lipids in the blood, the role of platelets in stimulating the prolif-

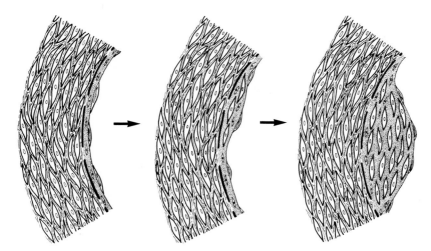

Fig. 65. Development of an atherosclerotic plaque (1). Left, normal artery showing tunica intima separated from tunica media by inner elastic membrane. Middle, early stages of plaque formation are marked by the appearance of smooth muscle cells in the intima where they normally never occur. Right, subsequent proliferation of smooth muscle cells in the intima results in the development of an atherosclerotic plaque impinging on the lumen. This may eventually acquire lipid deposits and develop areas of necrosis.

eration of smooth muscle cells, or neoplastic transformation, there is reason to predict that when the eventual explanation of atherosclerotic plaque formation is found it will have owed its origin to basic studies of growth and repair in blood vessels.

REFERENCES

1. Benditt, E.P. The origin of atherosclerosis. *Sci. Am.* **236**(2), 74–85 (1977).
2. Brem, H., and Folkman, J. Inhibition of tumor angiogenesis mediated by cartilage. *J. Exp. Med.* **141**, 427–439 (1975).
3. Cavallo, T., Sade, R., Folkman, J., and Cotran, R.S. Tumor angiogenesis. Rapid induction of endothelial mitoses demonstrated by autoradiography. *J. Cell Biol.* **54**, 408–420 (1972).
4. Chapman, W.B. The effect of the heart-beat upon the development of the vascular system in the chick. *Am. J. Anat.* **23**, 175–203 (1918).
5. Curtis, J., Conkle, D.M., Finch, W.T., Lanier, V.C., Jr., Younger, R.K., and Scott, H.W., Jr. The effects of experimental hypercholesterolemia on transposed arterial and venous autografts. *J. Surg. Res.* **18**, 163–167 (1975).
6. De Vries, H., and van den Berg, J. On the origin of poststenotic dilatations. *Cardiologia* **33**, 195–211 (1958).
7. Fishman, J.A., Ryan, G.B., and Karnovsky, M.J. Endothelial regeneration in the rat carotid artery and the significance of endothelial denudation in the pathogenesis of myointimal thickening. *Lab. Invest.* **32**, 339–351 (1975).
8. Florentin, R.A., Nam, S.C., Lee, K.T., and Thomas, W.A. Increased ^{3}H-thymidine incorporation into endothelial cells of swine fed cholesterol for 3 days. *Exp. Mol. Pathol.* **10**, 250–255 (1969).
9. Folkman, J. Tumor angiogenesis: A possible control point in tumor growth. *Ann. Intern. Med.* **82**, 96–100 (1975).
10. Folkman, J. The vascularization of tumors. *Sci. Am.* **234**(5), 58–73 (1976).
11. Gaynor, E. Increased mitotic activity in rabbit endothelium after endotoxin. An autoradiographic study. *Lab. Invest.* **24**, 318–320 (1971).
12. Gillman, T. A plea for arterial biology as a basis for understanding arterial disease. *In* "Biological Aspects of Occlusive Vascular Disease" (D.G. Chalmers and G.A. Gresham, eds.), pp. 3–23. University Press, Cambridge, Massachusetts, 1964.
13. Gimbrone, M.A., Jr., Leapman, S.B., Cotran, R.S., and Folkman, J. Tumor angiogenesis: Iris neovascularization at a distance from experimental intraocular tumors. *J. Nat. Cancer Inst.* **50**, 219–228 (1973).
14. Heymann, M.A., and Rudolph, A.M. Control of the ductus arteriosus. *Physiol. Rev.* **55**, 62–78 (1975).
15. Holman, E. Problems in the dynamics of blood flow. I. Conditions controlling collateral circulation in the presence of an arteriovenous fistula, following the ligation of an artery. *Surgery* **26**, 889–917 (1949).
16. Holman, E., and Taylor, G. Problems in the dynamics of blood flow. II. Pressure relations at site of an arteriovenous fistula. *Angiology* **3**, 415–430 (1952).
17. John, H.T., and Warren, R. The stimulus to collateral circulation. *Surgery* **49**, 14–25 (1961).
18. Kato, H., Kidd, L., and Olley, P.M. Effects of hypoxia on pulmonary vascular reactivity in pneumonectomized puppies and minipigs. *Circ. Res.* **28**, 397–402 (1971).
19. Kovacs, I.B., Mester, E., and Görög, P. Laser-induced stimulation of the vascularization of the healing wound. An ear chamber experiment. *Experientia* **30**, 341–343 (1974).

20. Lambert, P.B., Frank, H.A., Bellman, S., and Friedman, E. Observations on the recovery of continuity of divided arteries and veins. *Angiology* **14**, 121–133 (1963).
21. Leung, D.Y.M., Glagov, S., and Mathews, M.B. Cyclic stretching stimulates synthesis of matrix components by arterial smooth muscle cells in vitro. *Science* **191**, 475–477 (1976).
22. Maher, J.S. Morphologic and histochemical changes in postpartum uterine blood vessels. *Arch. Pathol.* **67**, 175–180 (1959).
23. Maurice, D.M., Zauberman, H., and Michaelson, I.C. The stimulus to neovascularization in the cornea. *Exp. Eye Res.* **5**, 168–184 (1966).
24. Mehrotra, R.M.L. An experimental study of the changes which occur in ligated arteries and veins. *J. Pathol. Bacteriol.* **65**, 307–313 (1953).
25. Meyer, W.W., and Lind, J. The ductus venosus and the mechanism of its closure. *Arch. Dis. Child.* **41**, 597–605 (1966).
26. Muller, W.H., Jr., Dammann, J.F., Jr., and Head, W.H., Jr. Changes in the pulmonary vessels produced by experimental pulmonary hypertension. *Surgery* **34**, 363–375 (1959).
27. Myers, M.B., and Cherry, G. Rate of revascularization in primary and disrupted wounds. *Surg. Gynecol. Obstet.* **132**, 1005–1008 (1971).
28. Naeye, R.L. Perinatal vascular changes associated with underdevelopment of the left heart. *Am. J. Pathol.* **41**, 287–293 (1962).
29. Poole, J.C.F., Cromwell, S.B., and Benditt, E.P. Behavior of smooth muscle cells and formation of extracellular structures in the reaction of arterial walls to injury. *Am. J. Pathol.* **62**, 391–414 (1971).
30. Rodbard, S. Vascular caliber. *Cardiology* **60**, 4–49 (1975).
31. Ross, R., and Glomset, J.A. Atherosclerosis and the arterial smooth muscle cell. *Science* **180**, 1332–1339 (1973).
32. Ross, R., Glomset, J., and Harker, L. Response to injury and atherogenesis. *Am. J. Pathol.* **86**, 675–684 (1977).
33. Schoefl, G.I., and Majno, G. Regeneration of blood vessels in wound healing. *Adv. Biol. Skin* **5**, 173–193 (1964).
34. Schwartz, S.M., and Benditt, E.P. Cell replication in the aortic endothelium: A new method for study of the problem. *Lab. Invest.* **28**, 699–707 (1973).
35. Schwartz, S.M., Stemerman, M.B., and Benditt, E.P. The aortic intima. II. Repair of the aortic lining after mechanical denudation. *Am. J. Pathol.* **81**, 15–42 (1975).
36. Spaet, T.H., Stemerman, M.B., Friedman, R.J., and Burns, E.R. Arteriosclerosis in the rabbit aorta: Long-term response to a single balloon injury. *Ann. N.Y. Acad. Sci.* **275**, 76–77 (1976).
37. Spiro, D., Lattes, R.G., and Wiener, J. The cellular pathology of experimental hypertension. I. Hyperplastic arteriolarsclerosis. *Am. J. Pathol.* **47**, 19–49 (1965).
38. Weibel, E.R. Early stages in the development of collateral circulation to the lung in the rat. *Circ. Res.* **8**, 353–376 (1960).
39. Winblad, J., Reemtsma, K., Vernhet, J.L., Laville, L.P., and Creech, O., Jr. Etiological mechanisms in the development of collateral circulation. *Surgery* **45**, 105–117 (1959).
40. Wright, H.P., Evans, M., and Green, R.P. Aortic endothelial mitosis and Evans blue uptake in cholesterol-fed subscorbutic guinea-pigs. *Atherosclerosis* **21**, 105–113 (1975).
41. Zauberman, H., Michaelson, I.C., Bergmann, F., and Maurice, D.M. Stimulation of neovascularization of the cornea by biogenic amines. *Exp. Eye Res.* **8**, 77–83 (1969).

7

Hypertension and Heart Growth

Coronary: L., *coronalis*, a crown (as the coronary arteries entwine the heart)

The heart has a remarkable capacity for growth, but is poorly endowed with regenerative ability, at least in higher vertebrates. Unlike skeletal muscle fibers, those in cardiac muscle are usually mononucleate and noninnervated. Metabolically, they do not share the capacity of skeletal muscle for anaerobic respiration, an attribute correlated with the generally smaller diameters of myocardial fibers designed to facilitate the exchange of metabolites.

The heart is an adaptable organ. In wild rats and rabbits, it is larger than in tame animals of the same species, but whether this is a genetic or physiological adaptation is admittedly not known. The cardiac muscle fibers of hummingbirds have gigantic mitochondria, while those of the sloth have small ones (5). The weight of the heart in relation to body size from rats to whales is remarkably constant, averaging about 0.45% over a range of absolute weights from less than 0.5 g to 116 kg. Yet in other animals more noted for their physical endurance, the relative weight of the heart is considerably greater. In the deer it measures 1.15%, while in the dog it is 0.8% (*13*). The relative weight of the greyhound heart averages 1.34%, but was 1.73% in the largest case on record!

The adaptability of the heart is also evident in ventricular size. In most mammals the left ventricle is about twice the dimensions of the right one. In the giraffe, whose heart must pump against a considerable head of pressure, the left ventricle is said to be about three times the size of the right (9). It cannot be assumed that such inequities are in fact genetic adaptations. Although they reflect the different work loads of the two ventricles, it is not possible to eliminate these differences experimentally to observe if the two ventricles might then become equal in size. It is worth noting, however, that

138

Fig. 66.   Cross sections of cardiac ventricles from (left to right) newborn, 8-day-, and 2-month-old infants. Left and right ventricles are of equal size prenatally. Following the postnatal development of functional inequities the right ventricular wall becomes thinner than that of the left ventricle (28).

such an experiment has been done by nature, for before birth both ventricles are pumping blood through the systemic circulation and are therefore subject to equal work loads. Accordingly, the prenatal heart is like "the double kernels of a nut," as William Harvey observed in 1628. Not until after birth when the left ventricle experiences a sudden increase in its work load does its growth rate exceed that of the right ventricle (Fig. 66). Actually, there is a drop in the burden on the right side as it takes over the pulmonary circulation following closure of the ductus arteriosus. The right ventricle ceases to grow postnatally, and in some cases may even atrophy until such time as the pulmonary circulation expands to the point of stimulating the resumption of right ventricular growth (28). However universal the postnatal ventricular inequities may be, the fact remains that they are not genetic adaptations, but physiological accommodations to the differential in their work loads.

DIFFERENTIATION VERSUS PROLIFERATION

Cardiac development is an interesting lesson in differentiation. The originally dissociated myoblasts of the embryo, once they have synthesized myofibrils, express the pulsatile activity which is so intrinsic in their species. The originally independent beats become synchronized as the fibers organize themselves into the histological architecture of cardiac muscle. Even while beating, such muscle fibers are still capable of cell division. Their rate of proliferation declines during development, but may persist in the rat until weaning. Cytokinesis presents a problem for such a highly specialized cell. It is achieved by the transverse disruption of fibrils in the vicinity of the dividing nucleus.

Owing to the small size of cardiac muscle fibers in immature animals, the concentration of DNA at birth is considerably higher than it is in the expanded sarcoplasmic mass of the adult heart. While the concentration of DNA may decline during maturation, the total DNA content of the heart is minimal at birth, increasing manifold during subsequent growth. As the rate of DNA synthesis decreases with development, RNA and protein synthesis continue unabated until the adult complement of contractile proteins has been attained. Even then, the mitotically static cardiac muscle continues to renew itself by the degradation and resynthesis of its sarcoplasmic proteins (21).

During postnatal development, increasing numbers of cardiac muscle fibers become binucleate, and in some animals 4, 8, or even 16 nuclei are

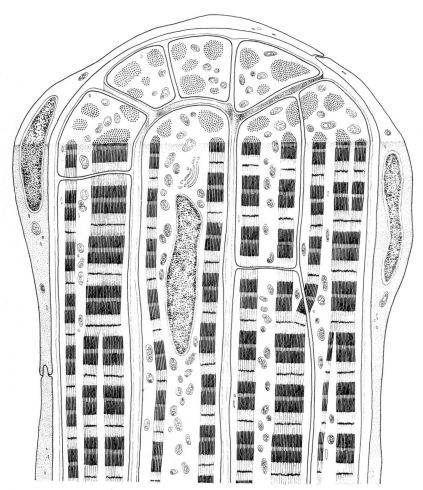

Fig. 67. Cut-away view of a trabecula from the spongy myocardium of an amphibian heart showing the bundle of muscle fibers, surrounded by endocardial epithelium (11, 31).

occasionally encountered. It is estimated that as many as 80% of the myocardial fibers in the adult rat heart are binucleate (*34*). In humans, the incidence of binucleate fibers rises from birth through childhood years, then declines to neonatal levels. This decline is believed to account for the age-related increase in polyploidy due to nuclear fusion (*30*).

The postnatal growth of the heart is achieved primarily by enlargement of its fibers. Concomitantly there is an increase in the blood supply to the heart. At birth there are about 4 muscle fibers per capillary (*19, 27*). In mature hearts the capillary:fiber ratio is 1:1, yet the number of capillaries per mm^2 remains constant. While some new fibers may be produced postnatally, it is hypertrophy of the muscle fibers that accounts for most of the postnatal growth. In infant mammals the diameters of heart muscle fibers may be only one-third to one-half that of adults. Thus, in infant rats, dogs, and humans they average 4–7 μm in width (*27*). At maturity, such fibers measure between 11 and 16 μm, a size range characteristic of the hearts of most other species—from bats to whales.

The hearts of lower vertebrates are different. In fishes, amphibians and reptiles the muscle fibers average 5 μm or less in diameter and their histological organization is unique (*11, 15, 31*). In cyclostomes and in smaller specimens of other cold-blooded vertebrates, the heart is avascular. Groups of several muscle fibers are arranged in trabeculae (Fig. 67) which are organized into a network of contractile tissue. The blood in the ventricular chamber is thus squeezed among the trabeculae in the course of cardiac contractions. Not until such animals grow larger do their hearts develop a less spongy musculature (Fig. 68) (*26*). This peripheral compact cardiac muscle becomes vascularized by the hypobranchial arteries originating on the

Fig. 68.   Left, spongy cardiac muscle in a sector of heart from a young specimen of a cold-blooded vertebrate. Blood percolates among the trabeculae nourishing the nonvascularized heart muscle. In older and larger specimens (right) there develops an outer layer of compact myocardium vascularized by the coronary circulation.

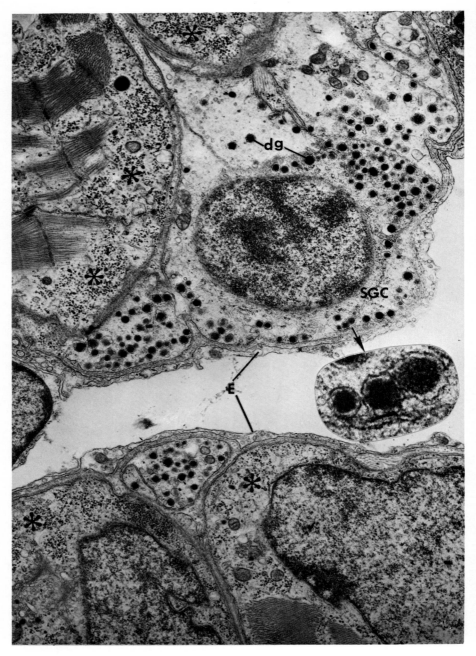

Fig. 69. Electron micrograph of the hagfish heart showing a specific granular cell (SGC), the dense granules (dg) of which are enlarged in the inset (arrow). Several cardiac muscle fibers are indicated (*) in portions of the two trabeculae lined with endothelial cells (E) which separate them from the intervening vascular channel (*18*).

efferent side of the gills (8). In this way, oxygenated blood is supplied to the muscle of the heart in contrast to the venous blood to which spongy cardiac muscle is exposed.

While heart muscle fibers are not innervated, the hearts themselves are. Such nerves are not necessary for the inherent beat of the heart, but regulate the speed of pulsatile activity. It is a curious thing, therefore, that the ventricles of cyclostomes are not innervated. Instead, they produce excessive quantities of adrenaline amounting to concentrations 50 times that of mammalian hearts. This is correlated with the existence of specific granular cells in such hearts, cells containing chromaffin-type granules believed to be the source of adrenaline (Fig. 69) (*18*).

It is well established that the hearts of birds and mammals grow to seemingly predetermined sizes commensurate with the body mass ultimately to be achieved. This is correlated with the switch from hyperplasia to hypertrophy of cardiac muscle fibers early in development, as a result of which there is an upper limit beyond which the heart muscle cannot enlarge without jeopardizing its physiological efficiency. In the case of lower vertebrates, particularly certain species of fishes, there seems to be no definite upper limit to the body size to which they can grow. If the heart is to keep pace with the continued growth of the body in animals of indeterminate size, it must not lose its capacity to augment the population of cardiac muscle fibers. Studies have shown that the diameters of such fibers remain much the same from small-fry to larger specimens, representing increases in body mass many thousands of times. It follows that if such hearts are to continue their growth they must somehow produce new muscle fibers. It is conceivable that the differentiated myocardial fibers in these animals are capable of cell division. Otherwise, new fibers might be recruited from a reserve population of undifferentiated cells. This interesting question is a problem for the future to resolve.

HYPERTROPHY OF THE OVERLOADED HEART

Although the hearts of birds and mammals are incapable of increasing their numbers of muscle fibers during the latter phases of their development, they make up for this by hypertrophy. Cardiac enlargement can be induced experimentally, or pathologically, by a variety of interventions, but the mechanism by which the muscle grows is much the same for all of them. First and foremost, there is an increase in fiber diameter from around 15 μm to over 20 μm. This is achieved by augmenting the number of myofibrils in the sarcoplasm and by the addition of myofilaments to the periphery of preexisting fibrils (*29*). The number of mitochondria also increases, and in extreme cases the incidence of binucleate cells and polyploid nuclei rises. The rates of RNA and protein synthesis increase, as do the activities of respiratory enzymes. Such cell division as may be stimulated is confined primarily to the connective tissue and vasculature. Although the number of capillaries may not proliferate to keep pace with the hypertrophy of the fibers, the overall diameters of the coronary vessels increase to enhance blood flow. Virtually all of these changes are reversible when the conditions responsible for hypertrophy are corrected or discontinued (2, 4, 35).

In its adaptation to overwork, the heart may go through several stages of compensation (22). The first stage involves the initial physiological responses of the heart to the stress of overwork prior to its morphological compensation. In due course, such a heart may meet the added demands and enter into a stable phase of hyperfunction when no further growth is required. If the demands upon the heart are too much to cope with, a phase of progressive exhaustion may supervene leading to permanent pathological changes and ultimately resulting in cardiac insufficiency. Presumably it is the inordinate enlargement of the fibers which contributes to the exhaustion of the heart. As the distance between the innermost parts of the muscle fiber and the nearest capillary increases, the contractile tissues suffer from metabolic insufficiency aggravated by their dependence on aerobic respiration.

It is instructive to note that certain stimuli of compensatory cardiac hypertrophy affect the entire heart while others promote either left or right ventricular enlargement. Studies in 1934 (33) on the hearts of ricksha-pullers in Canton revealed a high rate of cardiac hypertrophy correlated with the chronic exercise of their occupation. Subsequent experiments with animals on treadmills or following forced swimming have confirmed that exercise stimulates hypertrophy of the heart, and that this is manifested in both the left and right ventricles (35).

Not unrelated to this phenomenon is the similar response of the heart to anemia. This can be created experimentally by dietary deficiencies in iron and copper in rats fed a diet of milk, or more directly by repeated bleeding or by phenylhydrazine-induced hemolysis. Marked bilateral hypertrophy of the heart ensues (16), presumably triggered by the heightened cardiac output designed to compensate for the deficiency in oxygen transport. AV fistulas have similar effects related to the reduced circulatory efficiency under these conditions (20). Interestingly, even in the ice fish, an inhabitant of the supercooled waters of the Antarctic, the lack of red blood cells is correlated with hearts that are 2–3 times larger than those in other species possessing normal complements of red blood cells (14).

Both exercise and anemia may be assumed to elicit a compensatory response in the thyroid to raise the metabolic rate. It is no surprise, therefore, that the administration of thyroxine or its equivalent promotes cardiac enlargement (36). Even in the "frizzle fowl," a mutant chicken that loses its feathers, the heart becomes enlarged along with the hyperactive thyroid owing to the bird's attempt to keep warm by increasing its metabolic rate. Cardiac hypertrophy attributed to hyperthyroid conditions is evident in both ventricles, but more so in the right than the left side.

The most direct way to stimulate cardiac hypertrophy is by hypertension. This selectively causes left ventricular enlargement, the extent of which is proportional to the blood pressure. Aortic constriction either in the thorax or posterior to the diaphragm, by application of a ligature reducing the cross-sectional area of the aorta to a fraction of normal, forces the left ventricle to exert greater tension increasing its systolic blood pressure to overcome the stenosis. The degree of left ventricular hypertrophy may approach 50% in several weeks, particularly in young animals (4). This simulates the hypertrophy clinically associated with hypertension, or that established experimentally by application of a Goldblatt clamp to the renal artery, encapsulation of the kidney, or resection of one kidney in an animal treated with desoxycorticosterone acetate (DCA) plus 1% NaCl to drink (2). These techniques elicit high blood pressure by stimulating the juxtaglomerular apparatus to secrete renin. Possibly the cardiac hypertrophy caused by feeding diets rich in cholesterol or by inflicting lesions in the aortic valves may find their ultimate explanations in terms of hypertension.

Left ventricular hypertrophy induced by hypertension is reversible in several weeks following reduction in blood pressure. Indeed, it can be prevented altogether by hypophysectomy, a treatment that not only reduces the blood pressure below normal but brings about cardiac atrophy (1). Although replacement therapy with growth hormone may reduce the extent of such atrophy, it does not prevent it, nor does it affect the weights of intact hearts. Thyroxine, on the other hand, is more effective in reversing the effects of hypophysectomy, suggesting that the pituitary may favor cardiac growth via TSH.

Right ventricular hypertrophy is stimulated by a different set of conditions, not the least of which is unilateral pneumonectomy. This operation, involving the ligation of one pulmonary artery, drastically increases the vascular resistance on the other side. The pulmonary artery of the remaining lung ultimately expands, and its tunica media hypertrophies, but in the meantime the right ventricle is subjected to elevated systolic pressures. Partial constriction of the pulmonary artery elicits comparable responses leading to considerable right ventricular hypertrophy (17). It is undeniable, therefore, that the right ventricle, like the left one, is stimulated to enlarge in response to increases in the resistance against which it must pump.

Less mechanical interventions yield similar results. Oxygen deficiency, for example, whether by exposure to high altitude or reduction in the oxygen content at atmospheric pressure, is an effective way to promote right ventricular hypertrophy (Fig. 70) (10). Indeed, in young animals raised under hypoxic conditions the right ventricle may grow by hyperplasia as well as hypertrophy of its fibers. Cattle forced to graze in the mountains during their first summer run the risk of developing brisket disease, a developmental deformity in the chest associated with cardiac hypertrophy at high altitudes (Fig. 71) (12). The deleterious effects of oxygen deficiency may be translated to the right ventricle in terms of the associated rise in pulmonary arterial pressure. It is conceivable, however, that the stimulation of erythropoiesis by hypoxia (as well as pneumonectomy) may also play a role in stimulating right ventricular hypertrophy. Carbon monoxide, for example, while increasing red cell production also promotes cardiac enlargement (25). $CoCl_2$ likewise stimulates the production of erythropoietin which increases the rate of erythropoiesis and also causes hypertrophy of the right ventricle (32). It is possible, therefore, that the tissues of the heart might be responsive to oxygen deficiency *per se*, but if so this would be expected to affect the left ventricle as much as the right one.

Inasmuch as a variety of experimental interventions outlined above tend to increase the hematocrit while promoting right ventricular hypertrophy, this in itself could be responsible for the growth of the heart. To test this possibility, animals can be made polycythemic by hypertransfusion of red cells. This technique bypasses any interference with the respiratory mechanism, yet it has been shown to result in right ventricular hypertrophy (32). One wonders if the increased viscosity of polycythemic blood might explain this response, but the fact that the left ventricle does not enlarge under these circumstances argues against this explanation. On the other hand, when pregnant ewes are held at high altitude their hearts enlarge but those of their fetuses do not (23), perhaps because the fetal hearts are pro-

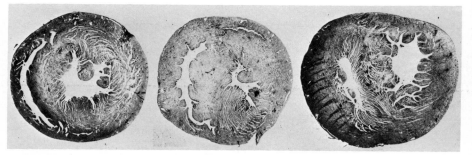

Fig. 70.   Cross sections through the ventricles of rat hearts illustrating a normal control (left), the effects of exposing adult animals to simulated high altitude (middle), and an animal born and raised in a low pressure chamber (right). The thickness of the right ventricular wall (on the left of each specimen) is increased in proportion to the duration of hypoxia (10).

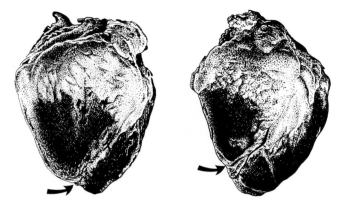

Fig. 71. Left, ventral view of a steer heart from an animal suffering from "brisket disease" brought on by having been exposed to high altitude. Right, heart of a control heifer. Note enlargement of the right ventricle of the affected heart, as indicated by the location of the apex (arrow) (*12*).

tected from the maternal polycythemia, if not the hypoxia itself. The absence of an effect in such cases may also be explained by the relative lack of pulmonary blood flow under prenatal conditions.

Throughout these studies of "halfhearted" hypertrophy, the opposite ventricle is not always unresponsive. For example, aortic constriction elicits its greatest response in the left ventricle, but a modest degree of right ventricular hypertrophy is sometimes observed. Similarly, the left ventricle is not infrequently found to hypertrophy following pulmonary arterial constriction, exposure to high altitude, or exposure to carbon monoxide, although to a considerably lesser extent than the right ventricle. How such vicarious responses are to be explained is not known, but it would seem that the synchronization between the two sides of the heart might result in increased, albeit passive, contraction of the less affected side sufficient to promote some growth.

GROWTH AND ATROPHY OF HEART GRAFTS

The technique of heart transplantation promises to yield valuable information concerning the control of cardiac growth. Fragments of fetal or newborn heart transplanted subcutaneously, or to the hamster cheek pouch, become revascularized and contractile. Whole hearts have even been grafted subcutaneously or to the mesenteries of newts and frogs where they may survive for many months (*3*). The pulsatile activity of such transplants is a useful

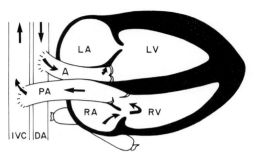

Fig. 72.   Diagram of auxiliary rat heart transplanted to the abdomen (6). Venae cavae and pulmonary veins are tied off. Donor aorta (A) is anastomosed end-to-side to that of the host (DA). Pulmonary artery (PA) of graft is sutured to host inferior vena cava (IVC). Blood enters graft aorta the "wrong" way and is diverted from the left ventricle (LV) and atrium (LA) into the coronary arteries. It is then returned to the right atrium (RA) and pumped to the right ventricle (RV) from which it exits via the pulmonary artery to the inferior vena cava of the host.

indication of the viability of such grafts, and its cessation accurately signals the ultimate rejection of allografts.

With the perfection of microvascular surgery, it has become possible to transplant hearts *in toto* to inbred strains of rats thus obviating immunological complications. Original attempts involved the end-to-end anastomosis of the host's dorsal aorta and inferior vena cava posterior to the level of the kidneys to the aorta and pulmonary artery of the transplant. Surprisingly, the hind quarters of such animals, while temporarily ischemic, usually became revascularized by collateral circulation. Later modifications favored the

Fig. 73.   Comparison of the response of the rat heart to auxiliary transplantation. Above is a normal (nontransplanted) heart as seen in transverse ventricular section. Below is a graft which has atrophied 2 months after operation (6).

end-to-side anastomosis of the donor aorta and pulmonary artery to the host's dorsal aorta and inferior vena cava, respectively, thus disturbing the host's circulation as little as possible (Fig. 72). These auxiliary heart transplants resume beating when their circulation is restored and may survive indefinitely. Certain pathologies occasionally develop as a result of faulty revascularization, not infrequently resulting in the ossification of large portions of the transplant.

The chief value of this technique in exploring the control of heart growth is its effect on the subsequent size of the transplant. These grafts are abnormally vascularized since the incoming blood, unable to enter the left ventricle from the aorta is forced into the coronary circulation. After perfusing the musculature, it is returned to the right atrium from which it is pumped to the right ventricle and out the pulmonary artery into the host's inferior vena cava. The left side of the transplanted heart is therefore bypassed and presumably subject to a work load far below that of untransplanted hearts.

host   graft

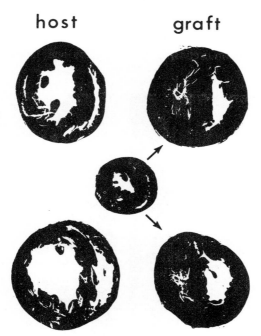

Fig. 74. Transverse sections through the midventricular regions of rat hearts illustrating the growth of infant hearts (center) 4 months following transplantation to the abdomens of adult (upper right) and infant (lower right) hosts (7). On the left are the respective host hearts for comparison. Although auxiliary infant hearts grow equally well in young or adult hosts, their right ventricles become hypertrophic.

Indeed, such hearts would appear to pump only enough blood for their own nourishment. Accordingly, many of them are found to have atrophied to weights as low as one third the mass at the time of grafting (Fig. 73) (6).

There is a minimum size below which microvascular surgery is not feasible, but it has nevertheless been possible to transplant hearts from weanling rats either into other infants or into adult hosts. Although such transplants would not necessarily be physiologically normal, they would be expected to reveal whether cardiac growth is totally dependent upon functional load or is at least in part the result of intrinsic potentials for growth expressed irrespective of the physiological burden. Experiments have shown that infant heart grafts, whether residing in infant or adult hosts, continue to grow nearly to adult dimensions regardless of host age or altered work loads (Fig. 74) (7). Such findings suggest that the immature heart is endowed with a potential for growth over and above that which may be stimulated by functional demands. It is worth noting, however, that in these grafts the left and right ventricles develop to the same proportions, a result that emphasizes that the size to which the heart or its components grow is not unresponsive to physiological work loads.

However capable of growth the mammalian heart may be, its capacity for repair and regeneration is conspicuous by its absence. Coronary infarcts that

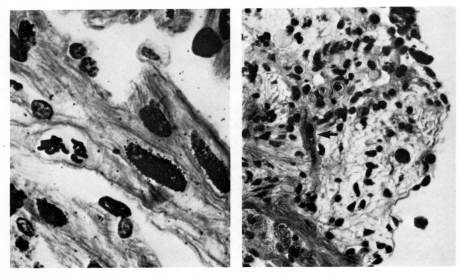

Fig. 75. Repair of the injured newt heart (24). Mitotic figures and labeled nuclei (left) are visible 20 days after wounding. By 30 days (right), connective tissue predominates in the wound area with occasional myocytes present (arrow).

result in the local destruction of myocardial fibers are repaired only by noncontractile scar tissue. Adult heart muscle fibers are incapable of proliferation, and in the absence of satellite cells there seems to be no pool of stem cells from which to recruit new ones. In immature mammals, injuries may sometimes be repaired by proliferation and differentiation if myoblasts are still present. Lower vertebrates fare better. Crush injuries to the hearts of frogs stimulate DNA synthesis in nearby muscle fibers. Indeed, in the newt it is possible to cut off the apex of the ventricle without lethal consequences. Proliferation and differentiation of new muscle fibers in adjacent tissues contribute to the regeneration of such hearts (Fig. 75) (24). This greater potential for repair may be correlated with the greater ability of hearts in lower vertebrates to grow throughout life by increasing their numbers of muscle fibers. Man's shortcomings suffer by comparison, but by retracing the path of evolution we may someday discover what was lost along the way.

REFERENCES

1. Beznak, M. The behaviour of the weight of the heart and the blood pressure of albino rats under different conditions. *J. Physiol. (London)* **124**, 44–63 (1954).
2. Beznak, M., Korecky, B., and Thomas, G. Regression of cardiac hypertrophies of various origin. *Can. J. Physiol. Pharmacol.* **47**, 579–586 (1969).
3. Cohen, N., and Rich, L.C. Exceptionally prolonged survival of allogeneic heart implants in untreated and previously skin grafted salamanders. *Am. Zool.* **10**, 536 (1970).
4. Cutilletta, A.F., Dowell, R.T., Rudnik, M., Arcilla, R.A., and Zak, R. Regression of myocardial hypertrophy. 1. Experimental model, changes in heart weight, nucleic acids and collagen. *J. Mol. Cell. Cardiol.* **7**, 767–781 (1975).
5. DiDio, L.J.A. Comparative study of the ultrastructure of the myocardium of the hummingbird and of the sloth. *J. Cell Biol.* **31**, 28A (1966).
6. Dittmer, J.E., and Goss, R.J. Size changes of auxiliary adult heart grafts in rats. *Cardiology* **58**, 355–363 (1973).
7. Dittmer, J.E., Goss, R.J., and Dinsmore, C.E. The growth of infant hearts grafted to young and adult rats. *Am. J. Anat.* **141**, 155–160 (1974).
8. Foxon, G.E.H. Problem of the double circulation in vertebrates. *Biol. Rev.* **30**, 196–228 (1955).
9. Goetz, R.H., and Keen, E.N. Some aspects of the cardiovascular system in the giraffe. *Angiology* **8**, 542–564 (1957).
10. Grandtner, M., Turek, Z., and Kreuzer, F. Cardiac hypertrophy in the first generation of rats native to simulated high altitude. Muscle fiber diameter and diffusion distance in the right and left ventricle. *Pfluegers Arch.* **350**, 241–248 (1974).
11. Gros, D., and Schrével, J. Ultrastructure comparée du muscle cardiaque ventriculaire de l'Ambystome et de sa larve, l'Axolotl. *J. Microsc. (Paris)* **9**, 765–784 (1970).
12. Grover, R.F., Reeves, J.T., Will, D.H., and Blount, S.G., Jr. Pulmonary vasoconstriction in steers at high altitude. *J. Appl. Physiol.* **18**, 567–574 (1963).
13. Herrmann, G.R. The heart of the racing greyhound: Hypertrophy of the heart. *Proc. Soc. Exp. Biol. Med.* **23**, 856–857 (1925–1926).

14. Holeton, G.F. Respiratory morphometrics of white and red blooded antarctic fish. *Comp. Biochem. Physiol. A* **54**, 215–220 (1976).
15. Kisch, B. The ultrastructure of the myocardium of fishes. *Exp. Med. Surg.* **24**, 220–227 (1966).
16. Korecky, B., and French, I.W. Nucleic acid synthesis in enlarged hearts of rats with nutritional anemia. *Circ. Res.* **21**, 635–640 (1967).
17. Laks, M.M., Morady, F., Garner, D., and Swan, H.J.C. Temporal changes in canine right ventricular volume, mass, cell size, and sarcomere length after banding the pulmonary artery. *Cardiovasc. Res.* **8**, 106–111 (1974).
18. Leak, L.V. Electron microscopy of cardiac tissue in a primitive vertebrate *Myxine glutinosa. J. Morphol.* **128**, 131–157 (1969).
19. Linzbach, A.J. Heart failure from the point of view of quantitative anatomy. *Am. J. Cardiol.* **5**, 370–382.
20. Marchetti, G.V., Merlo, L., Noseda, V., and Visiolo, O. Myocardial blood flow in experimental cardiac hypertrophy in dogs. *Cardiovasc. Res.* **7**, 519–527 (1973).
21. McCallister, B.D., and Brown, A.L. A biochemical and morphological study of protein synthesis in normal rat myocardium. *Cardiovasc. Res.* **3**, 79–87 (1969).
22. Meerson, F.Z. A mechanism of hypertrophy and wear of the myocardium. *Am. J. Cardiol.* **15**, 755–760 (1965).
23. Metcalfe, J., Meschia, G., Hellegers, A., Prystowsky, H., Huckabee, W., and Barron, D.H. Observations on the growth rates and organ weights of fetal sheep at altitude and sea level. *Q. J. Exp. Physiol.* **47**, 305–313 (1962).
24. Oberpriller, J.O., and Oberpriller, J.C. Response of the adult newt ventricle to injury. *J. Exp. Zool.* **187**, 249–259 (1974).
25. Penney, D., Benjamin, M., and Dunham, E. Effect of carbon monoxide on cardiac weight as compared with altitude effects. *J. Appl. Physiol.* **37**, 80–84 (1974).
26. Poupa, O., and Oštádal, B. Experimental cardiomegalies and "cardiomegalies" in free-living animals. *Ann. N.Y. Acad. Sci.* **156**, 445–468 (1969).
27. Rakušan, K., and Poupa, O. Changes in the diffusion distance in the rat heart muscle during development. *Physiol. Bohemoslov.* **12**, 220–227 (1963).
28. Recavarren, S., and Arias-Stella, J. Growth and development of the ventricular myocardium from birth to adult life. *Br. Heart J.* **26**, 187–192 (1964).
29. Richter, G.W., and Kellner, A. Hypertrophy of the human heart at the level of fine structure. An analysis and two postulates. *J. Cell Biol.* **18**, 195–206 (1963).
30. Schneider, R., and Pfitzer, P. Die Zahl der Kerne in isolierten Zellen des menschlichen Myokards. *Virchows Arch. B* **12**, 238–258 (1973).
31. Staley, N.A., and Benson, E.S. The ultrastructure of frog ventricular cardiac muscle and its relationship to mechanisms of excitation-contraction coupling. *J. Cell Biol.* **38**, 99–114 (1968).
32. Swigart, R.H. Polycythemia and right ventricular hypertrophy. *Circ. Res.* **17**, 30–38 (1965).
33. Tung, C.L., Hsieh, C.K., Bien, C.W., and Dieuaide, F.R. The hearts of ricksha-pullers. A study of the effect of chronic exertion on the cardiovascular system. *Am. Heart J.* **10**, 79–100 (1934).
34. Vahouny, G.V., Wei, R., Starkweather, R., and Davis, C. Preparation of beating heart cells from adult rats. *Science* **167**, 1616–1618 (1970).
35. Van Liere, E.J., Krames, B.B., and Northrup, D.W. Differences in cardiac hypertrophy in exercise and hypoxia. *Circ. Res.* **16**, 244–248 (1965).
36. Van Liere, E.J., and Sizemore, D.A. Regression of cardiac hypertrophy following experimental hyperthyroidism in rats. *Proc. Soc. Exp. Biol. Med.* **136**, 645–648 (1971).

8

Muscle: Atrophy versus Hypertrophy

Muscle: L., *musculus*, little mouse (which jumps the way muscles twitch)

Few tissues in the body rival the uniqueness of skeletal muscle. Its total mass exceeds that of all other organs. Its cells are multinucleate, sometimes containing hundreds of nuclei and measuring up to 100 μm in diameter and many centimeters in length. The concentration of protein in the muscle fiber is surpassed by few cells with the possible exceptions of erythrocytes, keratinocytes, or lens fibers. Yet in none of these other cells does the protein turn over, nor is it arranged in such exquisite geometry as in the cross-striated myofibrils into which actin and myosin are organized. The intimacy, both structural and functional, with which the skeletal muscle fiber is associated with nerve fibers as well as the tendons and bones on which they insert is unsurpassed by other cellular associations. Add to this the capacity of muscle to undergo hypertrophy or atrophy, not to mention dystrophy, and one can appreciate why so much research effort has focused on the many unique attributes of skeletal muscle.

As a model system in which to study cellular differentiation, muscle fibers have few equals. Their differentiated state can be accurately diagnosed both biochemically and ultrastructurally, and has been found to be initiated at the earliest fusion of myoblasts in the embryo. While cell division may not necessarily be incompatible with the presence of myofibrils, it does not occur in the multinucleate condition. Hence, muscle fibers grow exclusively by the recruitment of mononucleate myoblasts into their sarcoplasm, giving rise first to myotubes which undergo further enlargement by synthesizing quantities of contractile protein. The myofilaments of actin and myosin become arranged in the hexagonal arrays characteristic of the myofibrils into which they are organized. During maturation of the fiber, the nuclei take up peripheral positions just beneath the sarcolemma, while the sarcoplasm becomes populated with myofibrils. The latter grow at first by the accretion of new myofilaments on their surfaces until they acquire diameters of 1–2 μm. Thereafter, lateral growth ceases, but in some instances new fibrils may be formed by the longitudinal splitting of ones previously formed.

HOW MUSCLES GROW

If muscle fibers are to become large enough to extend from origin to insertion, their sarcoplasmic mass must grow to heroic dimensions. Such prodigious enlargement requires the maintenance of appropriate nucleocytoplasmic ratios. It is interesting to speculate why, in the evolution of such unusual cells, the nearly unprecedented mechanism of cell fusion should have been selected over the seemingly less complicated retention of mitotic potentials while eliminating cytokinesis. The latter mechanism is the way all other multinucleate cells are formed (with the exception of the zygote). It is axiomatic, however, that in a syncytium when one nucleus divides they all do so, reflecting the cytoplasmic origin of whatever triggers mitosis. Such a mechanism for nuclear proliferation would have been too efficient, for it would promote such a rapid growth of developing muscle fibers as to increase their nuclear populations by geometrical progression. This, however, would not have lent itself to the more finely tuned adjustments required by compensating muscle fibers in response to variations in functional loads. Clearly, it is better to acquire one nucleus at a time by myoblast fusion than to grow geometrically by doubling the nuclear population.

Muscles must grow in length as well as width if they are to keep pace with the elongation of the bones to which they are attached. If wire or ink markers are placed in the muscles of immature animals, they move apart during subsequent elongation suggesting that muscles grow by internal expansion (50). It is well known, however, that the sarcomeres of mammalian muscles do not normally depart from their typical lengths of 2–3 μm. If a muscle is stretched, however, by looping its tendon back on itself to shorten it, the sarcomeres may undergo a transient lengthening of up to 11% (59). This may represent an experimental confirmation of the lengthening of sarcomeres during the normal maturation of the muscle. Such modest shifts in the dimensions of sarcomeres, however, cannot account for the considerable degrees of elongation which most muscles undergo during growth. In various muscles of the mouse, for example, the number of sarcomeres per fiber may double or triple from birth to maturity (78). Autoradiographic labeling has shown that the new sarcomeres are added at the ends of the fibers. In view of the evidence from marker studies, that muscles as a whole grow by internal expansion, there is reason to believe that the component muscle fibers might slide past each other during growth. Although individual fibers extend from origin to insertion in small muscles, those in larger ones may reach only part way from one end of the muscle to the other, some anchored at the origin, some at the insertion, and others in intermediate positions unattached to either end of the muscle. These overlapping relays of fibers in whole muscles can be visualized by staining their motor endplates (Fig. 76) (50). Owing to the fact that endplates tend to be situated near the midpoints of muscle fibers, the number of transverse rows of terminals in register equals the number of overlapping fibers making up the length of the muscle.

The capacity of mammalian muscles to undergo hypertrophy is balanced by their usual inability to increase the numbers of fibers beyond early stages of development. There may be exceptions, however. Although exercise is a well-established method to increase the sizes of individual muscle fibers, it may also promote the longitudinal splitting of preexisting fibers. Perhaps this accounts for the larger number of fibers found in the biceps of athletes compared with nonathletic individuals. In contrast to birds and mammals, skeletal muscle fibers of fishes and amphibians have been shown to increase both in number and size during growth (49). The crayfish, on the other hand, retains the same number of fibers per muscle over a wide range of body

Fig. 76.   Demonstration of motor end plate distribution by staining whole muscles for acetylcholine (50). Left, rabbit tibialis anterior muscle showing multiple bands of end plates. Right, rat neck muscles in which single motor end plate bands are visible in the sternomastoids on either side of the sternohyoid muscle.

sizes. They increase the dimensions of their fibers, lengthening the sarco-meres severalfold from young to adult stages (7).

How much a muscle lengthens is commensurate with the job it must perform. Ranging from a few millimeters to many centimeters long, some muscles extend the full distance from origin to insertion while others may be attached to tendons of varying lengths. The latter situation is characteristic of appendages where it is of advantage to situate the bulk of the muscle prox-imally to insure distal slenderness. Whatever the arrangement, there is a limit beyond which a muscle cannot contract. The excursion distance through which the points of origin and insertion can be approximated does not exceed 44% of the relaxed position. It is noteworthy that this figure corresponds with the ratio of A band length to the distance between succes-sive Z bands in the sarcomere, which approximates 56% in the relaxed state.

In view of these relationships, the maximum efficiency of a muscle is achieved when there is no tendon interposed between its distal end and the point of insertion, in which case 100% of the muscle's contraction is trans-lated into pulling the insertion 44% closer to the origin. However, if a tendon is present which is as long as the muscle itself, then maximal contrac-tion of the muscle results in a shortening of the distance between origin and insertion by only about 22%, half the distance realized in the absence of a tendon. The question may be asked, What determines the relative lengths of muscle and tendon such that the resulting combination in maximal contrac-tion will effect an optimal approximation of the parts to which they are attached? To put this to the test, a simple and ingenious operation has been performed in young rabbits (14). This involved severing the crural ligament in the hind leg around which the tendon is looped in traversing the distance from the tibialis anterior muscle to its insertion on the metatarsals. Contrac-tion of the muscle normally pulls on the tendon to flex the foot, but after the operation there is more tendon and muscle than needed between the points of origin and insertion such that even maximal muscle contraction is in large measure wasted on taking up the slack. If such bunnies are to hop efficiently, they must adjust their muscle and tendon lengths accordingly. In unoper-ated animals, the tibialis anterior muscle extends about one half the distance from origin to insertion. In operated animals allowed to mature with the above handicap, the tibialis anterior grows extra long, attaining a length about two-thirds the distance between origin and insertion. The tendon is correspondingly shortened. By means of this developmental accommodation, the efficiency of the system is restored during the process of maturation (Fig. 77).

The implications of these findings cannot be exaggerated. They suggest that the anatomy of muscles and tendons is more profoundly influenced than hitherto suspected by the functional conditions under which they develop.

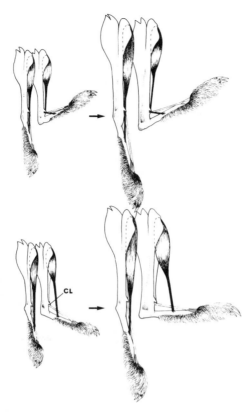

Fig. 77.   Experimental alteration of the muscle:tendon ratio (*14*). On the left are extended and flexed positions of the young rabbit foot; on the right are the same in the adult. In the normal rabbit (above) the length of the extended tibialis anterior equals that of its tendon (drawn in black). When the tendon is surgically shifted in front of the crural ligament (CL) in a young animal (below), flexion is limited. During subsequent maturation, however, its muscle grows extra long thereby increasing its pulling distance to permit nearly normal flexion of the foot.

Presumably, the work load of the muscle is translated into sarcomere synthesis at the ends of the fibers, with reciprocal effects on tendon elongation.

SATELLITE CELLS

Whatever conditions call forth the enlargement of muscle fibers, there must be a source of new nuclei to augment the population in the fiber. In the absence of mitoses in differentiated myonuclei, and of myoblasts in postembryonic stages, it was important to discover the source of mononucleate cells potentially capable of contributing their nuclei to growing muscle fibers. The discovery of satellite cells in 1961 (54) prompted much research in the intervening years, research that has yielded mounting evidence that these are indeed the reservoir of stem cells needed as a source of nuclei for the growth and regeneration of skeletal muscle. Originally disovered in frog muscle, satellite cells have since been found in newts, lizards, birds, and a wide variety of mammals, including man. Presumably present in all chordate skeletal muscle, they have not been described in cardiac muscle nor in the striated musculature of invertebrates. Satellite cells are typically sandwiched between the sarcolemma and its overlying baement lamella (Fig. 78). Lacking differentiated specializations in their cytoplasm, they are recognized by nuclear characteristics distinguishable under electron microscopy. Their num-

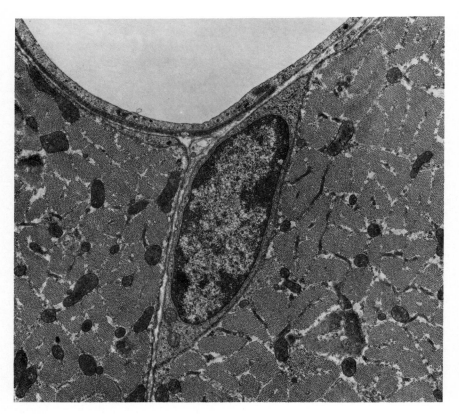

Fig. 78.   Satellite cell in the mouse lumbrical muscle (69). It is associated with the muscle fiber on the right.

bers are highest at birth in mammals when as many as one-third of the total
nuclear population in muscle fibers may be represented by satellite cells
(69). This proportion drops to only a few percent in adult animals, and to less
than 1% in old ones (60). When activated, satellite cells may be released
from the confines of the basement lamella to develop into myoblasts whose
numbers are amplified by proliferation before fusing with each other or with
differentiated muscle fibers. Proliferation of satellite cells may be stimulated
by overwork, denervation, or direct injury to muscle. In traumatized muscle, they provide the wherewithal for regeneration (54). Overwork, promoted by tenotomy of a muscle's synergists, more than doubles the population of satellite cells during the succeeding week, a process that coincides
with and contributes to hypertrophy of the overloaded muscle (37). It is not
clear why their numbers should increase after denervation, when muscle
fibers undergo atrophy. Whatever the reason, it would appear to be prompted

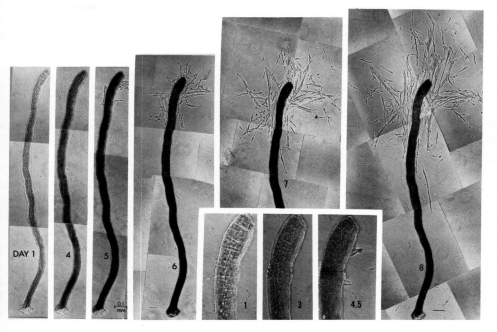

Fig. 79.   This is a 2-cm length of differentiated pectoral muscle fiber dissected from a Japanese
quail and explanted into tissue culture (43). The montages were photographed with phase
contrast at intervals as indicated in days. The myonuclei degenerate within the first day, but
occasional (satellite?) cells soon proliferate into the medium where their bipolar descendants
eventually fuse into multinucleate myotubes (days 7 and 8). The inset shows early stages when
the first cells become visible.

by mechanisms different from those associated with overwork. If denervation is combined with tenotomy of synergistic muscles, almost twice as
many satellite cells appear a week later as are seen in work hypertrophy
alone (37).

Perhaps the most graphic demonstration of satellite cells is to be seen *in
vitro*. When segments of adult muscle fibers are explanted, the myonuclei
promptly degenerate while the satellite cells survive and proliferate (6, 43).
They may go on to form clones of myoblasts inside the endomysium or in the
surrounding culture medium (Fig. 79). In either case, the mononucleate
bipolar cells fuse into myotubes similar to those typically produced in cultures of embryonic myoblasts.

FIBER TYPES

Some extrafusal muscles are adapted for slow, tonic contractions and their
fibers (type I) are of the small, red type. Others are adapted for quick, phasic
contractions and contain fibers (type II) that are large and white. Some
muscles are made up almost exclusively of slow fibers. The soleus of mammals and the avian pectoralis muscle and anterior latissimus dorsi are examples of this type. The gastrocnemius, plantaris, flexor digitorum longus, and
triceps brachialis are composed almost entirely of fast fibers. Most muscles
are a mixture of both slow and fast fibers, but each has its characteristic
proportions of the two. Almost every facet of structure and function is reflected in the fiber types. For example, in slow fibers the fibrils are irregularly arranged in a "Felderstruktur" pattern, while in fast fibers they are
more regularly organized in a "Fibrillenstruktur" arrangement. Slow fibers
are rich in myoglobin and fat, while fast ones are poor in these components.
Slow fibers are poor in glycogen, fast fibers contain much glycogen. Metabolically, slow fibers respire aerobically, fast fibers anaerobically. Accordingly,
slow fibers have greater mitochondrial activity with more oxidative enzymes,
while fast fibers have relatively less mitochondrial activity, and more glyco-
lytic enzymes. Slow fibers tend to be located in the interior of muscles and
have multiple innervation of the "en grappe" type. Fast fibers are usually
located peripherally, and have single innervations of the "en plaque" type. It
is to be emphasized, however, that the classification of muscle fibers into
these two types is an oversimplification (32). There are a number of intermediate types which may share attributes with both slow and fast fibers,
their classification depending upon the spectrum of enzymes to be his-
tochemically visualized. Muscle fibers in embryos are of indeterminate
character, only differentiating into adult types as they take on functional
activities during maturation (71).

In addition to the foregoing types of extrafusal fibers, there are two kinds of intrafusal fibers in skeletal muscle. These are the spindle fibers which act as proprioceptors enabling an animal to monitor changes in the tension of its muscles. Intrafusal fibers occur in groups of up to 10 or 12, first appearing prenatally but not completing their differentiation until after birth. When they first appear they are larger than the extrafusal fibers. Owing to their slower rates of growth, however, they become relatively smaller than other fibers in the adult (52). In the course of their maturation, intrafusal fibers differentiate into nuclear bag or nuclear chain fibers. In nuclear bag fibers the nuclei are clustered in the middle of the fiber interrupting the continuity of the myofibrils. These fibers have double innervation, each end receiving separate endplates. Nuclear chain fibers are half the diameter of the nuclear bag fibers and have a row of centrally located nuclei. Fine axons terminate in a network on the nonnuclear regions.

REACTIONS TO EXERCISE

Muscles are made for work, and it is to this that they are most responsive. Work comes in many forms, the most common of which is increased exercise. Experimentally, this is imposed by forcing animals to run on treadmills or to swim to exhaustion. Alternatively, a muscle can be made to overwork by tenotomy or denervation of its synergists. For example, the gastrocnemius, soleus, and plantaris make up the jumping complex of the hind leg. Inactivation of the gastrocnemius elicits a prompt reaction on the part of the soleus and plantaris. Similarly, removal of the tibialis anterior stimulates a compensatory response in its synergist, the extensor digitorum longus. Sometimes it is possible to remove part of one muscle, particularly one with two heads as in the case of the biceps, thus increasing the work load on the remnant. Comparable reactions may be obtained by incapacitating one leg to throw the burden on the muscles in the contralateral limb. Finally, muscular activity can be passively increased by electrical stimulation.

The reaction of muscle to exercise is manifold. Its hypertrophy is accompanied by hyperemia as the capillary:fiber ratio rises. The individual fibers also enlarge, the slow red ones more so than the fast white ones, perhaps because they are smaller to start with. The maximum size obtained by fibers is not exceeded. There is simply a shift in the bimodal distribution of fiber sizes from the smaller to the larger category (67). It is problematic whether the number of fibers increases with exercise unless it is by their longitudinal splitting. The mitochondria increase in number, size, and enzymatic activity. Creatine phosphokinase, succinic dehydrogenase, and hexokinase activities are elevated, as is ATPase. The glycogen content may rise 50%. Accompanying these changes is the prompt increase in amino acid uptake leading to more rapid synthesis of proteins and an increase in myofibril content (30). The concentration of protein in the muscle fiber remains unchanged, however.

There are two aspects of muscular work that must be taken into account in attempting to explain the hypertrophic response to increased work load. The muscle must contract with greater force, but it must also have something to pull against. The force of contraction is therefore enhanced by the tension of being stretched. Passive stretching, not contraction, has been shown to be the primary stimulus for hypertrophy in overworked muscles. For example, hypertrophy of the soleus in response to tenotomy of its synergist, the gastrocnemius, is prevented when its antagonist is denervated to keep it from being stretched (51). It is in this perspective that the effects of disuse are most logically to be interpreted.

IMMOBILIZATION

It is not easy to prevent a muscle from working. Restricted exercise is unsatisfactory in experimental animals which naturally resist forced inactivity. Immobilization of limbs is more effective, although one cannot entirely eliminate the voluntary isometric contractions of muscles. When a tendon is cut, the muscle is not prevented from contracting but neither is it stretched. Care must be taken, however, to prevent the reunion of the severed tendon, a reaction that occurs with inconvenient efficiency. Perhaps the most effective means of rendering a muscle nonfunctional is by denervation. Singly or in combination, these various methods of promoting muscular disuse have much to tell us about the physiological factors that promote the growth of muscle.

The position in which a limb is immobilized, whether in a plaster cast or by inserting a rod into the marrow cavities to two bones on either side of the joint, has a profound effect on the sizes of the muscles. If the flexors are relaxed the extensors are stretched, and vice versa. In general it is the relaxed muscles that atrophy when immobilized. Both red and white fibers atrophy at approximately the same rates, but the population densities of their myonuclei increase as their sarcoplasms diminish in mass. In young growing animals, the normal postnatal increase in the number of nuclei per fiber is reduced, an effect reversed when immobilization is discontinued. The number of sarcomeres in series is also reduced in immobilized adult muscles and their normal increase in the growing muscles of young animals is cut in half (79). The numbers of mitochondria decrease along with their cytochrome oxidase activities. There is also a decrease in the protein contents of the atrophying fibers, and a decline in the cholinesterase in the motor end plates. In contrast, acid hydrolase activity in the lysosomes increases. Although immobilization does not cause absolute atrophy of young growing muscles, it brings about a relative atrophy by reducing their rates of growth. When a cast is removed prior to maturity, catch-up growth to normal dimensions occurs, beginning 3–5 days after release (15). The reduced number of sarcomeres is restored to normal within a month (79).

Those muscles immobilized in lengthened positions are subject to chronic stretching. Accordingly, they tend to undergo hypertrophy, including extra elongation to the extent of augmenting their numbers of sarcomeres 20% (75). Such results testify to the fact that immobilization *per se* does not cause atrophy, but that it is the absence of the stretch stimulus that seems to be responsible. Since tenotomy simulates immobilization in a relaxed position, the ensuing atrophy may be attributed to the lack of tension. The possibility arises that this may also explain the atrophy of denervation, for in the absence of contraction, tension is normally not created.

DENERVATION ATROPHY

Muscles and their nerves are meaningless without each other. Together, their association poses some of biology's most fascinating problems, not the least of which is the nature of the neurotrophic influence responsible for maintaining the structural and functional integrity of skeletal muscle. To appreciate this problem it is necessary to review the normal innervation of muscle.

Each motor neuron sends out an axon which branches a number of times before establishing myoneural junctions with the fibers of the muscle it innervates. Thus, one nerve may innervate many muscle fibers, but in mammals each muscle fiber usually, but not always, receives only one axon. Those muscle fibers innervated by the sprouts of a single neuron make up a motor unit. The size of the motor unit is inversely proportional to the functional precision of the muscle. Hence, the ratio of neurons to muscle fibers in the extrinsic ocular muscles is approximately 1:5 in keeping with the fine adjustments needed in positioning the eye for steady vision. In contrast, the gluteus maximus, a muscle not noted for the acuity of its movements, has a nerve:muscle ratio of about 1:200.

The motor end plates which comprise the synapse between axon terminal and sarcolemma are of two general types (40). "En grappe" terminals, which may be widely distributed along a muscle fiber, tend to be associated with slow fibers. "En plaque" terminals, more consolidated in structure, are found on fast fibers. Although each muscle fiber usually has only a single motor end plate, exceptions to this rule are not uncommon, particularly among lower vertebrates. Single end plates may receive two or more axons derived from the same or different neurons. In other cases, more than one end plate may be present on a single fiber, sometimes from the axonal branches of one neuron, sometimes of polyneural origin. In higher vertebrates polyneural innervation of muscle fibers is typical of neonates (10). All but one of the endplates are eliminated after several weeks of age. Multiple innervation in infant birds and mammals may therefore be a throwback to our cold-blooded ancestors. The disappearance of extra terminals in infants resembles the normal turnover of end plates in adult mammals whereby older ones degenerate to be replaced with new ones established by collateral sprouts from the same axon (76).

The functional unit of the motor end plate is the acetylcholine (ACh) receptor. These are located in the sarcolemma where they mediate the effects of ACh from the neural component of the end plate. ACh receptors occur in densities up to $10^4/\mu m^2$ in the region of the myoneural junction, there being on the order of 10^7 receptors/end plate (Fig. 80) (38, 39). Outside the end plate region the ACh receptor density may be only $5/\mu m^2$. This accounts for the concentration of ACh sensitivity in the motor end plate and its near absence in extrajunctional sites on the muscle fiber. In fetal and neonatal animals ACh sensitivity is high throughout the length of the muscle fiber. Prior to maturation of the myoneural junction ACh receptors are synthesized everywhere in the sarcolemma, as evidenced by the ribosomal aggregates in subsarcolemmal locations throughout the length of the fiber (28). A similar distribution of ACh receptors is seen in cultured embryonic muscle even in the absence of mitosis or fusion of myoblasts (22,23), as well as in differentiating muscle fibers in regenerating salamander limbs (16). It is only when developing muscle fibers receive innervation that the extrajunctional ACh receptors disappear and the pattern of ACh sensitivity shifts predominately to the end plate region (Fig. 81).

Upon denervation the reverse occurs. The extrajunctional ACh receptor population increases more than tenfold in density (21). They are synthesized by subsarcolemmal ribosomes which appear after denervation (28). The appearance of ACh receptors in extrajunctional locations on the denervated muscle fiber is responsible for the marked hypersensitivity to ACh which develops along the entire length of the fiber. There is a concomitant decline in cholinesterase activity in the end plate regions, but a rise elsewhere in association with the spread of ACh receptors (46). Meanwhile, the diameters of the muscle fibers decrease steadily for many months. Although some fibers may eventually disappear, others persist with widths only one-fifth their normal diameters. The greatest atrophy is to be seen in the fast white fibers, perhaps because they are larger to begin with. As the sarcoplasm

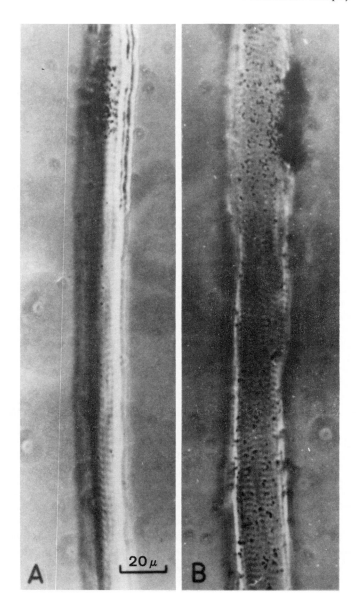

Fig. 80. Distribution of ACh receptors in normal (A) and denervated (B) rat diaphragm muscle fibers as seen in autoradiographs of preparations previously exposed to ^{125}I-α-bungarotoxin (38). In innervated fibers the ACh receptors are almost exclusively confined to the end plate region (A). After denervation, numerous extrajunctional bungarotoxin binding sites appear (B).

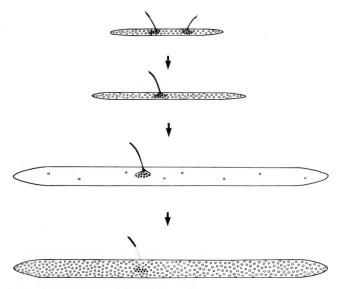

Fig. 81.   The distribution of acetylcholine receptors in developing muscle fibers (top to bottom). Solid circles are receptors associated with motor end plates. Open circles are extrajunctional ACh receptors. Muscle fibers of neonates (top) are multiply innervated and their ACh receptors are distributed throughout the sarcolemma. In older infants, all but one axon are withdrawn after which the ACh receptors become restricted mostly to the end plate. If a muscle is denervated it reacquires extrajunctional ACh receptors (bottom), the increased sensitivity of which results in fibrillation.

becomes depleted the myonuclei become more concentrated. With increasing rates of proteolysis myofibrils degenerate while protein synthesis undergoes little or no decrease.

Unlike the muscle fibers with which they are associated, satellite cells are not innervated. However, when the muscle fiber is denervated, their numbers increase severalfold in the next few weeks (37). Whether they proliferate or their numbers are augmented by the conversion of former myonuclei is not known. This response, like their reaction to muscle traumatization, suggests a possible sensitivity on the part of satellite cells to the state of their muscle fibers. Conceivably, it is the incipient degenerative changes in the muscle fiber, whether triggered by denervation (37), injury (64), overwork, or explanation *in vitro* (6, 43), which triggers the proliferation of satellite cells. Such a reaction has obvious advantages in terms of subsequent muscle repair, despite its futility in the absence of innervation.

Another response to denervation, less well established than that of satellite cells but of no less potential importance, is the proliferation of certain cells of undetermined type which are associated with the motor end plate

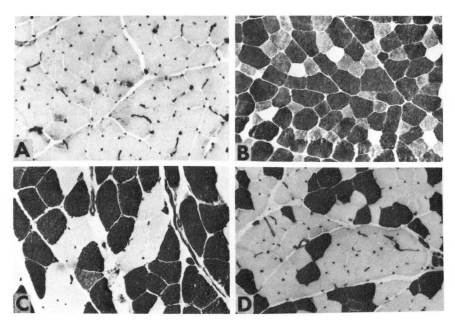

Fig. 84.   Effects of cross-reinnervation of skeletal muscles on the distribution of fiber types. The normal cat soleus muscle (A) is made up entirely of slow fibers with low ATPase activities, while the flexor hallucis longus (B) contains many more fast fibers (darkly stained for ATPase) than slow ones. One year after the nerves normally supplying each of these muscles were cut and sutured to the distal segments of the other, the cross-reinnervated soleus (C) has acquired numerous fast fibers while in the flexor hallucis longus (D) they have been largely replaced by slow fibers (*31*).

either case, however, the foreign nerve endings persist, whether functional or nonfunctional, when the original nerves return, thus resulting in hyperinnervated muscles. Experiments on mice and rats favor the functional persistence of the foreign innervation, sometimes with poor recovery by the original nerve when it grows back to its muscle. In such cases some muscle fibers may become doubly innervated by both foreign and original nerves (27). This in itself, however, suggests a certain preference for the original innervation inasmuch as extra nerves brought into an already innervated muscle do not form functional connections in the presence of the endogenous innervation.

It has been pointed out earlier that in most instances a muscle fiber is equipped with a single end plate. Although there are exceptions to this rule, the fact remains that the nerve-muscle relationship tends to be a rather exclusive one. To be otherwise would invite chaos.

The challenge is to determine how the presence of one nerve ending on a muscle fiber renders it unreceptive to the advances of others. The monogamy observed in adult fibers is not seen in young ones first establishing relations with axons. Immature muscle fibers typically become innervated by a number of nerve fibers, all but one of which eventually regress (10). It is conceivable that the multiple innervation of immature fibers may be correlated with the widespread sensitivity to ACh along their lengths (28). Nevertheless, experiments in rats have shown that hyperinnervation even of neonatal muscles does not occur when extra nerves are implanted into them, despite their extrajunctional ACh receptors (65).

Attempts to promote hyperinnervation of adult muscles results in the ramification of nerve fibers among the already innervated muscle fibers, but end plates fail to form. Such supernumerary nerves may not be without their effects, however, for they may induce a modest degree of muscular hypertrophy (33). Although extra nerves in a muscle do not make functional connections with the fibers, they readily do so as soon as the endogenous nerves are cut. This is reminiscent of the reinnervation of denervated muscle fibers by collateral sprouts from nearby residual nerves (1, 20).

Since surgical denervation readily permits the reinnervation of muscle by foreign nerves, it must alter whatever is normally responsible for precluding hyperinnervation. Experiments with pharmacological denervation have shed considerable light on this problem, for certain neurotoxins have been shown to permit the establishment of functional connections with extra nerves in already innervated muscles. Bupivacaine, for example, which anesthetizes the nerve by interfering with impulse conduction, permits hyperinnervation of muscles despite the morphological persistence of the original end plates (35). Cobra venom does likewise, as does botulinum toxin (19, 42). All of these agents, like surgical denervation, cause hypersensitivity to ACh. Indeed, botulinum promotes hyperinnervation from a muscle's own nerve by stimulating axonal overgrowth, the sprouts of which establish junctions elsewhere on the muscle fiber (19). The conclusion seems inescapable that the key to muscle innervation lies in the distribution of ACh receptors in the muscle fiber.

REGENERATION AND TRANSPLANTATION

Skeletal muscles, like most other tissues in the body, are capable of repairing injuries (12). These injuries can take many forms, including ischemia, burning, freezing, crushing, transection, and transplantation. Whatever the cause, the regenerative response may be continuous or discontinuous (3).

The former is observed following incision wounds in which regeneration proceeds by sprouting from the stumps of the severed fibers, followed by fusion to close the gap. The outgrowths may not always fuse with their original counterparts, but it is interesting to note that innervated components fuse only with uninnervated ones (77), thus avoiding the embarrassment of a double nerve supply. Furthermore, if the innervated half of one type of fiber fuses with the uninnervated half of another type, the latter converts to the former type.

Discontinuous regeneration of muscle fibers is the process by which mononucleate myoblasts proliferate and fuse to each other or to differentiated muscle fibers. Virtually all kinds of injury elicit this response by stimulating DNA synthesis within hours in the satellite cells, followed by their conversion to myoblasts. There is evidence that under such circumstances myonuclei from the differentiated muscle fiber may become partitioned off from the sarcoplasm and transformed into satellite cells (64). Such a process would serve two functions. It would aid in the replenishment of the satellite cell population and it would enable the myonuclei to be recycled while amplifying their numbers.

Transplantation cannot be achieved without a measure of trauma. Explants of muscle *in vitro* yield satellite cells which proliferate and fuse into myotubes. The same process occurs *in vivo* on a larger scale. Whether grafted whole or minced, there is widespread degeneration of fibers caused by the attending ischemia (12). Large central areas of the transplant are lost in this way, but peripheral fibers, closer to the blood supply and sooner to be revascularized, may survive. It is from this tissue that the muscle regenerates. The derivative myoblasts fuse into new myotubes which, if reinnervated, will continue to grow. Such regenerated muscles seldom attain their normal dimensions, but their fibers become parallel unless tension is prevented from developing.

The regeneration of transplanted muscles is enhanced by prior denervation. This treatment results in the survival of more fibers, and the earlier degeneration of those destined to be lost (13). Differentiation and contraction occur precociously. These effects may be attributed either to the greater capacity of denervated muscles for anaerobic respiration, better equipping them to survive ischemia, or to the larger population of satellite cells which develops after denervation and becomes available for regeneration.

The repair of minced muscle occurs no better in lower vertebrates than in higher ones, despite the ability of the former to regenerate entire appendages, musculature included. Even in the limbs of salamanders capable of this feat, completely excised muscles are not replaced any better than they are in mammals (11). Yet if such a limb is amputated proximal to the position of such a muscle, it is entirely restored in the course of limb regeneration. Hence, the limitations of tissue repair are entirely overcome by the process of epimorphic regeneration.

MYOPATHIES

The pathologies of muscle sometimes tell us as much as man-made experiments, particularly when neoplasias and dystrophies can be simulated in the laboratory. The cancers of muscle are particularly interesting since their growth depends upon the proliferation of undifferentiated cells. One of the most malignant forms of cancer is the rhabdomyosarcoma, a growth that can be induced by the direct injection of nickel sulfide (Ni_3S_2) in suspension into a skeletal muscle. After 3–4 months, there usually develops at the site of injection a rapidly growing rhabdomyosarcoma composed of many proliferating mononucleate cells interspersed between multinucleate fibers of abnormal configurations (4). The differentiation of fibers, however abnormal they may be, constantly removes mononucleate cells from the proliferating population. Continued growth is sustained by the constant replenishment of stem cells. Such tumors are not known to contract upon stimulation, but whether this is due to their lack of sufficient contractile apparatus or the absence of innervation remains to be determined.

Myasthenia gravis is an equally interesting condition, the etiology of which is now known. Its symptoms include atrophy, inflammation, and a progressive weakening of the muscle which eventually kills its victims. Enlargement of the thymus due to thymitis in this disease yielded the clue that finally revealed its autoimmune nature. Most patients contain antibodies in their serum against ACh receptors. This serum, injected into experimental animals, decreases the number of such receptors and inhibits the binding of α-bungarotoxin to them (2). Treatment may include the administration of anticholinesterases to enhance the action of ACh, or depression of antibody production by ACTH or prednisone. Remission of the disease by thymectomy may operate on the same principle (29).

The exploration of muscular dystrophy has been enhanced by genetic strains of experimental animals which develop conditions similar to, if not identical with, some of its forms which affect humans. It first manifests itself during maturation when central degeneration of muscle fibers occurs. This is accompanied by an increase in lysosomes and may lead eventually to a 50% decrease in the total number of fibers per muscle. Curiously, the remaining fibers may compensate by discontinuous regeneration and hypertrophy. The dystrophic muscle, therefore, may be a mixture of degenerated fibers interspersed with hypertrophic ones. The lesion would appear to be in the turnover of proteins, for there is an accelerated degradation of muscle protein indicating a defect in whatever physiological mechanisms regulate the rate of breakdown (44). The process of protein synthesis is evidently unaffected, for this is accelerated in muscular dystrophy as if to compensate for the rapid breakdown, a compensation insufficient to offset the net loss of protein.

The obvious symptoms of muscular dystrophy imply a myogenic etiology of the disease. This viewpoint is supported by the relatively limited regeneration of dystrophic muscle *in vitro* (61). Indeed, the regeneration of cultured dystrophic muscle fibers remains poor regardless of whether they are innervated by normal or dystrophic embryonic spinal cords. When limb buds are transplanted in exchange grafts between normal and dystrophic chick embryos, the muscles in the dystrophic limb retain the dystrophic characteristics despite their normal nerve supply (47). Cross-reinnervation experiments between parabiosed normal and dystrophic mice have also failed to alter the natures of either types of muscle (17). It is conceivable that muscular dystrophy might be attributed to the systemic distribution of humoral factors. To test this possibility, normal and dystrophic mice have been grafted together in parabiosis. The prolonged survival of the dystrophic partners and the decline in the severity of their disease favors the hypothesis that blood-borne factors may be involved (63). The fact that the normal parabiotic partners suffered no untoward effects from being joined with dystrophic ones suggests that some essential ingredient in normal blood may be lacking in muscular dystrophy. Apropos of this, it has been shown that when allophenic chimeras are created by dissociating the blastomeres of normal and dystrophic mouse embryos and when reassembling them into a single individual, the resulting mice are not dystrophic (62).

There is mounting evidence in favor of the neurogenic etiology of muscular dystrophy. Dystrophic muscles are supplied by fewer axons than normal, axons that tend to be abnormally myelinated (9). The motor end plates are hypertrophic and have less junctional folding than normal. Dystrophic muscles have fewer but larger motor units than do normal ones. This has been explained by the possible loss of neurons resulting in the reinnervation of their muscle fibers by axonal sprouts from neighboring motor units (55). To test the possible role of nerves in promoting muscular dystrophy, muscles have been grafted in various combinations between normal and dystrophic hosts (41). Normal grafts into normal animals degenerate at first but eventually regenerate nearly to normal. Normal muscles transplanted into dystrophic hosts generally undergo little or no regeneration, although they do not degenerate as much as do dystrophic muscles grafted into dystrophic hosts. The crucial experiment is the transplantation of dystrophic muscles into normal hosts. Although unanimity of results is lacking, such grafts tend to survive and regenerate, albeit not so well as do normal muscles. It is possible that the improvement in such dystrophic grafts might be accounted for by their innervation with normal nerves, just as the poor response of normal muscles in dystrophic hosts might be explained by a defect in the dystrophic nerves.

Unhappily, it is still premature to render a categorical verdict as to the basic nature of muscular dystrophy. There is too much compelling evidence in favor of both the myogenic and the neurogenic hypotheses, not to mention the possibility that a systemic factor, or lack of same, might be responsible for the disease. There is little doubt, however, that future research on both normal and pathological muscle will continue to shed light on our understanding of how this unique tissue grows.

REFERENCES

1. Aguilar, C.E., Bisby, M.A., Cooper, E., and Diamond, J. Evidence that axoplasmic transport of trophic factors is involved in the regulation of peripheral nerve fields in salamanders. *J. Physiol. (London)* **234**, 449–464 (1973).
2. Almon, R.R., Andrew, C.G , and Appel, S.H. Serum globulin in myasthenia gravis: Inhibition of α-bungarotoxin binding to acetylcholine receptors. *Science* **186**, 55–57 (1974).
3. Aloisi, M. Patterns of muscle regeneration. *In* "Regeneration of Striated Muscle and Myogenesis" (A. Mauro, S.A. Shafiq, and A.T. Milhorat, eds.), pp. 180–193. Excerpta Medica, Amsterdam, 1970.
4. Basrur, P.K., Sykes, A.K., and Gilman, J.P.W. Changes in mitochondrial ultrastructure in nickel sulfide-induced rhabdomyosarcoma. *Cancer* **25**, 1142–1152 (1970).
5. Bennett, M.R., and Pettigrew, A.G. The formation of synapses in reinnervated and cross-reinnervated striated muscle during development. *J. Physiol. (London)* **241**, 547–573 (1974).
6. Bischoff, R. Regeneration of single skeletal muscle fibers *in vitro*. *Anat. Rec.* **182**, 215–235 (1975).
7. Bittner, G.D., and Traut, D.L. Growth of crustacean muscles and muscle fibers. *J. Comp. Physiol., Sect. A,* **124**, 277–285 (1978).
8. Blunt, R.J., Jones, R., and Vrbová, G. Inhibition of cell division and the development of denervation hypersensitivity in skeletal muscle. *Pfluegers Arch.* **355**, 189–204 (1975).
9. Bradley, W.G., and Jenkison, M. Abnormalities of peripheral nerves in murine muscular dystrophy. *J. Neurol. Sci.* **18**, 227–247 (1973).
10. Brown, M.C., Jansen, J.K.S., and Van Essen, D. Polyneuronal innervation of skeletal muscle in newborn rats and its elimination during maturation. *J. Physiol. (London)* **261**, 387–422 (1976).
11. Carlson, B.M. The regeneration of a limb muscle in the axolotl from minced fragments. *Anat. Rec.* **166**, 423–436 (1970).
12. Carlson, B.M. The regeneration of skeletal muscle. A review. *Am. J. Anat.* **137**, 119–149 (1973).
13. Carlson, B.M., and Gutmann, E. Regeneration in free grafts of normal and denervated muscles in the rat: Morphology and histochemistry. *Anat. Rec.* **183**, 47–61 (1975).

14. Crawford, G.N.C. An experimental study of muscle growth in the rabbit. *J. Bone Jt. Surg.,* *Br. Vol.* **36** 294–303 (1954).
15. Crawford, G.N.C. The effect of temporary limitation of movement on the longitudinal growth of voluntary muscle. *J. Anat.* **111**, 143–150 (1972).
16. Dennis, M.J., and Ort, C.A. The distribution of acetylcholine receptors on muscle fibers of regenerating salamander limbs. *J. Physiol. (London)* **266**, 765–776 (1977).
17. Douglas, W.B. Sciatic cross-reinnervation of normal and dystrophic muscle in parabiotic mice: Isometric contractile responses of reinnervated tibialis anticus and triceps surae. *Exp. Neurol.* **48**, 647–663 (1975).
18. Duchen, L.W. Motor nerve growth induced by botulinum toxin as a regenerative phenomenon. *Proc. Roy. Soc. Med.* **65**, 196–197 (1972).
19. Duchen, L.W., Rogers, M., Stolkin, C., and Tonge, D.A. Suppression of botulinum toxin-induced axonal sprouting in skeletal muscle by implantation of an extra nerve. *J. Physiol. (London)* **248**, 1P–2P (1975).
20. Edds, M.V., Jr. Collateral nerve regeneration. *Q. Rev. Biol.* **28**, 260–276 (1953).
21. Fambrough, D.M. Acetylcholine receptors. Revised estimates of extrajunctional receptor density in denervated rat diaphragm. *J. Gen. Physiol.* **64**, 468–472 (1974).
22. Fambrough, D.M., and Devreotes, P. Synthesis and degradation of acetylcholine receptors in cultured chick skeletal muscle. *In* "Exploratory Concepts in Muscular Dystrophy II" (A.T. Milhorat, ed.), pp. 55–69. Excerpta Medica, Amsterdam, 1974.
23. Fambrough, D.M., and Rash, J.E. Development of acetylcholine sensitivity during myogenesis. *Dev. Biol.* **26**, 55–68 (1971).
24. Feng, T.-P., and Lu, D.-X. New lights on the phenomenon of transient hypertrophy in the denervated hemidiaphragm of the rat. *Sci. Sin.* **14**, 1772–1784 (1965).
25. Fenichel, G.M., Kibler, W.B., Olson, W.H., and Dettbarn, W.-D. Chronic inhibition of cholinesterase as a cause of myopathy. *Neurology* **22**, 1026–1033 (1972).
26. Fernandez, H.L., and Ramirez, B.U. Muscle fibrillation induced by blockage of axoplasmic transport in motor nerves. *Brain Res.* **79**, 385–395 (1974).
27. Frank, E., Jansen, J.K.S., Lømo, T., and Westgaard, R. Maintained function of foreign synapses on hyperinnervated skeletal muscle fibres of the rat. *Nature (London)* **247**, 375–376 (1974).
28. Gauthier, G.F., and Schaeffer, S.F. Ultrastructural and cytochemical manifestations of protein synthesis in the peripheral sarcoplasm of denervated and newborn skeletal muscle fibres. *J. Cell Sci.* **14**, 113–137 (1974).
29. Genkins, G., Papatestas, A., Horowitz, S., and Kornfeld, P. Studies in myasthenia gravis: Early thymectomy. *Am. J. Med.* **58**, 517–524 (1975).
30. Goldberg, A.L., Jablecki, C., and Li, J.B. Effects of use and disuse on amino acid transport and protein turnover in muscle. *Ann. N.Y. Acad. Sci.* **228**, 190–201 (1974).
31. Guth, L., Samaha, F.J., and Albers, R.W. The neural regulation of some phenotypic differences between the fiber types of mammalian skeletal muscle. *Exp. Neurol.* **26**, 126–135 (1970).
32. Guth, L., and Yellin, H. The dynamic nature of the so-called "fiber types" of mammalian skeletal muscle. *Exp. Neurol.* **31**, 277–300 (1971).
33. Guth, L., and Zalewski, A.A. Disposition of cholinesterase following implantation of nerve into innervated and denervated muscle. *Exp. Neurol.* **7**, 316–326 (1963).
34. Gwyn, D.G., and Aitken, J.T. The formation of new motor endplates in mammalian skeletal muscle. *J. Anat.* **100**, 112–126 (1966).
35. Hall-Craggs, E.C.B. Hyperinnervation of muscle following treatment with bupivacaine. *Anat. Rec.* **184**, 420–421 (1976).

36. Hall-Craggs, E.C.B., and Seyan, H.S. Histochemical changes in innervated and denervated skeletal muscle fibers following treatment with bupivacaine (Marcain). *Exp. Neurol.* **46**, 345–354 (1975).

37. Hanzlíková, V., Macková, E.V., and Hník, P. Satellite cells of the rat soleus muscle in the process of compensatory hypertrophy combined with denervation. *Cell Tissue Res.* **160**, 411–421 (1975).

38. Hartzell, H.C., and Fambrough, D.M. Acetylcholine receptors. Distribution and extrajunctional density in rat diaphragms after denervation correlated with acetylcholine sensitivity. *J. Gen. Physiol.* **60**, 248–262 (1972).

39. Hartzell, H.C., and Fambrough, D.M. Acetylcholine receptor production and incorporation into membranes of developing muscle fibers. *Dev. Biol.* **30**, 153–165 (1973).

40. Hess, A. Vertebrate slow muscle fibers. *Physiol. Rev.* **50**, 40–62 (1970).

41. Hironaka, T., and Miyata, Y. Transplantation of skeletal muscle in normal and dystrophic mice. *Exp. Neurol.* **47**, 1–15 (1975).

42. Jansen, J.K.S., Lømo, T., Nicolaysen, K., and Westgaard, R.H. Hyperinnervation of skeletal muscle fibers: Dependence on muscle activity. *Science* **181**, 559–561 (1973).

43. Konigsberg, U.R., Lipton, B.H., and Konigsberg, I.R. The regenerative response of single mature muscle fibers isolated *in vitro*. *Dev. Biol.* **45**, 260–275 (1975).

44. Kruh, J., Dreyfus, J.C., Schapira, G., and Gey, G. Abnormalities of muscle protein metabolism in mice with muscular dystrophy. *J. Clin. Invest.* **39**, 1180–1184 (1960).

45. Landmesser, L. Contractile and electrical responses of vagus-innervated frog sartorius. *J. Physiol. (London)* **213**, 707–725 (1971).

46. Linkhart, T.A., and Wilson, B.W. Role of muscle contraction in trophic regulation of chick muscle acetylcholinesterase activity. *Exp. Neurol.* **48**, 557–568 (1975).

47. Linkhart, T.A., Yee, G.W., and Wilson, B.W. Myogenic defect in acetylcholinesterase regulation in muscular dystrophy of the chicken. *Science* **187**, 549–550 (1975).

48. Lømo, T., and Rosenthal, J. Control of ACh sensitivity by muscle activity in the rat. *J. Physiol. (London)* **221**, 493–513 (1972).

49. Luquet, P., and Durand, G. Évolution de la teneur en acides nucléiques de la musculature épaxiale au cours de la croissance chez la truite arc-en-ciel (*Salmo gairdnerii*); Rôles respectifs de la multiplication et du grandissement cellulaire. *Ann. Biol. Anim., Biochim. Biophys.* **10**, 481–492 (1970).

50. Mackay, B., and Harrop, T.J. An experimental study of the longitudinal growth of skeletal muscle in the rat. *Acta Anat.* **72**, 38–49 (1969).

51. Mackova, E., and Hnik, P. Compensatory muscle hypertrophy in the rat induced by tenotomy of synergistic muscles. *Experientia* **27**, 1039–1040 (1971).

52. Maier, A., and Eldred, E. Postnatal growth of the extra- and intrafusal fibers in the soleus and medial gastrocnemius muscles of the cat. *Am. J. Anat.* **141**, 161–177 (1974).

53. Mark, R.F., and Marotte, L.R. The mechanism of selective reinnervation of fish eye muscles. III. Functional, electrophysiological and anatomical analysis of recovery from section of the IIIrd and IVth nerves. *Brain Res.* **46**, 131–148 (1972).

54. Mauro, A. Satellite cell of skeletal muscle fibers. *J. Biophys. Biochem. Cytol.* **9**, 493–495 (1961).

55. McComas, A.J., Sica, R.E.P., and Upton, A.R.M. Quantitative motor unit studies in human muscular dystrophy. *In* "Exploratory Concepts in Muscular Dystrophy II" (A.T. Milhorat, ed.), pp. 560–563. Excerpta Medica, Amsterdam, 1974.

56. Melichna, J., and Gutmann, E. Stimulation and immobilization effects on contractile and histochemical properties of denervated muscle. *Pfluegers Arch.* **352**, 165–178 (1974).

57. Midrio, M., Bouquet, F., Durighello, M., and Princi, T. Role of muscular disuse in the genesis of fibrillation is denervated muscle. *Experientia* **29**, 58–59 (1973).

58. Muchnik, S., Ruarte, A.C., and Kotsias, B.A. Effects of actinomycin D on fibrillation activity in denervated skeletal muscles of the rat. *Life Sci.* **13**, 1763–1770 (1973).

59. Muhl, Z.F., and Grimm, A.F. Tendon shortening in striated muscle. *Experientia* **31**, 1053–1054 (1975).

60. Ontell, M. Muscle satellite cells: A validated technique for light miscroscopic identification and a quantitative study of changes in their population following denervation. *Anat. Rec.* **178**, 211–227 (1974).

61. Paul, C.V., and Powell, J.A. Organ culture studies of coupled fetal cord and adult muscle from normal and dystrophic mice. *J. Neurol. Sci.* **21**, 365–379 (1974).

62. Peterson, A.C. Chimaera mouse study shows absence of disease in genetically dystrophic muscle. *Nature (London)* **248**, 561–564 (1974).

63. Pope, R.S., Murphy, E.D., and West, W.T. Histopathology of dystrophic and normal mice after timed periods of parabiosis. *Anat. Rec.* **151**, 151–158 (1965).

64. Reznik, M. Origin of myoblasts during skeletal muscle regeneration. Electron microscopic observations. *Lab. Invest.* **20**, 353–363 (1969).

65. Robbins, N. Neurotrophic interactions: A brief review and some experiments. I. Neuronal regulation of muscle membrane properties: A mini-review. *In* "Exploratory Concepts in Muscular Dystrophy II" (A.T. Milhorat, ed.), pp. 398–405. Excerpta Medica, Amsterdam, 1974.

66. Robert, E.D., and Oester, Y.T. Electrodiagnosis of nerve-impulse deprived skeletal muscle. *J. Appl. Physiol.* **28**, 439–443 (1970).

67. Rowe, R.W.D., and Goldspink, G. Surgically induced hypertrophy in skeletal muscles of the laboratory mouse. *Anat. Rec.* **161**, 69–75 (1968).

68. Schiaffino, S. Hypertrophy of skeletal muscle induced by tendon shortening. *Experientia* **30**, 1163–1164 (1974).

69. Schultz, E. A quantative study of the satellite cell population in postnatal mouse lumbrical muscle. *Anat. Rec.* **180**, 589–595 (1974).

70. Scott, S.A. Persistence of foreign innervation on reinnervated goldfish extraocular muscles. *Science* **189**, 644–646 (1975).

71. Shafiq, S.A., Asiedu, S.A., and Milhorat, A.T. Effect of neonatal neurectomy on differentiation of fiber types in rat skeletal muscle. *Exp. Neurol.* **35**, 529–540 (1972).

72. Sola, O.M., Christensen, D.L., and Martin, A.W. Hypertrophy and hyperplasia of adult chicken anterior latissimus dorsi muscles following stretch with and without denervation. *Exp. Neurol.* **41**, 76–100 (1973).

73. Stewart, D.M. Effect of age on the response of four muscles of the rat to denervation. *Am. J. Physiol.* **214**, 1139–1146 (1968).

74. Stewart, D.M., Sola, O.M., and Martin, A.W. Hypertrophy as a response to denervation in skeletal muscle. *Z. Vergl. Physiol.* **76**, 146–167 (1972).

75. Tabary, J.C., Tabary, C., Tardieu, C., Tardieu, G., and Goldspink, G. Physiological and structural changes in the cat's soleus muscle due to immobilization at different lengths by plaster casts. *J. Physiol. (London)* **224**, 231–244 (1972).

76. Tuffery, A.R. Growth and degeneration of motor end-plates in normal cat hind limb muscles. *J. Anat.* **110**, 221–247 (1971).

77. Webb, P. The effect of innervation, denervation, and muscle type on the reunion of skeletal muscle. *Br. J. Surg.* **60**, 180–182 (1973).

78. Williams, P.E., and Goldspink, G. Longitudinal growth of striated muscle fibres. *J. Cell Sci.* **9**, 751–767 (1971).

79. Williams, P.E., and Goldspink, G. The effect of immobilization on the longitudinal growth of striated muscle fibres. *J. Anat.* **116**, 45–55 (1973).

80. Williams, P.E., and Goldspink, G. The effect of denervation and dystrophy on the adaptation of sarcomere number to the functional length of the muscle in young and adult mice. *J. Anat.* **122**, 455–465 (1976).

81. Zalewski, A.A. Effects of reinnervation on denervated skeletal muscle by axons of motor, sensory, and sympathetic neurons. *Am. J. Physiol.* **219**, 1675–1679 (1971).

9

Adaptive Plasticity of the Nervous System

Retina: L., *rete*, net

Neurons are improbable cells. With the possible exception of certain eggs, no other cell in the body surpasses in sheer cytoplasmic volume that of nerve cells with axons or dendrites more than a meter long in larger species. Unlike their rivals in size, the multinucleate skeletal muscle fibers, the vast cytoplasmic domain of the neuron is presided over by a single nucleus. The outsized dimensions of nerve cells are attained during the long period of development when they must continue to grow to keep pace with the rest of the body following their premature loss of mitotic competence in prenatal stages. What they lack in proliferative abilities, however, they make up in their capacities for regeneration, an attribute of peripheral nerves which for unaccountable reasons is not shared by the central nervous system.

The complexity of the nervous system is amplified by the dependence of nerves on the cells with which they must maintain synaptic connections. Whether linked with sense organs or effectors, or in synapse with other nerve cells, the structural and functional integrity of the neuron cannot be maintained in the absence of end organ connections any more than muscles and sense cells can escape the atrophy that ensues denervation. Equally impressive are the symbiotic relationships between nerves and the variety of their associated glial elements, not the least of which are the Schwann cells. The anatomical intimacy with which these satellites are bound up with nerves attests to the existence of physiological and developmental affiliations thus far only suspected.

The number and sizes of fibers in a peripheral nerve reflect the effects of the end organs. As the peripheral field in a growing animal enlarges, the number of nerve fibers supplying it increases, as do their diameters. Similarly, the nerves of large species contain more fibers than do homologous nerves in small ones, although such phylogenetic increases in the absolute numbers of fibers per nerve represent relative decreases in proportion to body weight. The number of fibers per nerve is said to decline in old age.

Although the number of neurons and nerve fibers is presumably determined by genetic factors, the sizes to which they grow are very much affected by their end organ connections. For example, if a motor nerve is prevented from reaching its muscle the fiber diameters remain small (3). Immobilization of a limb causes the dimensions of its nerve fibers to decrease (27), an effect similar to that seen in nerves supplying tenotomized muscles. Conversely, functional overload of muscles by increased exercise or the denervation of synergistic muscles brings about an increase in nerve fiber diameters, but no change in their numbers (75). Sensory nerves show similar changes when their peripheral fields are enlarged. Thus, if some of the cells in a spinal ganglion of a rat are destroyed as a result of cutting their peripheral fibers, the remaining ones which must grow back into the entire peripheral field undergo enlargement (13). A comparable situation prevails in the regenerating lizard tail (68). Since no new spinal ganglia are differentiated in the process of regeneration, the regenerate itself becomes innervated by fibers derived from the three pairs of sensory ganglia immediately anterior to the level of amputation. Required to innervate a considerably expanded peripheral field, the cells in these ganglia increase in volume. Since this reaction is limited to the originally smaller neurons, there are more cells of larger dimensions, with no increase in overall maximum size.

SCHWANN CELLS

Peripheral nerves are intimately associated with their Schwann cells, from which the myelin sheath is derived. Myelination is initiated when the axonal diameter approaches 1 μm (54). Prior to this, there is a multiplication of Schwann cells associated with the nerves of young animals, a number of axons being imbedded in the cytoplasm of each one (Fig. 85) (20). Once myelination commences, however, the Schwann cells cease to divide and each one becomes exclusively associated with a single nerve fiber. The myelin sheath is laid down as a series of lamellae winding around the circumference of the axon (88). During nerve elongation, myelination proceeds in a proximal-distal direction. The thickness of the myelin sheath increases less rapidly than does the diameter of the axon. Therefore, the relative thickness of the myelin sheath decreases with the growth of the nerve fiber.

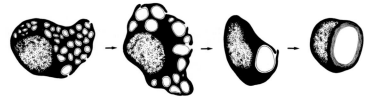

Fig. 85. Developmental relationship between Schwann cell (black cytoplasm, stippled nucleus) and axons (open cross sections) (*20*). With increasing age (left to right) the number of axons associated with each Schwann cell decreases as their diameters increase. Schwann cells cease dividing when their ratio to the number of axons is 1:1, after which they begin to lay down a myelin sheath (right).

Not all peripheral nerves are myelinated, although even those which are not have Schwann cells. Whether or not myelination occurs would appear to be determined by the nature of the nerve rather than that of the Schwann cells (*1*). If a myelinated fiber is cut and its proximal stump united with the distal portion of an unmyelinated nerve, many of the regenerating fibers become myelinated (Fig. 86). Unmyelinated fibers growing into pathways

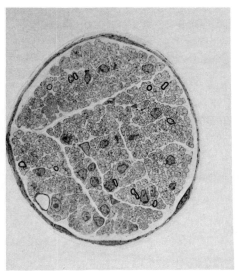
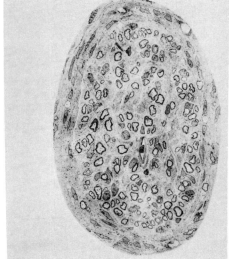

Fig. 86. Myelination of an unmyelinated nerve segment grafted to a myelinated nerve (*1*). Left, control mouse cervical sympathetic trunk (CST) in which less than a dozen of the approximately 2000 fibers are myelinated (1000×). Right, segment of CST one month after its transplantation between the severed stumps of the normally myelinated sural nerve (700×). Note the striking increase in the number of myelinated axons due to the regeneration of normally myelinated fibers into a population of susceptible Schwann cells.

formerly occupied by myelinated nerves do not become myelinated. This may be taken to indicate that the differentiation of myelin sheaths from Schwann cell cytoplasm is induced by the nerve fibers themselves.

The proliferation of Schwann cells would seem also to be under the control of their associated nerves. Although the presence of a myelin sheath precludes Schwann cell proliferation, the loss of myelin restores mitotic potential. This is dramatically illustrated in the Wallerian degeneration of cut nerves. The Schwann cells distal to the level of transection now lose their myelin sheaths and undergo a burst of mitotic activity (45). During the next few weeks the population of such cells may increase considerably, more so in relation to large fibers than small ones. Repeated crushing of a nerve stimulates a progressive rise in the Schwann cell population downstream (83). The dedifferentiated Schwann cells perform a phagocytic function in assisting in the removal of debris from the degeneration of severed axons and fragments of myelin sheaths. In contrast to myelinated nerves, severed unmyelinated ones exhibit an increase in the proliferation of Schwann cells which is primarily confined to the site of injury (73). This would appear to be more a wound healing response to local trauma than a reaction to the Wallerian degeneration of axons.

Each Schwann cell occupies a length of nerve fiber between successive nodes of Ranvier. An internode is roughly 100 times as long as its fiber diameter. The number of internodes, and therefore Schwann cells, does not increase once the nerve fiber has become myelinated. Thus, internode lengths increase with the circumferential and longitudinal expansion of nerve fibers in growing animals (86, 78). If an adult nerve is cut and allowed to regenerate, the ensuing proliferation of distal Schwann cells, followed by their redifferentiation to envelop the regenerated fibers with new myelin sheaths, results in the production of shorter internode lengths than in the

Fig. 87.   Internode dimensions of myelin sheaths in relation to axonal dimensions (86). In infant rabbit (top) the internodes are short and the thin myelin sheath corresponds to the narrow axon. Adult rabbits (middle) have longer and larger axons with increased lengths of internodes and thickness of myelin sheaths. Regenerated adult nerves (bottom) are enveloped by shorter internodes owing to the proliferation of Schwann cells in nerves severed from their cell bodies.

original nerve because there are several times as many Schwann cells as before (Fig. 87) (86). However, if the nerve of a *young* animal is allowed to regenerate, the resulting adult internode lengths are longer than in regenerated adult nerves because of the continued increase in body size after the completion of nerve regeneration.

NEURAL SPECIFICITY

The specificity with which each end organ is innervated by the "right" nerve is something which has intrigued neuroscientists for years, the more so for the apparent lack of direction exhibited by regenerating peripheral nerve fibers. By cross union with the distal segments of other nerves, they can readily be made to grow down foreign pathways. Even sensory fibers will regenerate along motor pathways, although they fail to establish synaptic connections with the muscle fibers to which they are led. Various motor fibers will innervate incorrect muscles following cross-reinnervation, although the foreign innervation may not persist if the original nerves are allowed to grow back. Nevertheless, such nerves retain their intrinsic physiological attributes and may be responsible for reversing the fast and slow muscle fibers to their own types (see Chap. 8). A comparable situation obtains in the case of sense organs. By cross union of nerve fibers, for example, it is possible to cause taste buds to be reinnervated from a variety of sources (see Chap. 2). Some, but not all, of these different nerves have been shown to promote taste bud regeneration when they reach the appropriate epidermis. Thus, it is not the nerve which is altered by such experimental manipulations, but the response of the end organ to the type of fiber by which it becomes innervated.

This lack of neural specificity is exhibited in the phenomenon of collateral innervation. If a peripheral field is partially denervated, the remaining nerves nearby sprout side branches which grow into the adjacent territory deprived of nerves (2). Both sensory and motor fibers exhibit this phenomenon, but it has been most thoroughly studied in the case of partially denervated muscles (24). Here, axonal sprouts originating from the nodes of Ranvier of intact nerves grow into the denervated portion of the muscle where they establish new motor end plates on muscle fibers separated from their original axons. In this way, the motor units of the surviving neurons are expanded, a phenomenon illustrated by the partial denervation of salamander hind limbs (2). These are normally supplied by the 15th, 16th, and 17th spinal nerves, which lead to the anterior, middle, and posterior regions of the leg, respectively. When the 16th nerve is cut, the territories of the 15th and 17th nerves expand into the central region of the leg. Thus, the boundaries of a region supplied by a given nerve are established by confrontation with the peripheral fields of adjacent ones, but the original territory may expand if the one next to it is vacated. It may be concluded, therefore, that the tendency for collateral branching is normally held in abeyance by the presence of neighboring nerves. What attribute it is that inhibits collateral sprouting is not understood, except that it is not the physical presence of axons which is responsible. If a pharmacological blockade is applied to the 16th spinal nerve of a newt, while not actually transecting the nerve, then the 15th and 17th nerves still send collaterals into the territory normally supplied by the 16th nerve. The application of colchicine, which blocks axoplasmic flow, or xylocaine, which interferes with the transmission of impulses, are both effective in eliminating whatever influence is responsible for holding in check the tendencies of adjacent nerves to expand their peripheral fields (2).

PERIPHERAL NERVE REGENERATION

The capacity of nerve fibers to regenerate has much in common with the process by which nerve fibers normally find their way to their end organs in the embryo or develop collateral sprouts in the adult. It is in the nature of such fibers to continue growing until they establish peripheral connections. If prevented from doing so, they may simply grow in aimless profusion, giving rise to neuromas such as frequently develop in the stumps of amputated limbs. On the other hand, it has been shown that nerves can regenerate across greater distances than were spanned by the original fibers if given the opportunity to do so. For example, if the two arms of a newt are amputated distally and their stumps grafted end-to-end, the nerves of one will grow into the other if the latter is denervated. Under these circumstances, regenerating nerve fibers grow in the wrong direction along old pathways, sometimes nearly doubling their original lengths (31).

Whether nerves are severed by cutting or crushing, Wallerian degeneration of the axons separated from the nucleated portion of the nerve promptly ensues. Axon cylinders are destroyed, myelin sheaths disintegrate, and Schwann cells proliferate in the distal stumps (45). There may be limited retrograde degeneration of the proximal stumps, even accompanied by a modest proliferative response among the associated Schwann cells. This reaction is soon reversed, however, by the outgrowth of surviving fibers. When the nerve is simply crushed, an operation that interrupts the continuity of axons but leaves the connective tissue sheaths more or less intact, the proximal nerve fibers more easily find their way back to their end organs than in cases of nerve transection when regenerating fibers are apt to grow astray. Although there is much variation, mammalian peripheral nerve fibers

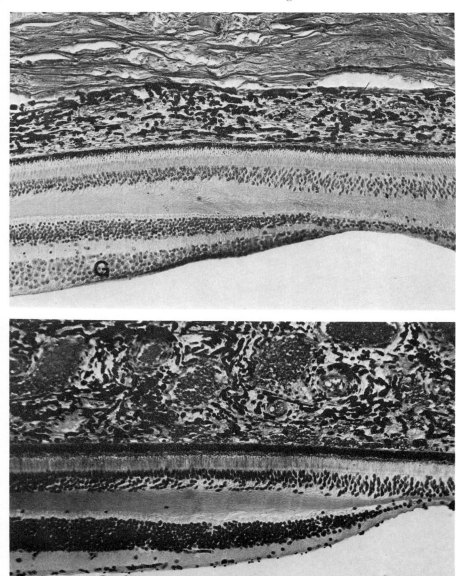

Fig. 93.   Chimpanzee retinas from an animal exposed from birth to 1½ hours of light per day (above), and one allowed only a few minutes of light a day (below) (15). Note the thick layer of retinal ganglion cells (G) on the vitreous surface in the former case, compared with the sparse population of such cells in the latter.

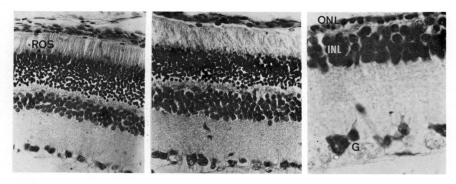

Fig. 94. Influence of continuous illumination on the albino rat retina (66). Left, section through a normal retina showing orderly arrangement of rod outer segments (ROS). Middle, after 4 days of constant exposure to light the rod outer segments are in disarray. Right, a retina following 14 days of illumination. Only a single layer of pycnotic nuclei remains of the outer nuclear layer (ONL) of photoreceptors. The inner nuclear layer (INL) and retinal ganglion cells (G) are not affected.

during the third day when they separate from the inner segments of the cells. The nuclei of the rod cells are pyknotic by two weeks and completely disappear in a month (66). Continuous illumination eventually destroys all but about 2% of the rods and 40% of the cones in the rat retina (48). The primary lesion is in the outer segments of the photoreceptors, which is consistent with the fact that no degeneration takes place in infant rats until they reach about four weeks of age, before which time their outer segments are not fully mature (67).

In retinal dystrophy, the lesion is in the retinal pigment cells (25), although it is the photoreceptor that eventually suffers. Studies of dystrophic strains of rats have confirmed the absence of phagocytosis by the retinal pigment cells, leading to a buildup of rod outer segment material (Fig. 95). The onset of symptoms in growing animals does not occur until the rod outer segment approaches mature lengths, after which further growth is interrupted as the escalation of proteins ceases (38). The onset of the disease is earlier in unpigmented strains of rats (49), and its progress is delayed if animals are kept in the dark (22).

In the normal retina there is a delicate balance between the growth of the rod outer segment and its phagocytosis by the retinal pigment cells. There may be a physiological feedback from the latter to the former, for in the absence of phagocytosis, as in retinal dystrophy, further elongation of the rod outer segments is arrested. Without synthesis of new rod outer segment material the very survival of the rod cell is jeopardized. Such cells may eventually disappear in retinal dystrophy, as they do following vitamin A deficiency (21) or exposure to continuous illumination (66). The latter two

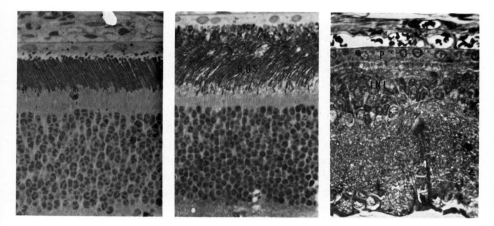

Fig. 95.   Comparison of normal and dystrophic rat retinas (22). Left, a normal retina in a 22-day-old rat. Middle, a dystrophic retina in a 22-day-old rat showing incipient deterioration of the rod outer segments (ROS). Right, a 1-year-old dystrophic retina in which the photoreceptor cells have disappeared altogether, leaving the pigment epithelium (P) adjacent to the bipolar cells of the inner nuclear layer (INL).

conditions seem to have much in common, for both involve a relative deple-tion of rhodopsin. Vitamin A deficiency prevents the synthesis of rhodopsin in the first place. Continuous bright illumination promotes its breakdown faster than it can be regenerated. Despite the seemingly similar effects of these two influences, they do not operate synergistically. Indeed, vitamin A deficiency protects the retina from the deleterious effects of light (64), suggesting that it may be rhodopsin or one of its byproducts that is responsi-ble for light-induced retinal damage.

CNS REGENERATION

The extent to which nerve cells in the central nervous system fail to survive injuries to their fibers is not known, although it is probably not so prevalent as to explain why the brain and spinal cord of higher vertebrates are so much more deficient in regenerative abilities than are the peripheral nerves. There are many conceivable explanations for the poor regenerative ability of the CNS. The deficiency may reside in the neurons themselves. It is not uncommon to observe the initiation of regeneration from severed nerve fibers in the CNS, but these tend later to be retracted without estab-lishing functional synaptic connections. Another possibility is that the failure of mammalian CNS regeneration might be attributed to the production of antibodies against autoantigens from the CNS which, because of the blood–brain barrier, do not ordinarily escape from the brain unless an injury is inflicted. Treatment of rats with immunosuppressants favors recovery in some cases following spinal cord transection (29). Encouraging as these approaches may appear to be, however, the results should be interpreted with considerable caution.

The same reservations apply to investigations of extraneuronal factors affecting CNS regeneration. Such studies are predicated on the possibility that neurons in the brain and spinal cord may be fully competent to regenerate but prevented from doing so by the milieu in which they are imbedded. Like other tissues in the body, the CNS is capable of forming scar tissue at the site of a lesion. Experimental attempts to hold this reaction in abeyance by injection of trypsin or the application of a Millipore filter around the severed spinal cord have met with some success in suppressing scar formation and promoting regeneration (5, 30). Some years ago there was reason to suspect that the outgrowth of severed axons in the CNS might be hindered by the ground substance, the consistency of which could be altered by treatment with bacterial pyrogens, factors otherwise responsible for the production of fevers in infections (89). Experimental results were in some cases quite impressive, for the recovery of animals with transected spinal cords was little short of remarkable. Unhappily, the results were only temporary, and no treatment has since been discovered to prevent this loss of functional recovery.

Apropos of the hypothesis that the CNS might simply be an unfavorable environment for nerve fiber regeneration, it is logical to determine to what extent peripheral nerves known for their ability to regenerate can grow into the substance of the CNS. This has been studied in two ways. One has been to cut the dorsal roots of spinal nerves, the cell bodies of which are located in the dorsal ganglia. Such axons can in fact grow back into the spinal cord (63). Similarly, if the severed ventral motor nerve is anastomosed to the dorsal root, it too can regenerate into the spinal cord and establish synaptic connections (4). A reciprocal approach to the problem has been to transplant fragments of skin or skeletal muscle into the substance of the brain to determine if nerve fibers in the CNS can grow into peripheral tissues. They can in fact penetrate these tissues, although the evidence that functional connections are established is problematic (62). Thus, regenerative ability in the mammalian CNS is not altogether lacking, but it is clearly deficient, perhaps for more complicated reasons than have been heretofore suggested.

The nature of this deficiency may be put into perspective by comparison with lower vertebrates, the central nervous systems of which are remarkable for their capacities to perform many feats of growth mammals cannot achieve (89). In embryonic or newly hatched fishes, morphological evidence of spinal

the growth rate during the period of lactation. In such animals, less protein develops in the cerebral hemispheres, the cells of which are smaller than normal (*14*). The number of cells is not affected. Interference with the growth of the brain during these early developmental periods produces lasting and irreversible effects on later development. Thus, a malnourished infant may grow more slowly, but the rate of its development *per se* does not slow down accordingly (*26*). However capable the body as a whole may be of catch-up growth upon refeeding after a period of malnutrition, certain crucial aspects of neural development cannot be compensated once the corresponding ages and stages of development have been outgrown.

This aspect of mammalian growth may signify one of the chief differences between higher and lower vertebrates. Fishes do not necessarily outgrow their capacity to augment the population of neurons in their brains. Young fishes exhibit a higher rate of mitotic activity in their brains than do older ones, but even in mature specimens there is still to be found a residual capacity for DNA synthesis (*71*). In view of the indeterminate body size of many species of fishes, it is inevitable that the brain must have evolved a mechanism to keep pace with the enlargement of the rest of the body. It does so by retaining the proliferative potential of its ependymal cells from which new neurons can be recruited. Whether this occurs in response to feedback from a continuously growing body, or is itself the cause of somatic enlargement, is an interesting problem for conjecture. Nevertheless, the possibility cannot be denied that the superior regenerative ability in the brains of lower vertebrates compared with birds and mammals may be causally related to their retention of growth potentials in the CNS throughout life.

REFERENCES

1. Aguayo, A.J., Epps, J., Charron, L., Bray, G.M. Multipotentiality of Schwann cells in cross-anastomosed and grafted myelinated and unmyelinated nerves: Quantitative microscopy and radioautography. *Brain Res.* **104**, 1–20 (1976).
2. Aguilar, C.E., Bisby, M.A., Cooper, E., and Diamond, J. Evidence that axoplasmic transport of trophic factors is involved in the regulation of peripheral nerve fields in salamanders. *J. Physiol. (London)* **234**, 445–464 (1973).
3. Aitken, J.T., and Thomas, P.K. Retrograde changes in fibre size following nerve section. *J. Anat.* **96**, 121–129 (1962).
4. Barnes, C.D., and Worrall, N. Reinnervation of spinal cord by cholinergic neurons. *J. Neurophysiol.* **31**, 689–694 (1968).
5. Bassett, C.A.L., Campbell, J.B., and Husby, J. Peripheral nerve and spinal cord regeneration: Factors leading to success of a tubulation technique employing Millipore. *Exp. Neurol.* **1**, 386–406 (1959).

6. Bennett, E.L., Rosenzweig, M.R., and Diamond, M.C. Rat brain: Effects of environmental enrichment on wet and dry weights. *Science* **163**, 825–826 (1969).

7. Bernstein, J.J. The regenerative capacity of the telencephalon of the goldfish and rat. *Exp. Neurol.* **17**, 44–56 (1967).

8. Bittner, G.D., and Mann, D.W. Differential survival of isolated portions of crayfish axons. *Cell Tissue Res.* **169**, 301–311 (1976).

9. Blaxter, J.H.S., and Jones, M.P. The development of the retina and retinomotor responses in the herring. *J. Mar. Biol. Assoc. U.K.* **47**, 677–697 (1967).

10. Bondy, S.C., and Margolis, F.L. Effects of unilateral visual deprivation on the developing avian brain. *Exp. Neurol.* **25**, 447–459 (1969).

11. Butler, E.G., and Ward, M.B. Reconstitution of the spinal cord after ablation in adult *Triturus*. *Dev. Biol.* **15**, 464–486 (1967).

12. Cankovic, J.G. Contribution to the study of regenerative-degeneration qualities of the fasciculi optici in mammals under experimental conditions. *Acta Anat.* **70**, 117–123 (1969).

13. Cavanaugh, M.W. Quantitative effects of the peripheral innervation area on nerves and spinal ganglion cells. *J. Comp. Neurol.* **94**, 181–219 (1951).

14. Chase, H.P., Lindsley, W.F.B., and O'Brien, D. Undernutrition affects development of the cerebellum in rats. *Nature (London)* **221**, 554–555 (1969).

15. Chow, K.L., Riesen, A.H., and Newell, F.W. Degeneration of retinal ganglion cells in infant chimpanzees reared in darkness. *J. Comp. Neurol.* **107**, 27–40 (1957).

16. Coulombre, A.J. The role of intraocular pressure in the development of the chick eye. I. Control of eye size. *J. Exp. Zool.* **133**, 211–225 (1956).

17. Coulombre, A.J., and Coulombre, J.L. Lens development. I. Role of the lens in eye growth. *J. Exp. Zool.* **156**, 39–48 (1964).

18. Coulombre, J.L., and Coulombre, A.J. Influence of mouse neural retina on regeneration of chick neural retina from chick embryonic pigmented epithelium. *Nature (London)* **228**, 559–560 (1970).

19. Currie, J., and Cowan, W.M. Some observations on the early development of the optic tectum in the frog (*Rana pipiens*), with special reference to the effects of early eye removal on mitotic activity in the larval tectum. *J. Comp. Neurol.* **156**, 123–142 (1974).

20. Diner, O. Les cellules de Schwann en mitose et leurs rapports avec les axones au cours du development du nerf sciatique chez le Rat. *C.R. Acad. Sci. Paris* **261**, 1731–1734 (1965).

21. Dowling, J.E. Night blindness. *Sci. Am.* **215**(4), 78–84 (1966).

22. Dowling, J.E., and Sidman, R.L. Inherited retinal dystrophy in the rat. *J. Cell Biol.* **14**, 73–110 (1972).

23. Eccles, J.C. The plasticity of the mammalian central nervous system with special reference to new growths in response to lesions. *Naturwissenschaften* **63**, 8–15 (1976).

24. Edds, M.V., Jr. Collateral nerve regeneration. *Q. Rev. Biol.* **28**, 260–276 (1953).

25. Edwards, R.B., and Szamier, R.B. Defective phagocytosis of isolated rod outer segments by RCS rat retinal pigment epithelium in culture. *Science* **197**, 1001–1003 (1977).

26. Eichenwald, H.F., and Fry, P.C. Nutrition and learning. *Science* **163**, 644–648 (1969).

27. Eisen, A.A., Carpenter, S., Karpati, G., and Bellavance, A. The effect of muscle hyper- and hypoactivity upon fibre diameters of intact and regenerating nerves. *J. Neurol. Sci.* **20**, 457–469 (1973).

28. Fankhauser, G., Vernon, J.A., Frank, W.H., and Slack, W.V. Effect of size and number of brain cells on learning in larvae of the salamander *Triturus vividescens*. *Science* **122**, 692–693 (1955).

29. Feringa, E.R., Johnson, R.D., and Wendt, J.S. Spinal cord regeneration in rats after immuno-suppressive treatment. *Arch. Neurol.* **32**, 676–683 (1975).

30. Freeman, L.W., MacDougall, J., Turbes, C.C., and Bowman, D.E. The treatment of

experimental lesions of the spinal cord of dogs with trypsin. *J. Neurosurg.* **17,** 259–265 (1960).

31. Goshgarian, H.G. The regeneration of brachial nerves of contralateral origin into denervated fused newt forelimbs. *J. Exp. Zool.* **197,** 347–356 (1976).
32. Goss, R.J. "Principles of Regeneration." Academic Press, New York, 1969.
33. Graziadei, P.P.C. Cell dynamics in the olfactory mucosa. *Tissue Cell* **5,** 113–131 (1973).
34. Guillery, R.W. Binocular competition in the control of geniculate cell growth. *J. Comp. Neurol.* **144,** 117–127 (1972).
35. Gyllensten, L., Malmfors, T., and Norrlin, M.-L. Effect of visual deprivation on the optic centers of growing and adult mice. *J. Comp. Neurol.* **124,** 149–160 (1965).
36. Hasegawa, M. Restitution of the eye from the iris after removal of the retina and lens together with the eye-coats in the newt, *Triturus pyrrhogaster. Embryologia* **8,** 362–386 (1965).
37. Herron, W.L., Jr., and Riegel, B.W. Production rate and removal of rod outer segment material in vitamin A deficiency. *Invest. Ophthalmol.* **13,** 46–53 (1974).
38. Herron, W.L., Riegel, B.W., Myers, O.E., and Rubin, M.L. Retinal dystrophy in the rat—A pigment epithelial disease. *Invest. Ophthalmol.* **8,** 595–604 (1969).
39. Hess, A. The effects of eye removal on the development of cholinesterase in the superior colliculus. *J. Exp. Zool.* **144,** 11–19 (1960).
40. Hickey, T.L. Translaminar growth of axons in the kitten dorsal lateral geniculate nucleus following removal of one eye. *J. Comp. Neurol.* **161,** 359–382 (1975).
41. Hoy, R.R., Bittner, G.D., and Kennedy, D. Regeneration in crustacean motoneurons: Evidence for axonal fusion. *Science* **156,** 251–252 (1967).
42. Hughes, W.F. Influence of tectal ablation on the morphogenesis of retinal ganglion cells in the chick. *Anat. Rec.* **172,** 333 (1972).
43. Ingoglia, N.A., Weis, P., and Mycek, J. Axonal transport of RNA during regeneration of the optic nerves of goldfish. *J. Neurobiol.* **6,** 549–563 (1975).
44. Johns, P.R., and Easter, S.S., Jr. The adult goldfish retina grows by adding new cells. *Neurosci. Abstr.* **1,** 95 (1975).
45. Joseph, J. Further studies in changes of nuclear population in degenerating non-myelinated and finely myelinated fibres. *J. Anat.* **82,** 146–152 (1948).
46. Keefe, J.R. An analysis of urodelian retinal regeneration. I. Studies of the cellular source of retinal regeneration in *Notophthalmus viridescens* utilizing ^{3}H-thymidine and colchicine. *J. Exp. Zool.* **184,** 185–206 (1973).
47. Keefe, J.R. An analysis of urodelian retinal regeneration: IV. Studies of the cellular source of retinal regeneration in *Triturus cristatus carnifex* using ^{3}H-thymidine. *J. Exp. Zool.* **184,** 239–257 (1973).
48. LaVail, M.M. Survival of some photoreceptor cells in albino rats following long-term exposure to continuous light. *Invest. Ophthalmol.* **15,** 64–70 (1976).
49. LaVail, M.M., Sidman, R.L., and Gerhardt, C.O. Congenic strains of RCS rats with inherited retinal dystrophy. *J. Hered.* **66,** 242–244 (1975).
50. Levine, R. Regeneration of the retina in the adult newt, *Triturus cristatus,* following surgical division of the eye by a limbal incision. *J. Exp. Zool.* **192,** 363–379 (1975).
51. Lopashov, G.V., and Sologub, A.A. Artificial metaplasia of pigmented epithelium into retina in tadpoles and adult frogs. *J. Embryol. Exp. Morphol.* **28,** 521–546 (1972).
52. Lyall, A.H. The growth of the trout retina. *Q. J. Microscop. Sci.* **98,** 101–110 (1957).
53. Maraini, G., Carta, F., and Franguelli, R. Metabolic changes in the retina and the optic centres following monocular light deprivation in the new-born rat. *Exp. Eye Res.* **8,** 55–59 (1969).
54. Matthews, A. An electron microscopic study of the relationship between axon diameter and

the initiation of myelin production in the peripheral nervous system. *Anat. Rec.* **161**, 337–351 (1968).

55. Matulionis, D.H. Light and electron microscopic study of the degeneration and early regeneration of olfactory epithelium in the mouse. *Am. J. Anat.* **145**, 79–100 (1976).

56. Montero, V.M., and Guillery, R.W. Degeneration in the dorsal lateral geniculate nucleus of the rat following interruption of the retinal or cortical connections. *J. Comp. Neurol.* **134**, 211–242 (1968).

57. Moulton, D.G. Dynamics of cell populations in the olfactory epithelium. *Ann. N.Y. Acad. Sci.* **237**, 52–61 (1974).

58. Moulton, D.G., Celebi, G., and Fink, R.P. Olfaction in mammals—two aspects: Proliferation of cells in the olfactory epithelium and sensitivity to odours. *In* "Ciba Foundation Symposium on Taste and Smell in Vertebrates" (G.E.W. Wolstenholme and J. Knight, eds.), pp. 227–250. Churchill, London, 1970.

59. Murray, M. Regeneration of retinal axons into the goldfish optic tectum. *J. Comp. Neurol.* **168**, 175–196 (1976).

60. Murray, M., and Goldberger, M.E. Restitution of function and collateral sprouting in the cat spinal cord: The partially hemisected animal. *J. Comp. Neurol.* **158**, 19–36 (1974).

61. Murray, M., Grafstein, B. Changes in the morphology and amino acid incorporation of regenerating goldfish optic neurons. *Exp. Neurol.* **23**, 544–560 (1969).

62. Nathaniel, E.J.H., and Clemente, C.D. Growth of nerve fibers into skin and muscle grafts in rat brains. *Exp. Neurol.* **1**, 65–81 (1959).

63. Nathaniel, E.J.H., and Nathaniel, D.R. Regeneration of dorsal root fibers into the adult rat spinal cord. *Exp. Neurol.* **40**, 333–350 (1973).

64. Noell, W.K., and Albrecht, R. Irreversible effects of visible light on the retina: Role of vitamin A. *Science* **172**, 76–80 (1971).

65. Noell, W.K., Delmelle, M.C., and Albrecht, R. Vitamin A deficiency effect on retina: Dependence on light. *Science* **172**, 72–76 (1971).

66. O'Steen, W.K., and Anderson, K.V. Photically evoked responses in the visual system of rats exposed to continuous light. *Exp. Neurol.* **30**, 525–534 (1971).

67. O'Steen, W.K., Anderson, K.V., and Shear, C.R. Photoreceptor degeneration in albino rats: Dependency on age. *Invest. Ophthalmol.* **13**, 334–339 (1974).

68. Pannese, E. Detection of neurofilaments in the perikaryon of hypertrophic nerve cells. *J. Cell Biol.* **13**, 457–461 (1962).

69. Rensch, B. Increase of learning capability with increase of brain size. *Am. Nat.* **90**, 81–95 (1956).

70. Richter, W. Regenerative Vorgänge nach einseitiger Entfernung des caudalen Endhirnabschnittes einschliesslich des telo-diencephalen Grenzbereiches bei *Ambystoma mexicanum. J. Hirnforsch.* **10**, 515–534 (1968).

71. Richter, W., and Kranz, D. Altersabhängigkeit der Aktivität der Matrixzonen im Gehirn von *Xiphophorus helleri* (Teleostei). Autoradiographische Untersuchungen. *J. Hirnforsch.* **13**, 109–116 (1971).

72. Riesen, A.H. Effects of stimulus deprivation on the development and atrophy of the visual sensory system. *Am. J. Orthopsychol.* **30**, 23–36 (1960).

73. Romine, J.S., Bray, G.M., and Aguayo, A.J. Schwann cell multiplication after crush injury of unmyelinated fibers. *Arch. Neurol.* **33**, 49–54 (1976).

74. Rovainen, C.M. Regeneration of Müller and Mauthner axons after spinal transection in larval lampreys. *J. Comp. Neurol.* **168**, 545–554 (1976).

75. Samorajski, T., and Rolsten, C. Nerve fiber hypertrophy in posterior tibial nerves of mice in response to voluntary running activity during aging. *J. Comp. Neurol.* **159**, 553–558 (1975).

76. Sara, V.R., Lazarus, L., Stuart, M.C., and King, T. Fetal brain growth: Selective action by growth hormone. *Science* **186**, 446–447 (1974).

77. Scheff, S., Benardo, L., and Cotman, C. Progressive brain damage accelerates axon sprouting in the adult rat. *Science* **197**, 795–797 (1977).

78. Schlaepfer, W.W., and Myers, F.K. Relationship of myelin internode elongation and growth in the rat sural nerve. *J. Comp. Neurol.* **147**, 255–266 (1972).

79. Schonbach, C., and Schonbach, J. Protein renewal in the photoreceptor outer segments of the pigeon retina. *Experientia* **28**, 836–837 (1972).

80. Sotelo, C., and Palay, S.L. Altered axons and axon terminals in the lateral vestibular nucleus of the rat. Possible example of axonal remodeling. *Lab. Invest.* **25**, 653–671 (1971).

81. Stone, L.S. Polarization of the retina and development of vision. *J. Exp. Zool.* **145**, 85–96 (1960).

82. Straznicky, K., and Gaze, R.M. The growth of the retina in *Xenopus laevis:* An autoradiographic study. *J. Embryol. Exp. Morphol.* **26**, 67–79 (1971).

83. Thomas, P.K. The cellular response to nerve injury. 3. The effect of repeated crush injuries. *J. Anat.* **106**, 463–470 (1970).

84. Turner, J.E., and Singer, M. The ultrastructure of Wallerian degeneration in the severed optic nerve of the newt (*Triturus viridescens*). *Anat. Rec.* **181**, 267–286 (1975).

85. Van Marthens, E., and Zamenhof, S. Deoxyribonucleic acid of neonatal rat cerebrum increased by operative restriction of litter size. *Exp. Neurol.* **23**, 214–219 (1969).

86. Vizoso, A.D., and Young, J.Z. Internode length and fibre diameter in developing and regenerating nerves. *J. Anat.* **82**, 110–134 (1948).

87. Vrensen, G., and de Groot, D. The effect of dark rearing and its recovery on synaptic terminals in the visual cortex of rabbits. A quantitative electron microscopic study. *Brain Res.* **78**, 263–278 (1974).

88. Webster, D. The geometry of peripheral myelin sheaths during their formation and growth in rat sciatic nerves. *J. Cell Biol.* **48**, 348–367 (1971).

89. Windle, W.F. Regeneration of axons in the vertebrate central nervous system. *Physiol. Rev.* **36**, 427–440 (1956).

90. Young, R.W. Visual cells and the concept of renewal. *Invest. Ophthalmol.* **15**, 700–725 (1976).

91. Young, R.W., and Bok, D. Participation of the retinal pigment epithelium in the rod outer segment renewal process. *J. Cell Biol.* **42**, 392–403 (1969).

92. Zamenhof, S., Mosley, J., and Schuller, E. Stimulation of the proliferation of cortical neurons by prenatal treatment with growth hormone. *Science* **152**, 1396–1397 (1966).

93. Zamenhof, S., van Marthens, E., and Grauel, L. Prenatal cerebral development: Effect of restricted diet, reversal by growth hormone. *Science* **174**, 954–955 (1971).

94. Zamenhof, S., van Marthens, E., and Grauel, L. DNA (cell number) and protein in neonatal rat brain: Alteration by timing of maternal dietary protein restriction. *J. Nutr.* **101**, 1265–1270 (1971).

10

Intraocular Control of Lens Development

Pupil: L., *pupa*, doll (as oneself is seen reflected in another's eye)

It is the function of the lens "to see without being seen." Suspended in intraocular fluid, this transparent jewel is one of the body's most unique organs. It is neither vascularized nor innervated and is even devoid of fibroblasts. The histological purity of the lens mas made it an ideal tissue in which to study antigenic development by immunological and electrophoretic methods (3).

Embryologically, the lens is derived from the ectoderm under the inductive influence of the optic cup. Developing first as a sphere of cells, the posterior moiety then differentiates into lens fibers while the anterior cells remain epithelial (Fig. 96). The fibrous and epithelial poles of the lens meet at the equator where, in a circular growth zone, proliferation provides the cells destined to differentiate into lens fibers (9). Here the incompatibility between mitosis and cellular differentiation is unequivocal, for once a daughter cell is committed to becoming a lens fiber all other potentials are abandoned (16). Quantities of crystallin proteins are laid down and all cytological organelles are gradually lost until even the nucleus itself disappears. Though still capable of limited metabolic activity, these lens fibers are condemned to live in limbo, prisoners for life of their own specializations.

CRYSTALLIN PROTEINS

The fibrous proteins of the lens, or crystallins, have been useful in studies of both ontogeny and phylogeny. The uncontaminated nature of crystallin preparations permits their precise analysis by a variety of separation methods, chief among which has been the agar gel diffusion technique.

210

Equator

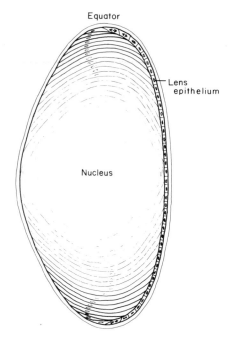

Lens
epithelium

Nucleus

Fig. 96. Sagittal section through a mammalian lens. Proliferation of epithelial cells at the equator yields cells destined to differentiate into lens fibers. Such cells become buried by subsequent fibers laid down over them. They lose their nuclei and cytoplasmic organelles in the course of differentiation, eventually becoming filled with fibrils of crystallin proteins. The nucleus of the lens represents those tissues deposited early in development.

Antiserum against lens is allowed to interact with antigen preparations as they diffuse toward each other through agar at rates dependent upon their molecular sizes. When their paths cross, reaction bands between the different crystallins and their corresponding antibodies can be spatially visualized, the effect sometimes being further resolved by electrophoretic separation. Alternatively, if the antibodies are coupled with fluorescent dyes and applied to histological sections, the crystallin proteins, complexed to the fluorescent antibodies, can be localized using ultraviolet microscopy.

In the chick embryo the earliest appearance of crystallin proteins is associated with the induction of the lens by the optic cup (*15*). By 31–50 hours of incubation (11–17 somites) α-crystallins are first synthesized. Not long afterward (60 hours), β-crystallins are found. By 7 days of incubation, γ-crystallins become detectable in the developing chick lens. Interestingly, some of the same antigenic proteins are later to be found in the iris, the pigmented retinal epithelium, and the cornea, all three of which, in certain amphibians, possess the potential for lens regeneration.

By making antibodies against the lens of one species and comparing their cross-reactions with lenses from its own and other species, it is possible to identify such antigens as may or may not be shared in common, thereby revealing phylogenetic affinities between animals (11). Not unexpectedly, the incidence of cross-reactions is more or less correlated with taxonomic proximities. There is considerable identity between the lenses of different mammals, but only about half of their antigens are present in amphibians and even fewer in fishes. The lens of the squid, having evolved independently, exhibits no immunological identity with vertebrate lenses. In general, it would appear that during the course of vertebrate lens evolution a succession of antigenic proteins have been added with little or no loss of their antecedents. For all its apparent simplicity, therefore, the molecular complexity of the lens has been compounded in its evolution from fish to mammal.

As structural proteins, the crystallins of the lens are unique. Once a lens epithelial cell at the equator goes through its last division and embarks upon the pathway of differentiation, it begins synthesis of the crystallin proteins. Which ones it makes vary with age. Lens fibers laid down postnatally synthesize a different set of crystallins from the ones made in the embryonic lens. Those fibers that differentiate in the embryo end up in the center, or nucleus, of the adult lens. Later generations of fibers are laid down around them. Consequently, the antigenic composition of the adult lens varies from nucleus to cortex (18).

Protein synthesis depends upon mRNA, and in most cells, lens epithelium included, this is short-lived and rapidly turned over. Consequently, if its production is blocked by actinomycin D, there is a prompt cessation of protein synthesis. In lens fibers, however, protein synthesis can go on for 24 hours after treatment with actinomycin D, suggesting the presence of long-lived mRNA to mediate the production of proteins in a cell whose nucleus is inactivated and destined to disappear (20). Once the full complement of structural proteins has been laid down in the lens fiber and all of the cytoplasmic organelles, including the ribosomes, have disappeared, there is no provision for further protein synthesis. Nor is there degradation of crystallins once they have been produced. The lens fiber survives indefinitely as a living fossil dating back to the embryonic stage when it was first laid down.

CELL PROLIFERATION IN THE LENS EPITHELIUM

The growth of the lens depends upon the proliferative capacity of its epithelium (12). Existing as a veneer on the anterior hemisphere of the lens, this epithelial monolayer is normally not mitotically active except at the equator where it is contiguous with the differentiated lens fibers. Here the potential for growth is never lost, although the rate at which new fibers are produced may drop to only one per day in the lenses of adult rats and rabbits (*10*). Nevertheless, the cumulative effect of even very slow rates of cellular differentiation can result in unwanted enlargement of lenses in nongrowing adult mammals. It has been suggested that this slow but persistent accretion may contribute to some of the age-related changes in the focusing ability of the human lens.

Notwithstanding the paucity of proliferation among adult lens epithelial cells, this tissue never loses its mitotic compentence even in the nonequatorial regions covering the anterior surface of the lens. Here it is possible to stimulate DNA synthesis and cell division by a number of experimental methods. For example, if atropine, a parasympathetic antagonist, is topically applied to the rabbit eye the lens epithelial cells are stimulated to proliferate (*2*). Conversely, when parasympathomimetic drugs (DFP, pilocarpine) are applied, mitosis is inhibited. These results are interpreted to indicate that the proliferative activity of the lens epithelium may be subject to mechanical forces exerted by the contraction of zonular fibers which insert on the lens and assist in focusing. Mitotic activity can be inhibited by exposure to aqueous extracts of lenses, an effect perhaps not unrelated to the inhibitory action of γ-crystallins on lens mitoses *in vitro* (*30*).

The maintenance of lenses in culture would seem to be a simple matter, for the intraocular lens is itself suspended in a lavage of aqueous humor. The physiological delicacy of the lens, however, necessitates utmost care in the transfer from *in vivo* to *in vitro* conditions. Presumably associated with the osmotic upheaval of being abruptly bathed in a culture medium not necessarily duplicating the composition of the aqueous humor, the epithelium of the explanted lens typically undergoes a burst of mitotic activity during the first several days. Although such lenses may survive for a month or two, they seldom exhibit further cell division, except when serum is added to the culture medium (*1*). For reasons not well understood, this initiates a round of mitoses throughout the lens epithelium.

The simplest way to stimulate cell division in the lens epithelium is to inflict a wound. A needle puncture is sufficient to set off the wound healing response in cultured lenses as well as those *in vivo*. If this is done to the intraocular rabbit lens, there is triggered an episode of DNA synthesis followed by mitosis in which the temporal sequence of events can be spatially visualized in the surrounding lens epithelium (Fig. 97) (*13*). This is made evident by autoradiography of whole mounts of the injured lens epithelium stripped from the lens at various intervals after injury and exposure to ^{3}H-thymidine. Within 24 hours after wounding, there can be seen a band of labeled cells synthesizing DNA immediately peripheral to the wound itself.

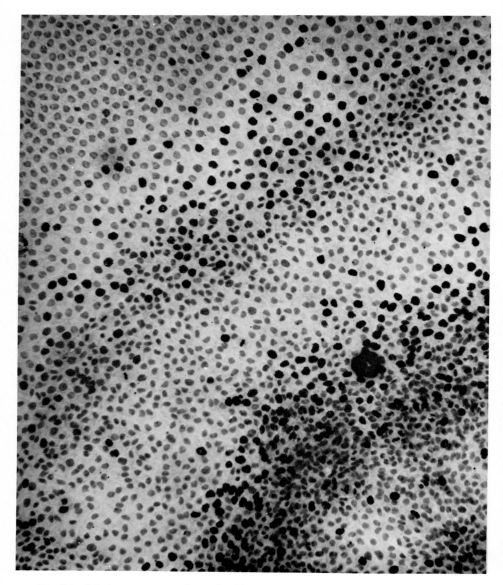

Fig. 97. Whole mount autoradiograph of lens epithelium from a rabbit eye 48 hours after wounding (out of field toward lower right). The eye was injected with ³H-thymidine 2 hours before fixation. Two waves of radioactive nuclei and mitotic figures are visible. Note that the as yet unaffected cells (upper left) are more densely packed than those about to synthesize DNA (13).

Some hours later, these same cells are found to be in mitosis while those farther out are now synthesizing DNA. By two days, these concentric rings of DNA synthesis and mitosis are some distance out from the wound and a second wave has begun centrally. Although these ripples of proliferation might be assumed to be triggered by the centrifugal diffusion of a mitotic stimulant, they are more likely the result of a chain reaction propagated outward from the wound site. Although the nature of this reaction remains undetermined, it is perhaps not unrelated to the migratory activities of the epithelium in effecting wound closure. Peripheral to, and therefore prior to, the advent of DNA synthesis one can note a decrease in the population density of the epithelial cells such as might be expected where centripetal migration would deplete the surrounding epithelium. It is not inconceivable that this epithelium might be sensitive to the distances between its cells and programmed to respond mitotically whenever the cellular concentration drops below normal, whether in response to the expansion of the lens or the need to resurface a lesion.

SIZE AND POLARITY

If the eye is to be an optical instrument of the precision required by most animals, the growth of its parts must be harmoniously integrated. To probe the developmental factors responsible for coordinating the growth rates and ultimate sizes of the parts of the eye some ingenious experiments have been carried out with provocative results. Foremost among these studies is Harrison's classic investigation of how the size of the lens is regulated to that of the eyeball, and vice versa (14). This he studied by exchanging presumptive lens ectoderm between embryos of potentially large and small species of salamanders. These interspecific grafts were induced by the optic cups of their hosts to differentiate into lenses and the resulting eyes were allowed to develop until the differential in body size between the two species had become established. The results showed that the original mismatch between graft and host was accommodated by alterations in the sizes of both the lens and the rest of the eyeball. Hence, the lens developing from ectoderm of the smaller species grew larger than normal, but did not attain the full dimensions characteristic of the host lens. The difference was made up by a reduction in the size of the rest of the eye cup, resulting in an eye of intermediate size, the parts of which were proportional. In the reciprocal experiment the growth of the lens from the larger species was reduced while that of the rest of the eye was enhanced, again giving rise to a harmonious organ of intermediate dimensions. These findings emphasize the fact that the growth of the lens and of the rest of the eye adjust to each other rather than to the dimensions of the body as a whole, which is the only mechanism by which disharmonious growth could be avoided. Clearly, it is more important to possess a functional eye of the wrong size than a disproportionate one commensurate with the size of the body.

The interplay between parts in the development of the eye inferred by Harrison's experiments has subsequently been explored in the developing eye of the chick embryo. When the wall of the developing eye is pierced with a short length of capillary tubing the intraocular fluids can be partially drained off thereby preventing the normal buildup of hydrostatic pressure within the eye. During subsequent development, these intubated eyeballs, including their lenses, fail to attain normal dimensions, although their retinas continue to grow (Fig. 89) (4). These findings, like those of Harrison's, suggest that the growth of the lens is coordinated with that of the rest of the eyeball, the retina to the contrary notwithstanding. If the lens is extirpated from an embryonic eye, subsequent ocular development is retarded (except for the retina) (5). In this case, the mass of the vitreous is reduced presumably accounting for the diminished expansion of the eyeball. Though the source of the vitreous is unknown, its production would appear to be in some way dependent upon the lens.

The lens is not altogether independent in its development from the rest of the eye. When an extra lens is implanted into the developing eye of a chick embryo the two lenses continue their development, but 9 days later their combined volumes approximate that for a single normal lens (Fig. 98), implying that the mass of lens produced is under the influence of some other part of the eye (7). There is reason to believe that the neural retina may provide the necessary control of lens growth. It may also determine its polarity.

If an embryonic lens is removed and replaced in reverse orientation, with the lens epithelium facing the retina and the lens fibers on the corneal side, a reversal of polarity occurs during subsequent development (Fig. 98) (6). That is, the former lens epithelial cells displaced to the posterior aspect undergo differentiation into lens fibers. Even when the original lens is replaced by a piece of lens epithelium, the fibers into which it differentiates are polarized toward the retina. Comparable results have been obtained in the regeneration of lenses in newt eyes following lentectomy. If a wedge of dorsal iris from which the new lens normally regenerates is excised and replaced inside out, the lens to which it gives rise develops a polarity consistent with the rest of the eye rather than the iris epithelium from which it was derived (25). Reversal of a partially regenerated lens likewise results in the development of lens fibers from the epithelium and the establishment of a new epithelium over the already formed fibers (23). These findings bear witness to the possibility that the differentiation of lens fibers is profoundly influenced by the presence of the retina. They leave unanswered, however, the problem of

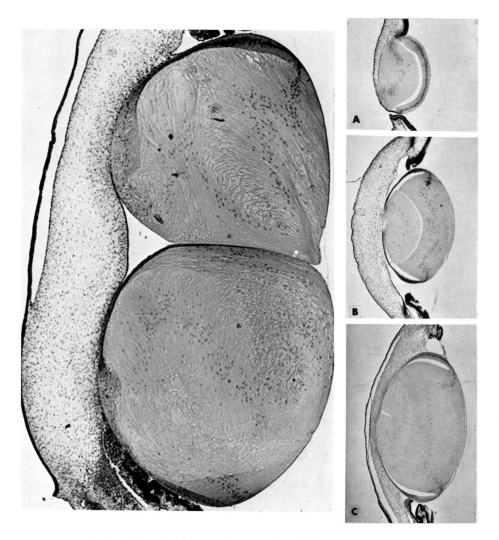

Fig. 98.   Left, double lens development (7). In a 5-day-old chick embryo the ocular lens was replaced with two lenses from donors of the same age. On the fourteenth day of incubation, the combined volumes of the two lenses equaled that of a single normal lens.

Right, reversal of lens polarity (6). At 5 days of incubation the anterior-posterior polarity of the lens was surgically reversed. After 2, 6, and 10 days (A–C) the former lens epithelium has differentiated into lens fibers on the retinal side, while the former mass of fibers has become covered with a new layer of lens epithelium on the corneal side. From Coulombre, J.L., and Coulombre, A.J., *Science* **142**, 1489–1490, 13 December 1963. Copyright 1963 by the American Association for the Advancement of Science.

how such a retinal influence can be transmitted across a relatively massive quantity of vitreous to exert its influence on the developing lens. Apropos of lens polarity, it is interesting to note that when pituitary glands are implanted into newt eyes they can induce secondary lenses from the dorsal iris, the polarities of which are oriented to the pituitary (*19*).

LENS REGENERATION

One of nature's strangest curiosities is the phenomenon of lens regeneration which occurs in certain amphibiains. Salamanders of the genus *Notophthalmus* (*Triturus*) give rise to perfect and complete lens regenerates from the epithelial cells of the dorsal iris (Fig. 99). Lens regeneration is also possible in *Xenopus* tadpoles where the corneal epidermis rounds up into a new lens, its basement membrane forming the lens capsule (*31*). This ability is lost at metamorphosis and can be abolished in the larval eye by implanting a pellet of thyroxine. Conversely, fragments of cornea from postmetamorphic *Xenopus* can regenerate lenses when grafted to larvae. Lenses do not regenerate in birds or mammals, not even in the lentectomized chick embryo.

Following lentectomy in the newt, one of the first events to occur is the invasion of the iris by macrophages (*33*). Appearing during the first postoperative day and reaching maximum numbers after 3–5 days, these cells are responsible for phagocytosis of pigment granules extruded from the epithe-

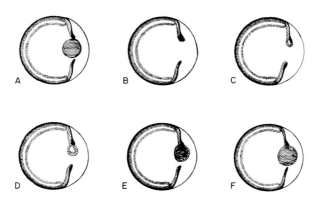

Fig. 99.   Stages of lens regeneration in the lentectomized larval salamander eye. A, normal eye before lentectomy. Following loss of the original lens the dorsal iris becomes slightly swollen (B) whereupon its marginal cells begin to round up about a week postoperatively (C). Several days later (D) a small lens vesicle is present, the posterior cells of which are in the early stages of fiber differentiation. Further development is achieved by continued enlargement of the lens (E) until it finally detaches from the iris (F).

lial cells of the iris. Meanwhile the rate of RNA synthesis in the cells of the iris epithelium increases as early as two days after lentectomy. DNA synthesis begins by the fourth day, reaching a peak after one week (Fig. 100) (8, 22, 35). The nuclei enlarge, the number of ribosomes increases and the mitochondria become abnormally large and distorted. These activated cells then round up into a lens vesicle, whereupon those situated posteriorly commence their differentiation into lens fibers. After 10 days they synthesize α-crystallins, and during the second and third weeks after lentectomy, respectively, β- and γ-crystallins can be detected (17). By the fourth week the regenerating lens becomes detached from the dorsal iris from which it developed. It continues to grow until the size of the original lens has been attained. Despite the fact that this regenerate was produced from a tissue different from the one that gave rise to the original lens in the embryo, it is identical to the original in every respect, including the antigenic proteins in its fibers.

It is perplexing that regenerative ability should be limited to the dorsal iris of the newt eye. Actually, there is a gradient of decreasing potentialities on

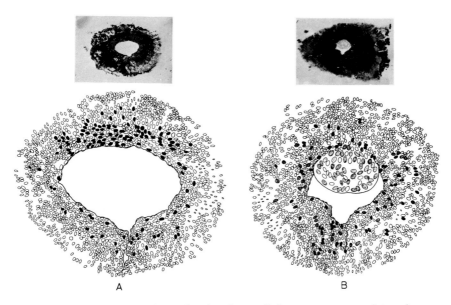

A   B

Fig. 100.   Autoradiographic evidence that dorsal iris cells become incorporated into the regenerated lens in the newt (8). A, photograph and drawing of an iris 6 days after lentectomy and 24 hours following intraperitoneal injection of ^{3}H-thymidine. The labeled nuclei are black, and are concentrated in the dorsal iris. B, iris and regenerating lens 15 days after lentectomy and 10 days after labeling. Most of the labeled nuclei from the dorsal iris are now part of the lens, their radioactivity having been diluted by mitotic divisions.

either side of the midpoint which can be demonstrated by removing the dorsal sector, in which case two lenses may regenerate from the dorsal corners of the remaining iris provided the cuts were made no farther down than the 3 and 9 o'clock positions. The iris itself regenerates in due course following iridectomy, the regenerate retaining the capacity to produce a lens. If an artificial pupil is created in the dorsal iris, a lens may regenerate from its dorsal rim providing the original lens has been removed. Similarly, apertures created as far back as the retina on the dorsal circumference of the eye can give rise to lens-like regenerates derived from the pigmented retinal epithelium (26). Regeneration of lenses is also possible from transplanted pieces of dorsal iris (Fig. 101). Grafted to another eye, a lens may be regenerated provided the host eye has been previously lentectomized (36). Even if transplanted to the eye of a genus of salamander (e.g., *Ambystoma*) normally incapable of regenerating lenses, such iris grafts are still able to produce lens regenerates (in the absence of the host lens) (21). Since iris grafts from

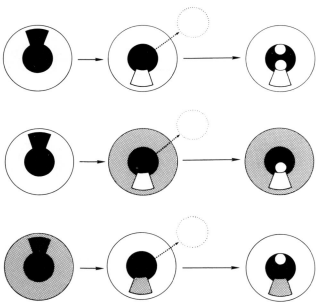

Fig. 101. Lens regeneration from transplanted segments of dorsal iris. When grafted to the lentectomized eye of a newt normally capable of regeneration (*Notophthalmus*) (top row) lenses regenerate simultaneously from both graft and host dorsal iris. If a newt iris is transplanted to an *Ambystoma* eye normally unable to regenerate (shaded), a lens develops only from the grafted newt iris (middle row). The reciprocal experiment (bottom row) results in lens regeneration only from the host iris, not the graft. Regenerative ability therefore resides in the cells of the dorsal iris and is not a property of the intraocular environment.

nonregenerating species in the eyes of newts capable of regeneration are still unable to produce lenses, it must be concluded that the presence or absence of regenerative ability is resident in the cells of the dorsal iris, not the intraocular milieu.

This is not to say, however, that intraocular factors are not important. The neural retina seems to exert a promoting influence on lens regeneration, for retinectomy retards regeneration of a lens in the same eye (28). Moreover, when the dorsal iris is transplanted to an extraocular site in the body, its capacity for lens regeneration is markedly enhanced if part of the retina is left attached. What this influence may be is a matter for conjecture, but it could be related to neurotrophic influences necessary for limb regeneration. Experiments have shown that if the iris is implanted into a limb blastema its incidence of lens formation is several times greater than when grafter subcutaneously, but there is s sharp drop in the number of lens vesicles formed if such limbs are denervated (24).

Many attempts have been made to demonstrate lens regeneration from irises maintained in tissue culture. No regeneration is possible in such explants when cultured immediately following lentectomy. Irises may survive for up to 4 weeks *in vitro* without giving rise to lenses, but they retain the capacity to do so if grafted back into a lentectomized eye (29).

Although no lens forms in culture, the cells of the dorsal iris may exhibit an increase in the synthesis of RNA and DNA, a response that mimics the early reactions to lentectomy *in vivo*. If explanation of the dorsal iris is postponed a week, depigmentation of its cells may occur *in vitro*. When a young lens regenerate is already present, it may continue to differentiate in culture, especially if retinal tissue is present (34).

What prevents the dorsal iris from regenerating a lens except following lentectomy? It is conceivable that the original lens might produce inhibitors of lens regeneration, the cells of the dorsal iris being induced to initiate lens differentiation by their withdrawal. Attempts to demonstrate the existence of such substances by repeated injections of aqueous humor from intact eyes into lentectomized ones have failed to confirm the existence of factors capable of specifically inhibiting lens regeneration. Another possibility is that the lens may somehow counteract the influence from the neural retina which seems to be important for lens growth.

Self-inhibition by the lens can be explored by surgical manipulations. For example, if a piece of dorsal iris with a still attached regenerating lens is transplanted to a lentectomized eye, lens regeneration from the host iris is not necessarily inhibited unless it is in contact with the transplanted regenerate (32). Proximity and position, therefore, seem to be important factors in deciding whether or not lens regeneration will occur from the iris. In another approach to the problem, two eyeballs have been grafted side by

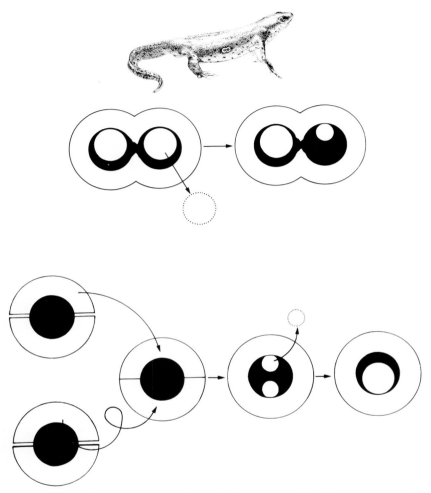

Fig. 102. Above, lens regeneration in "parabiotic" newt eyes (27). When both eyeballs are grafted together in the orbit or on the flank (as shown here), each replaces its lens lost in operation. When one regenerate is removed it is still replaced from the dorsal iris despite the confluency of ocular fluid. Below, the double dorsal eye experiment in which two dorsal halves of newt eyeballs are grafted together. Each then simultaneously regenerates a lens to replace the one discarded in the operation. However, if one of the regenerates is later extirpated it is not replaced owing to suppression by the remaining lens (28).

side with confluency between their intraocular fluids. Lentectomy of one of these "parabiotic" eyeballs was followed by normal regeneration (Fig. 102) (27). Presumably the lens of its partner eye was too far away to exert inhibitory influences. However, if the dorsal halves of two eyes are joined together to give rise to a single sphere, both dorsal irises regenerate lenses simultaneously. If one of these lenses is subsequently removed, it is not replaced presumably owing to inhibitory influences from the partner lens already occupying the pupil (Fig. 102) (28). Again, proximity between lens and iris seems to be the dominant factor determining whether or not regeneration will occur.

Much still remains to be learned about the regulation of lens growth. Yet the relative simplicity of this organ coupled with its isolation from the vascular and nervous systems, would seem to require that if its growth is controlled chemically (rather than by physical influences) the intraocular fluid must act as a vehicle for the conveyance of such substances, be they inhibitors or stimulants. It is anticipated that the eventual identification and characterization of agents that control the growth of the lens will be made possible by analysis of the aqueous humor in which they are likely to be dissolved.

REFERENCES

1. Bagchi, M., Harding, C., Unakar, N., and Reddan, J. Experimental modification of the temporal aspects of the serum-induced cell cycle in the cultured adult rabbit lens. *Ophthalmic Res.* **2**, 133–142 (1971).
2. Bito, L.Z., Davson, H., and Snider, N. The effects of autonomic drugs on mitosis and DNA synthesis in the lens epithelium and on the composition of the aqueous humour. *Exp. Eye Res.* **4**, 54–61 (1965).
3. Bloemendal, H. The vertebrate lens. A useful system for the study of fundamental biological processes on a molecular level. *Science* **197**, 127–138 (1977).
4. Coulombre, A.J. The role of intraocular pressure in the development of the chick eye. I. Control of eye size. *J. Exp. Zool.* **133**, 211–225 (1956).
5. Coulombre, A.J., and Coulombre, J.L. Lens development. I. Role of the lens in eye growth. *J. Exp. Zool.* **156**, 39–48 (1964).
6. Coulombre, J.L., and Coulombre, A.J. Lens development: Fiber elongation and lens orientation. *Science* **142**, 1489–1490 (1963).
7. Coulombre, J.L., and Coulombre, A.J. Lens development. IV. Size, shape, and orientation. *Invest. Ophthalmol.* **8**, 251–257 (1969).
8. Eguchi, G., and Shingai, R. Cellular analysis on localization of lens forming potency in the newt iris epithelium. *Dev. Growth Differt.* **13**, 337–349 (1971).
9. Hanna, C., and Keatts, H.C. Chicken lens development: Epithelial cell production and migration. *Exp. Eye Res.* **5**, 111–115 (1966).
10. Hanna, C., and Keatts, H.C. Cell migration in the adult rat and rabbit lens. *Exp. Eye Res.* **7**, 244–246 (1968).

11. Halbert, S.P., Locatcher-Khorazo, D., Swick, L., Witner, R., Seegal, B., and Fitzgerald, P. Homologous immunological studies of ocular lens. I. *In vitro* observations. *J. Exp. Med.* **105**, 439–452 (1957).

12. Harding, C.V., Reddan, J.R., Unakar, N.J., and Bagchi, M. The control of cell division in the ocular lens. *Int. Rev. Cytol.* **31**, 215–300 (1971).

13. Harding, C.V., and Srinivasan, B.D. A propagated stimulation of DNA synthesis and cell division. *Exp. Cell Res.* **25**, 326–340 (1961).

14. Harrison, R.G. Correlation in the development and growth of the eye studied by means of heteroplastic transplantation. *Arch. Entwicklungsmech. Org.* **120**, 1–55 (1929).

15. Maisel, H., and Langman, J. Lens proteins in various tissues of the chick eye and in the lens of animals throughout the vertebrate series. *Anat. Rec.* **140**, 183–193 (1961).

16. Modak, S.P., Morris, G., and Yamada, T. DNA synthesis and mitotic activity during early development of chick lens. *Dev. Biol.* **17**, 544–561 (1968).

17. Ogawa, T. The similarity between antigens in the embryonic lens and in the lens regenerate of the newt. *Embryologia* **9**, 295–305 (1967).

18. Papaconstantinou, J. Biochemistry of bovine lens proteins. II. The γ-crystallins of adult bovine, calf and embryonic lenses. *Biochim. Biophys. Acta* **107**, 81–90 (1965).

19. Powell, J.A., and Segil, N. Secondary lens formation caused by implantation of pituitary into the eyes of the newt, *Notophthalmus*. *Dev. Biol.* **52**, 128–140 (1976).

20. Reeder, R., and Bell, E. Short- and long-lived messenger RNA in embryonic chick lens. *Science* **150**, 71–72 (1965).

21. Reyer, R.W. Lens regeneration from homoplastic and heteroplastic implants of dorsal iris into the eye chamber of *Triturus viridescens* and *Amblystoma punctatum*. *J. Exp. Zool.* **133**, 145–190 (1956).

22. Reyer, R.W. DNA Synthesis and the incorporation of labeled iris cells into the lens during lens regeneration in adult newts. *Dev. Biol.* **24**, 533–558 (1971).

23. Reyer, R.W. Repolarization of reversed, regenerating lenses in adult newts, *Notophthalmus viridescens*. *Exp. Eye Res.* **24**, 501–509 (1977).

24. Reyer, R.W., Woolfitt, R.A., and Withersty, L.T. Stimulation of lens regeneration from the newt dorsal iris when implanted into the blastema of the regenerating limb. *Dev. Biol.* **32**, 258–281 (1973).

25. Stone, L.S. Further experiments on lens regeneration in eyes of the adult newt *Triturus v. viridescens*. *Anat. Rec.* **120**, 599–624 (1954).

26. Stone, L.S. Regeneration of the iris and lens from retina pigment cells in adult newt eyes. *J. Exp. Zool.* **129**, 505–534 (1955).

27. Stone, L.S. Experiments dealing with the role played by the aqueous humor and retina in lens regeneration of adult newts. *J. Exp. Zool.* **153**, 197–210 (1963).

28. Stone, L.S. Experiments dealing with the inhibition and release of lens regeneration in eyes of adult newts. *J. Exp. Zool.* **161**, 83–107 (1966).

29. Stone, L.S., and Gallagher, S.B. Lens regeneration restored to iris membranes when grafted to neural retina environment after cultivation *in vitro*. *J. Exp. Zool.* **139**, 247–262 (1958).

30. Voaden, M.J. A chalone in the rabbit lens? *Exp. Eye Res.* **7**, 326–331 (1968).

31. Waggoner, P.R. Lens differentiation from the cornea following lens extirpation or cornea transplantation in *Xenopus laevis*. *J. Exp. Zool.* **186**, 97–109 (1973).

32. Williams, L.A., and Higginbotham, L.T. The role of a normal lens in Wolffian lens regeneration. *J. Exp. Zool.* **191**, 233–251 (1975).

33. Yamada, T., and Dumont, J.N. Macrophage activity in Wolffian lens regeneration. *J. Morphol.* **136**, 367–384 (1972).

34. Yamada, T., Reese, D.H., and McDevitt, D.S. Transformation of iris into lens *in vitro* and its dependency on neural retina. *Differentiation* **1**, 65–82 (1973).
35. Yamada, T., and Roesel, M.E. Activation of DNA replication in the iris epithelium by lens removal. *J. Exp. Zool.* **171**, 425–432 (1969).
36. Zalik, S.E., and Scott, V. Development of ^{3}H-thymidine-labelled iris in the optic chamber of lentectomized newts. *Exp. Cell Res.* **66**, 446–448 (1971).

11

Lactation and Mammary Growth

Amazon: Gr., *a* + *mazos*, without breast

So unique are mammary glands that an entire class of vertebrates has been named for them (*11*). Even the egg-laying platypus secretes milk for the nourishment of its young, milk which in the absence of nipples is licked from the surface of the mother's abdomen. In higher mammals the number of mammary glands is roughly proportional to litter size, ranging from a single pair in guinea pigs, goats, and humans to as many as a dozen in rats. Their remarkable capacity for postpartum growth followed by regression after weaning is well known. They are also responsive to litter size and the suckling stimulus. One might expect, therefore, that the growth and differentiation of mammary glands would be subject both to humoral and neural stimuli.

Mammary glands initiate their development as invaginations from the epidermis. During infancy there develops a branching series of ducts which slowly expands in area with little or no differentiation. Not until puberty do the secretory alveoli differentiate on the terminal branches under the influence of estrogen. The failure of this differentiation in males may be due in part to the lack of estrogen, but may also be actively suppressed by testosterone. Male hormones injected into pregnant rats initiate regression of the early rudiment of the fetal mammary glands (*8*). *In vitro*, testosterone suppresses DNA synthesis and the production of the milk protein, casein (*25*).

Ovariectomy of weanling rats has been shown to prevent the sharp increase in mammary growth that otherwise occurs with sexual maturity (*1*). In adult animals, ovariectomy is associated with a decrease in the incidence of mammary cancer and will retard the growth of mammary tumors already present. Estrogen administration, on the other hand, tends to promote cancer of the mammary glands.

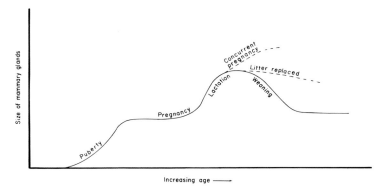

Fig. 103. Physiological factors affecting growth of the mammary glands in rats. Initial enlargement occurs at puberty, to be amplified during subsequent pregnancy and lactation. If an animal is pregnant during lactation her mammary glands become extra large. After weaning the glands normally regress, a reaction that is suppressed if the first litter is replaced at weaning by a new one from another rat.

Other hormones play equally important roles in mammary development. Without insulin, for example, DNA synthesis is suppressed in mammary epithelium. Insulin is also necessary for prolactin to induce the differentiation of the mammary epithelial cells following their proliferation, as well as the promotion of casein synthesis (12).

Although maturation of the mammary glands takes place at puberty (17), a further growth spurt occurs during pregnancy when the effects of progesterone are felt (Fig. 103) (20). In the mouse there is a peak of cellular proliferation on the fourth day of pregnancy due to the secretion of progesterone by the corpora lutea. A second peak occurs on the twelfth day when progesterone secretion from the placenta is maximal (18).

THE SUCKLING STIMULUS

The final stages of mammary development occur after parturition. The initiation of lactation depends upon the presence of suckling infants. It fails to occur if the fetuses have been removed prenatally or at birth. Under normal conditions, however, the relatively modest growth of the mammary glands during pregnancy is far surpassed by the burst of proliferation that occurs during the first week or so of lactation in the rat. As in so many other organs, the growth of the lactating mammary glands is characterized by a hyperplastic phase when the rate of DNA synthesis rises, followed by a period of cellular hypertrophy when RNA synthesis dominates the developmental process. Conversely, the DNA and RNA contents of the mammary glands decline during postlactational involution to minimal levels three weeks after weaning (Fig. 103) (22).

This natural regression of the mammary glands at the end of lactation can be prevented, or at least postponed, if the mother mates during her postpartum estrus and is therefore pregnant again during lactation (24). Concurrent pregnancy has been shown to act synergistically with lactation, leading to a rise in DNA and RNA contents of the mammary glands at 18 days after birth of the first litter. Furthermore, if newborn litters are substituted for old ones at the time of weaning the normal regression of the mammary glands can be delayed, but not prevented altogether (21). Compelled to serve two or more litters in a row, the mammary glands of rats remain functional but at reduced efficiency. Although the DNA content of such glands remains normal throughout the course of suckling two or three successive litters, the content of RNA declines to less than two-thirds of the level at the end of the first period of lactation. This is reflected in decreased rates of milk secretion leading to slower growth rates in the foster litters. It would seem that the functional capacity of the mammary glands is not inexhaustible, but that lactation and maintenance of the glands are very much a function of the suckling stimulus. This is consistent with dairy animals in which daily milking is necessary for continued lactation (2).

Although the results of removing some of an animal's mammary glands may vary from one species to another, compensatory growth of the remaining gland(s) tends to occur only in pregnant or lactating animals (10). In the lactating goat, for example, the loss of one of the two mammary glands results in a doubling of the milk output from the remaining gland. This is achieved in part by the growth of the gland but also by an increase in the milk yield on a per gram basis. To what extent this represents a response to the operation of intrinsic regulatory mechanisms, or an adaptation to the demands for more milk by suckling offspring, can only be determined by experimentation.

The most direct approach to this problem is to alter litter size. In general, the number of cells per mammary gland, the DNA content, and the milk yield are greater when large litters are being suckled (15, 19). For example, a litter of 12 rat pups has been shown to elicit the production of 16.4 gm milk/day from the mother on her fourteenth day of lactation, while only 5.6 gm milk/day are produced with a litter of two (9). This means that each pup in the litter of 12 consumes only 1.4 gm milk/day, while twice this amount is available to each pup in the litter of two. These results are tantamount to the established fact that milking frequency increases the yield in dairy animals (2). What remains to be determined is whether it is the nervous stimulation of suckling or the removal of milk that is responsible for the capacity of mammary glands to adjust their size and productivity to the demands for more milk. It is worth noting that in marsupials the teats to which the pouch young are attached become enlarged while unsuckled ones remain small.

If the teats (galactophores) of some of a rat's mammary glands are ligated while allowing others to be suckled normally, the intact glands are overburdened but the ligated ones continue to be suckled without the expulsion of milk. There is at first a marked buildup of unreleased secretory products in the ligated glands. Eventually, however, they grow smaller and have decreased contents of DNA and RNA (23). The remaining glands become enlarged, with increasing DNA and RNA contents, in proportion to litter size (14, 23). Such glands are significantly larger than those in unoperated rats, again emphasizing the importance of litter size on mammary gland enlargement. The decrease in size of galactophore-ligated glands can be overcome if the suckling stimulus is sufficiently intense. The DNA contents of ligated glands is maintained at normal levels if the litter size is not less than 12 pups, indicating that although milk cannot be ejected from ligated glands their tendency to involute can be counteracted by sufficiently strong suckling stimuli (14). These results would seem to indicate that the nervous stimulation of suckling may play an important role in regulating mammary gland growth. Such activity, however, is unquestionably linked to hormonal factors, for suckling has been shown to stimulate the secretion of growth hormone and prolactin from the anterior pituitary, while pituitary prolactin concentration is decreased in mothers separated from their offspring (7). A comparable situation may prevail in the case of pregnant rats which typically lick their nipples during the course of gestation. If this self-licking is prevented by fitting the rat with a rubber collar around her neck, mammary gland growth is significantly retarded (13). Thus, licking in the absence of milk removal is responsible for about half of the mammary gland growth that takes place prior to birth.

The most direct way to test the effect of nerves on gland growth is by denervation. When this is achieved by spinal section, the secretion of prolactin and growth hormone by the pituitary, which is normally stimulated by suckling, is decreased in the absence of sensory nerves. Denervation of mammary glands in the goat has been achieved by the transplantation of one udder (Fig. 104). Autografts can be made in a one-step operation by cutting the nerves but leaving the blood vessels intact in a pedicle while transferring the gland to the nearby thigh or flank. In other cases it has been possible to make free grafts of an udder to the neck by surgical anastomosis of its vessels to those in the graft site. Despite such impressive feats of surgery, the transplanted glands remain capable of producing milk in the absence of innervation (10). To the extent that their milk yield is decreased, that in the remaining intact gland compensates. These studies prove that nerves are not

Fig. 104.   Graft of an udder to the neck of a goat where its blood vessels were anastomosed to the carotid artery and jugular vein (*10*). In the absence of intrinsic innervation, this udder continued to yield as much milk as did the one left *in situ*.

indispensable for the structural and functional maintenance of the mammary gland.

AGING OF GRAFTED GLANDS

Mammary transplantation has been an important technique in another line of investigation, namely, research on aging. In view of the prodigious expansion of the mammary duct system in preparation for lactation, a phenomenon repeated with each birth, there is reason to suspect that the size of the mammary gland is potentially indeterminate, an attribute that lends itself to studies of growth regulation. When explanted *in vitro*, mammary gland tissue dedifferentiates, but is capable of proliferation and differentiation if stimulated with insulin and prolactin (*16*). When transplanted *in vivo* to genetically compatible hosts, small fragments of mammary duct prove capable of extensive growth and branching to give rise to complete glands (Fig. 105). This regeneration is most successful following implantation into fat pads in which mammary glands are normally imbedded. Poor growth occurs if the original gland is still present in the fat pad (*6*). Therefore, hosts are prepared by complete extirpation of their mammary gland rudiments

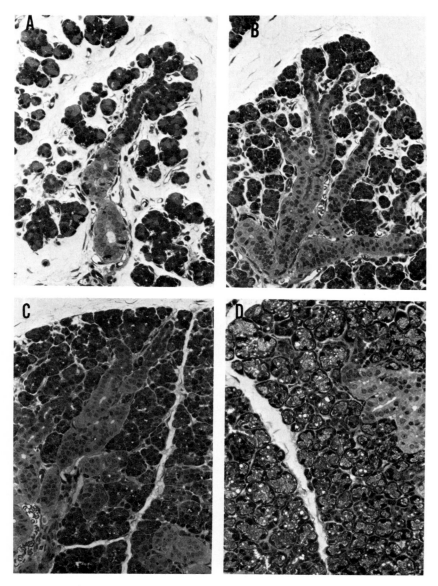

Fig. 108. Development of the rat submandibular salivary gland at 2, 7, 14, and 42 days of age (A–D) showing proliferation and maturation of terminal tubules (A, B) into secretory acini (C, D) (8).

supervened, becoming the dominant form of growth after weaning. Thus, a phase of DNA replication early in postnatal development gives way to a period dominated by RNA and protein synthesis. It is perhaps no coincidence that this switch from hyperplasia to hypertrophy upon maturation of the salivary glands occurs at about the time when solid foods begin to be consumed.

A still later change occurs at the time of sexual maturation when differences between the male and female salivary glands appear. In general, granular tubules are more predominant in the male than in the female salivary glands, while acini are more numerous in females. During pregnancy, there is an increase in the diameters of the tubules of female salivary glands. Castration of the male brings about a decline in the weight of the glands, more so in adults than in immature animals (24). Testosterone administration causes masculinization of the female glands, including increased size and cellular proliferation. Why it is that the salivary glands are in fact secondary sex organs remains something of a mystery.

They are also responsive to other hormones. Hypophysectomy causes atrophy of the salivary glands, including decreases in tubule diameter, number of secretory granules, amylase production, and sodium concentration in the saliva. These effects can be partly reversed by the administration of growth hormone (21). Thyroidectomy brings about a decrease in the size of the salivary glands, an effect reversed by TSH, thyroxine, or triiodothyronine (7). In diabetes, the weight of the salivary glands also decreases, while insulin stimulates growth.

Not all hormones exert a positive influence on salivary gland growth and activity. Parathyroidectomy, for example, is followed by a doubling in the size of the salivary glands due primarily to cell enlargement (5). The decrease in plasma calcium concentration accompanying parathyroidectomy is perhaps in part attributable to the increase in calcium ion content in the salivary glands themselves. It would seem that the parathyroid gland may have as intimate a relationship with the salivary glands as with the other excretory organs, the kidneys (see Chap. 17).

When the duct of an excretory gland is ligated the results can be disastrous. Salivary glands are no exception. The cells of the acini degenerate after 2–3 days, and may have completely disappeared by the end of a month. Concomitantly, the synthesis of secretory products is arrested during the first week and the weight of the gland as a whole drops to half normal in four days, falling to 20% after four weeks (14). The ducts react differently. Their cells lose their polarity, but due to their proliferation there is an increase in the duct system. This reaction is probably a reparative response to the distension of ligated ducts which cannot get rid of their secretory contents.

If the ligature on the salivary duct is removed during the first week, the effects may be reversible (6). New acinar cells regenerate, and at the end of a week they may be producing secretory granules again. If the ligation is a permanent one, compensation may occur in the other nonligated salivary glands (10, 21). This compensation is both physiological, involving increased secretion of saliva, and morphological in terms of gland enlargement. These responses are similar to those occurring after partial ablation of salivary glands. When part of a single gland is lost, there is a wound-healing response that may involve limited production of new acini from residual duct cells (12). The more common reaction, however, is for remaining glands to compensate for the loss or incapacitation of others. Although there is little or no compensatory growth in a salivary gland following extirpation of its opposite partner, considerable growth may be elicited if several other glands are removed or their ducts ligated (21, 24). This would suggest that there may be little specificity between the different kinds of salivary glands despite their structural and functional distinctions.

There are three pairs of salivary glands all contributing saliva to the oral cavity. The submandibular gland is a large, consolidated oval organ located ventrally in the neck of the rat. The sublingual gland is considerably smaller and is closely apposed to the submandibular, so much so that care is needed to separate them. The parotid glands are situated subcutaneously behind the ears. Their configuration is more diffuse than the others, making their complete dissection difficult.

ADAPTATIONS TO FUNCTIONAL ALTERATIONS

Saliva may be serous or mucous. The parotid secretion is serous; the submandibular gland secretes a mixed saliva that is more serous than mucous; the sublingual saliva is almost completely mucous. The key to the function and growth of the salivary glands lies in the autonomic nerves supplying them. All three kinds of salivary glands receive parasympathetic innervation, mostly via the chorda tympani. Sympathetic nerves from the superior cervical ganglion innervate the submandibular and parotid glands, but not the sublingual. Both sympathetic and parasympathetic denervation lead to atrophy of the glands they innervate. There may be a transient increase in weight presumably owing to the accumulation of secretory products, but after the first few days there is a steady decline in gland size, mitotic activity, and synthesis of DNA, RNA, and salivary proteins, including amylase (16). When the salivary glands of infants are denervated, their differentiation is not prevented although the sizes to which they grow are below normal (17).

Autonomic stimulation induces both salivation and growth of the salivary glands, a phenomenon dramatically illustrated by the effects of isoproterenol. This sympathomimetic amine is a β-adrenergic stimulator that exerts its effects on the submandibular and parotid glands but not on the sublingual gland which normally lacks sympathetic innervation. Intraperitoneal injection of isoproterenol elicits copious salivation within 5–10 minutes, soon depleting the salivary glands of their secretory products. This is followed by a period in which new secretory material is synthesized. There are increases in the glyogen concentration, monoamine oxidase activity, amino acid uptake, and protein synthesis, including the production of salivary amylase. These changes are accompanied by increases in the rates of DNA and RNA synthesis leading to heightened mitotic activity in 2–4 days (2, 4). If isoproterenol administration is continued, chronic stimulation of the glands leads to their three- to fivefold enlargement during the next two weeks. In such glands the tubule diameters are decreased but the acini are enlarged, as are their cells due in part to the polyploidy accompanying this kind of growth. If isoproterenol is given to infant rats, salivary gland maturation is accelerated by enhanced DNA synthesis after two days of age and acinar cell stimulation by two weeks after birth (3, 9, 19). The effects of isoproterenol cannot be prevented by sympathectomy since the drug acts directly on the β-adrenergic receptors at the cellular level (21). Propranolol, a β-adrenergic antagonist, blocks the action of isoproterenol.

The effects produced by autonomic stimulation can be induced indirectly by a variety of experimental interventions, each of which illustrates the versatility of these glands in responding to an interesting assortment of functional stimuli. Not surprisingly, the salivary glands are affected by the nature of the diet. If rats are fed a diet bulked out with inert cellulose, the resulting rise in food consumption increases salivary secretion and causes submandibular hypertrophy (22). Conversely, animals deprived of the need to chew by being fed a liquid diet (e.g., Metrecal) undergo atrophy of their salivary glands, especially the parotid (16, 22). This atrophy is characterized by decreases in cell size, mitosis, and DNA and RNA contents of the acini, not the ducts. Animals held on a liquid diet and then refed solid food promptly undergo a burst of mitotic activity with increases in cell size (18). The stimulatory effect of a bulk diet on salivary gland growth is partially prevented by sympathectomy as well as parasympathectomy, being totally abolished by complete autonomic denervation (20).

Presumably the effects of diet are mediated at least in part by way of the teeth. If the incisors of a rat are repeatedly amputated every few days, the salivary glands hypertrophy (15). This curious effect is produced only by amputation of the lower, not upper, incisors. Amputation of the molars or ulceration of the oral mucosa is ineffectual. Furthermore, the response is unilateral. If only one lower incisor is amputated, hypertrophy is confined to the ipsilateral glands. The effect is mediated primarily via the sympathetic innervation, but to a lesser extent through the parasympathetic nerves. The response is seen in the enlargement of acini by cell hyperplasia and hypertrophy, while the tubules remain normal or even decrease in size.

An even more surprising method for stimulating salivary gland growth is to feed diets supplemented with such proteolytic enzymes as trypsin, protease, pancreatin, or papain. This treatment promotes hypertrophy of the submandibular glands, but not of the sublingual or parotid glands. Hypertrophy is due to cell enlargement and increased acinar size. Further experiments have demonstrated the lack of effect when the proteolytic enzymes are administered by stomach tube (23). This indicates that the site of action is somewhere in the oral cavity, a suspicion confirmed by the effects of sectioning the glossopharyngeal nerves. This proves that proteolytic enzymes must stimulate the taste buds in order to promote submandibular hypertrophy, which they do by transmission of impulses via the glossopharyngeal nerves to the autonomic nervous system. The response of the submandibular gland to proteolytic enzymes is partially prevented by sympathectomy and totally abolished by sectioning all autonomic innervation.

A final method for stimulating salivary gland growth is one that does not operate directly on the oral cavity. If rats are exposed to a hot environment (e.g., 34°C) they react by profuse salivation, licking their fur in an attempt to keep cool by evaporation. Clearly, the salivary glands are reacting in a manner analogous to that of sweat glands at high temperatures. Although the sublingual glands do not become enlarged under these conditions, the submandibular glands increase 50% in mass within three weeks, an effect primarily dependent upon their parasympathetic innervation and only to a lesser extent on sympathetic nerves (11).

In general, therefore, there are several pathways by which the salivary glands can be made to undergo hypertrophy (Fig. 109). Isoproterenol mimics the action of the sympathetic nervous system to stimulate salivation and growth. Proteolytic enzymes in the diet exert their influences exclusively through the sympathetic innervation. Repeated amputation of the lower incisors acts primarily via the sympathetic nerves but partly through the parasympathetic innervation. Compensatory growth of the remaining salivary glands following excision of the others is mediated by way of both types of autonomic innervation. Bulk diets also operate in this way. Exposure to hot environments stimulates salivary gland growth primarily via the parasympathetic nerves, but in part through the sympathetic innervation. Other influences operate independently of nerves. The hypertrophic effects of growth hormone or parathyroidectomy take place even in the totally denervated salivary glands. Salivary glands are therefore responsive to a rich

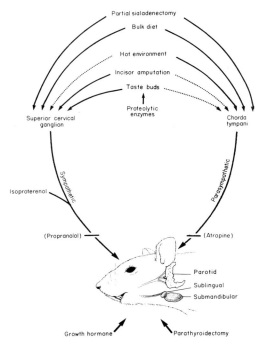

Fig. 109. Outline of various experimental interventions that promote hypertrophy of the rat salivary glands. Solid arrows indicate the principal autonomic nerves by which the physiological effects are mediated. Broken arrows are less important neural pathways as revealed by selective denervation studies. Inhibitory influences are in parentheses. Growth hormone and parathyroidectomy stimulate salivary gland hypertrophy independent of autonomic innervation.

variety of physiological stimuli which act by way of neural as well as humoral pathways to increase function while stimulating growth.

REFERENCES

1. Ball, W.D. Development of the rat salivary glands. III. Mesenchymal specificity in the morphogenesis of the embryonic submaxillary and sublingual glands of the rat. *J. Exp. Zool.* **188**, 277–288 (1974).
2. Barka, T. Effect of isoproterenol on amino acid transport into rat salivary glands. *Exp. Cell Res.* **64**, 371–379 (1971).
3. Barka, T., Chang, W.W.L., and Van der Noen, H. Stimulation of DNA synthesis by isoproterenol in rat submandibular gland during postnatal growth. *Cell Tissue Kinet* **6**, 135–146 (1973).
4. Baserga, R., and Heffler, S. Stimulation of DNA synthesis by isoproterenol and its inhibition by actinomycin D. *Exp. Cell Res.* **46**, 571–580 (1967).

5. Bazerque, P.M., Meiss, A., and Moroni, M.N. Autonomous denervation and salivary glands hypertrophy after parathyroidectomy. *Acta Physiol. Lat. Am.* **21**, 107–111 (1971).

6. Bhaskar, S.N., Lilly, G.E., and Bhussry, B. Regeneration of the salivary glands in the rabbit. *J. Dent. Res.* **45**, 37–41 (1966).

7. Bray, G.A. Effects of thyroid status and adrenergic blocking drugs on isoproterenol-induced enlargement of the salivary glands. *Proc. Soc. Exp. Biol. Med.* **124**, 1073–1076 (1967).

8. Chang, W.W.L. Cell population changes during acinus formation in the postnatal rat submandibular gland. *Anat. Rec.* **178**, 187–201 (1974).

9. Chang, W.W.L., and Barka, T. Stimulation of acinar cell proliferation by isoproterenol in the postnatal rat submandibular gland. *Anat. Rec.* **178**, 203–209 (1974).

10. Elmér, M., and Ohlin, P. Compensatory hypertrophy of the rat's submaxillary gland. *Acta Physiol. Scand.* **76**, 396–398 (1969).

11. Elmér, M., and Ohlin, P. Salivary glands of the rat in a hot environment. *Acta Physiol. Scand.* **79**, 129–132 (1970).

12. Hanks, C.T., and Chaudhry, A.P. Regeneration of rat submandibular gland following partial extirpation. A light and electron microscopic study. *Am. J. Anat.* **130**, 195–207 (1971).

13. Jacoby, F., and Leeson, C.R. The post-natal development of the rat submaxillary gland. *J. Anat.* **93**, 201–216 (1959).

14. Ohlin, P., and Perec, C. Secretory responses and choline acetylase of the rat's submaxillary gland after duct ligation. *Experientia* **23**, 248–249 (1967).

15. Ohlin, P., and Perec, C.J. Choline acetylase in normal and denervated submaxillary glands of rats after repeated teeth amputations. *Acta Physiol. Scand.* **69**, 134–139 (1967).

16. Schneyer, C.A. Mitotic activity and cell number of denervated perotid of adult rat. *Proc. Soc. Exp. Biol. Med.* **142**, 542–547 (1973).

17. Schneyer, C.A., and Hall, H.D. Effects of denervation on development of function and structure of immature rat parotid. *Am. J. Physiol.* **212**, 871–876 (1967).

18. Schneyer, C.A., and Hall, H.D. Parasympathetic regulation of mitosis induced in rat parotid by dietary change. *Am. J. Physiol.* **229**, 1614–1617 (1975).

19. Srinivasan, R., Chang, W.W.L., van der Noen, H., and Barka, T. The effect of isoproterenol on the postnatal differentiation and growth of the rat submandibular gland. *Anat. Rec.* **177**, 243–254 (1973).

20. Wells, H. Functional and pharmacological studies on the regulation of salivary gland growth. *In* "Secretory Mechanisms of Salivary Glands" (L.H. Schneyer and C.A. Schneyer, eds.), pp. 178–190. Academic Press, New York, 1967.

21. Wells, H., and Peronace, A.A.V. Synergistic autonomic nervous regulation of accelerated salivary gland growth in rats. *Am. J. Physiol.* **207**, 313–318 (1964).

22. Wells, H., and Peronace, A.A.V. Functional hypertrophy and atrophy of the salivary glands of rats. *Am. J. Physiol.* **212**, 247–251 (1967).

23. Wells, H., Peronace, A.A.V., and Stark, L.W. Taste receptors and sialadenotrophic action of proteolytic enzymes in rats. *Am. J. Physiol.* **208**, 877–881 (1965).

24. Zebrowski, E.J. Effects of age and castration on compensatory hypertrophy and sialic acid levels of the male rat submandibular gland. *Arch. Oral Biol.* **17**, 447–454 (1972).

13

The Exocrine Pancreas

Trypsin: Gr., *tryein*, to wear down

The need for two kinds of enzymes in the vertebrate digestive tract, one with acid and the other with alkaline pH optima, has necessitated the subdivision of the gut into stomach and intestines, respectively. The digestive enzymes of the latter are secreted by one of the body's most versatile exocrine glands. Phylogenetically, the pancreas does not exist as such in cyclostomes in which digestive glands are embedded in the walls of the intestines. In teleost fishes the pancreas is found to have evolved as an outgrowth from the small intestine to which it is connected by a network of ducts. Not infrequently, however, the secretory acini of the fish pancreas are to be found intermingled with the liver. Moreover, their islets of Langerhans may be consolidated in some species into a discrete organ, the principal islet. Hence, the characteristic association between the endocrine and exocrine pancreas in higher forms would appear to be an accident of histology, although the possibility of some underlying physiological strategy in this relationship cannot be denied.

The functional unit of the pancreas is the secretory acinus, a bulb of cells located at the termination of each duct. It is the function of each cell in the acinus to synthesize and secrete enzymes capable of digesting fats (lipase), carbohydrates (amylase), and proteins (trypsin, chymotrypsin). In view of the prodigious output of enzymes by these cells, there are few if any other organs of the body which surpass the pancreas in their rates of protein synthesis. Each cell in the secretory acinus is richly endowed with endoplasmic reticulum, a characteristic responsible for the basophilia typical of the basal ends of these cells. The enzyme precursors thus synthesized are packaged by the Golgi apparatus into zymogen granules which crowd the apical ends of the cells. As an expanding organ, even the most highly specialized of these cells are capable of mitosis. Although cell division is rare in the normal adult pancreas, it is not incompatible with the differentiated state of the secretory cells, mitotic figures being found sharing the same cytoplasm with zymogen granules (23). Pancreatic cells are capable of synthesizing DNA and enzymes simultaneously, and even in the process of mitosis they continue to incorporate amino acids (24).

If the pancreatic duct is ligated the exocrine component of the pancreas degenerates, the islets of Langerhans being spared (26). Curiously, degeneration of the ligated pancreas is often accompanied by a transient proliferative response, perhaps representing a futile attempt at regeneration. A more successful regenerative response is observed following partial resection of the pancreas (14). Two days after operation there is a severalfold increase in proliferation, the magnitude of which is proportional to the degree of ablation. This compensatory hyperplasia is not confined to the site of operation, but is distributed throughout the remaining glandular tissue. Nevertheless, restoration of the original mass of the pancreas is usually incomplete.

ETHIONINE-INDUCED DEGENERATION

The capacity of the pancreas for regeneration and its susceptibility to degeneration are associated with the high rate of protein synthesis that is the hallmark of pancreatic function. Interference with this function results in severe damage to the secretory cells, suggesting that their morphological integrity depends upon the preservation of their physiological competence. Deficiencies in methionine, an essential amino acid of particular importance for the formation of disulfide linkages in protein synthesis, decrease the production of pancreatic enzymes (16). More dramatic effects are achieved by the administration of ethionine, an analogue of methionine. Daily injections into rats have deleterious effects on the exocrine pancreas, reducing the production of amylase, lipase, and chymotrypsinogen, and causing extensive degeneration of the organ, virtually obliterating it after about two weeks (Fig. 110). Protection is afforded by the simultaneous administration of methionine. Cytologically, the effects of ethionine are characterized by disruption of the endoplasmic reticulum, loss of cytoplasmic basophilia, and the development of endoplasmic whorls in the cytoplasm (4, 8). The nuclei become dense, the mitochondria swollen, and there is a decrease in the Golgi apparatus and the numbers of zymogen granules (Fig. 111). During the first few days, as degeneration of the pancreas proceeds, there is a concomitant decrease in its enzymes (19). The organ as a whole shrinks to a fraction of its original mass and in fact becomes rather difficult to locate. Histologically, the destruction of the pancreas is limited to the secretory

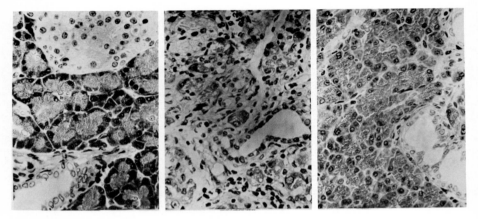

Fig. 110.   Effects of daily intraperitoneal injections of ethionine (35 gm/kg) on the rat pancreas as seen in light micrographs. Left, control pancreas showing intact acini (islet at top). Middle, disintegration of secretory acini with loss of most zymogen granulation after 8 days of ethionine treatment. Right, regenerated pancreas following cessation of treatment. (Courtesy of David F. Gardner, M.D., University of Pennsylvania School of Medicine.)

acinar cells, the ducts and islets of Langerhans being preserved. Upon cessation of treatment with ethionine, the pancreatic remnant undergoes remarkable recovery (5). The destructive changes are reversed and the cells proliferate and redifferentiate. By 2–3 weeks after the last ethionine injection, the pancreas is restored nearly to normal (Fig. 111) (9).

If the vulnerability of the pancreas to ethionine is attributable to the protein synthesis in which it is engaged, then other compounds that interfere with this process should mimic the effects of ethionine. Studies have shown that the antibiotic puromycin, which acts by inhibiting protein synthesis, also causes necrosis of pancreatic acini (15). Actinomycin D, which blocks RNA synthesis, elicits many of the cytological manifestations of ethionine, including endoplasmic whorls and depletion of zymogen granules (21), effects that lead to declines in the levels of pancreatic enzyme activity (18).

The fetal pancreas, while possessing zymogen granules, is relatively nonfunctional in the absence of intestinal digestion. It might be predicted, therefore, that it should be less susceptible to the deleterious effects of ethionine. If pregnant rats are injected with this analogue, the maternal pancreatic tissue degenerates as expected, but that in the fetuses is less seriously disrupted (7). Even when ethionine is injected directly into the fetus, its pancreas is not so seriously affected as in postnatal animals. This evidence is consistent with the interpretation that the vulnerability of the

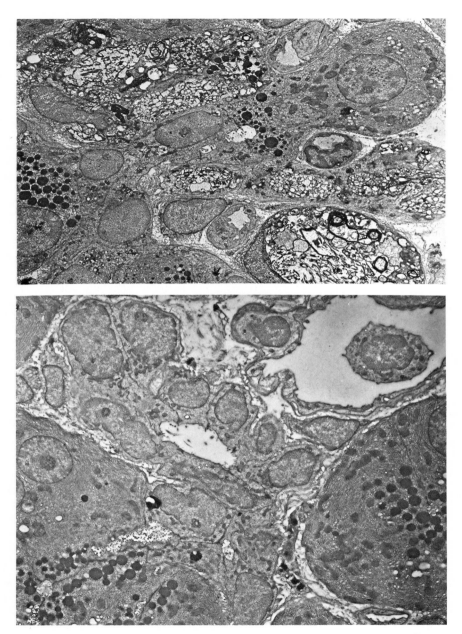

Fig. 111. Electron micrographs of degenerating and regenerating rat pancreas following ethionine treatment (8, 9). After 10 days of ethionine administration on a protein-free diet, the acinar cells become filled with cellular debris, and zymogen granules are relatively sparse (above). Regenerative changes are seen below 18 days after cessation of ethionine treatment. The cells have become reorganized into acini, and zymogen granules are again produced in quantity.

pancreas to such agents lies in the role of protein synthesis in preserving the integrity of this organ.

THE SOYBEAN EFFECT

A clue to the control of pancreatic growth may be found in the effects of soybean diet on the function and size of the pancreas. Animals fed raw soybean undergo pancreatic hypertrophy due to proliferation and enlargement of the cells (3). A comparable response is seen in animals fed lima beans or egg white, but only if they are uncooked. These and other substances contain heat-labile trypsin inhibitors which stimulate pancreatic growth by inactivating trypsin, thereby interfering with such feedback mechanisms as may be responsible for the regulation of enzyme synthesis and secretion by the pancreas. Trypsin inhibitors are therefore valuable tools with which to analyze the physiological regulation of pancreatic growth and activity. In general, chickens and rats fed diets rich in raw soybean secrete pancreatic juice rich in trypsin at the expense of the other enzymes, an adaptation that may eventually overcome the effects of the trypsin inhibitor (12, 13). Parenthetically, it is worth noting that if rats are fed raw soybeans while being treated with ethionine, the latter dominates the former and the pancreas degenerates (2).

The growth response of the pancreas in compensating for the presence of trypsin inhibitors suggests that the composition of the diet might influence the repertoire of enzymes secreted. Physiological studies have confirmed that the kinds of enzymes produced by the pancreas are correlated with the availability of substrates to be digested. High carbohydrate diets stimulate the secretion of increased amounts of amylase while protein rich diets have a similar effect on trypsin and chymotrypsin secretion (1, 10, 25). How the composition of the intestinal contents might be monitored to bring about an adaptive response by the pancreas is not known, nor is the pathway of communication between the intestine and the pancreas understood. The hormones secretin and pancreozymin, both secreted by the intestine, are important factors in the stimulation of pancreatic secretion. Pancreozymin, now known to be the same hormone as cholecystokinin (20), causes acinar cell hypertrophy (22) and hyperplasia (17), while increasing their enzyme contents (6). Indeed, if the proximal 30% of the small intestine is resected, removing the presumed source of hormones, the pancreatic hypertrophy otherwise stimulated by soybean trypsin inhibitor is suppressed (11). However, these hormones are not known to be selective in their effects on which enzymes are to be secreted. Clearly, there is a problem of quality control whereby pancreatic enzyme production must be correlated with specific digestive requirements.

There is no evidence that different pancreatic cells secrete different enzymes, nor is there reason to believe that zymogen granules are anything but mixed bags of enzyme precursors. Hence, the control of enzyme production must operate in the synthetic machinery of the cell itself, presumably by the selective repression or derepression of genes coded for the synthesis of specific enzymes. A solution to the problem of how pancreatic growth is controlled depends upon our understanding of how the pancreas adapts its enzyme synthesis to the composition of the diet.

REFERENCES

1. Ben Abdeljlil, A., and Desnuelle, P. Sur l'adaptation des enzymes exocrines du pancreas a la composition due regime. *Biochim. Biophys. Acta* **81**, 136–149 (1964).
2. Beswick, I.P., Bouchier, I.A.D., and Pirola, R.C. The effect of oral trypsin inhibitor on pancreatic damage in the rat. *Br. J. Exp. Pathol.* **52**, 244–251 (1971).
3. Beswick, I.P., Pirola, R.C., and Bouchier, I.A.D. The cause of pancreatic enlargement in rats fed raw soybean. *Br. J. Exp. Pathol.* **52**, 252–255 (1971).
4. Fitzgerald, P.J., and Herman, L. Degeneration and regeneration of the pancreas. *Bull. N.Y. Acad. Med.* **41**, 804–810 (1965).
5. Fitzgerald, P.J., Herman, L., Carol, B., Rogue, A., Marsh, W.H., Rosenstock, L., Richards, C., and Perl, D. Pancreatic acinar cell regeneration. I. Cytologic, cytochemical and pancreatic weight changes. *Am. J. Pathol.* **52**, 983–1011 (1968).
6. Fölsch, U.R., and Wormsley, K.G. The pancreatic secretion of enzymes in rats treated with soybean diet. *Scand. J. Gastroenterol.* **9**, 679–683 (1974).
7. Gardner, D.F., and Goss, R.J. Effect of ethionine on fetal pancreatic development in the rat. *Proc. Soc. Exp. Biol. Med.* **127**, 1130–1134 (1968).
8. Herman, L., and Fitzgerald, P.J. The degenerative changes in pancreatic acinar cells caused by DL-ethionine. *J. Cell Biol.* **12**, 277–296 (1962).
9. Herman, L., and Fitzgerald, P.J. Restitution of pancreatic acinar cells following ethionine. *J. Cell Biol.* **12**, 297–312 (1962).
10. Howard, F., and Yudkin, J. Effect of dietary change upon the amylase and trypsin activities of the rat pancreas. *Br. J. Nutr.* **17**, 281–294 (1963).
11. Ihse, I. Abolishment of oral trypsin inhibitor stimulation of the rat exocrine pancreas after duodeno-jejunal resection. *Scand. J. Gastroenterol.* **11**, 11–15 (1976).
12. Konijn, A.M., Birk, Y., and Guggenheim, K. Pancreatic enzyme pattern in rats as affected by dietary soybean flour. *J. Nutr.* **100**, 361–368 (1970).
13. Lepkovsky, S., Furuta, F., Koike, T., Hasegawa, N., Dimick, M.K., Krause, K., and Barnes, F.J. The effect of raw soya beans upon the digestion of proteins and upon the functions of the pancreas of intact chickens and of chickens with ileostomies. *Br. J. Nutr.* **19**, 41–56 (1965).
14. Lehv, M., and Fitzgerald, P.J. Pancreatic acinar cell regeneration. IV. Regeneration after surgical resection. *Am. J. Pathol.* **53**, 513–535 (1968).
15. Longnecker, D.S., Shinozuka, H., and Farber, E. Molecular pathology of *in-vivo* inhibi-

tion of protein synthesis: Electron microscopy of rat pancreatic acinar cells in puromycin-induced necrosis. *Am. J. Pathol.* **52**, 891–915 (1968).

16. Lyman, R.L., and Wilcox, S.S. Effect of acute amino acid deficiencies on carcass composition and pancreatic function in the force-fed rat. I. Deficiencies of histidine, methionine, phenylalanine, and threonine. *J. Nutr.* **79**, 28–36 (1963).

17. Mainz, D.L., Black, O., and Webster, P.D. Hormonal control of pancreatic growth. *J. Clin. Invest.* **52**, 2300–2304 (1973).

18. Marsh, W.H., and Fitzgerald, P.J. Pancreas acinar cell regeneration. X. Effect of actinomycin D on enzymatic activities. *Am. J. Pathol.* **64**, 357–371 (1971).

19. Marsh, W.H., Goldsmith, S., Crocco, J., and Fitzgerald, P.J. Pancreatic acinar cell regeneration. II. Enzymatic, nucleic acid, and protein changes. *Am. J. Pathol.* **52**, 1013–1038 (1968).

20. Niess, E., Ivy, C.A., and Nesheim, M.C. Stimulation of gallbladder emptying and pancreatic secretion in chicks by soybean whey protein. *Proc. Soc. Exp. Biol. Med.* **140**, 291–296 (1972).

21. Rodriguez, T.G. Ultrastructural changes in the mouse exocrine pancreas induced by prolonged treatment with actinomycin D. *J. Ultrastr. Res.* **19**, 116–129 (1967).

22. Rothman, S.S., and Wells, H. Enhancement of pancreatic enzyme synthesis by pancreozymin. *Am. J. Physiol.* **213**, 215–218 (1967).

23. Sesso, A., Abrahamsohn, P.A., and Tsanaclis, A. Acinar cell proliferation in the rat pancreas during early postnatal growth. *Acta Physiol. Lat. Am.* **23**, 37–50 (1973).

24. Sesso, A., Abrahamsohn, P.A., Freymüller, E., and Valeri, V. Incorporation d'amino-acides tritiés, par les cellules acineuses du pancréas, observée aux microscopes optique et électronique, avant et pendant la mitose chez le Rat jeune. *C.R. Acad. Sci. Paris* **266**, 1668–1670 (1968).

25. Snook, J.T. Dietary regulation of pancreatic enzyme synthesis, secretion and inactivation in the rat. *J. Nutr.* **87**, 297–305 (1965).

26. Zeligs, J.D., Janoff, A., and Dumont, A.E. The course and nature of acinar cell death following pancreatic ligation in the guinea pig. *Am. J. Pathol.* **80**, 203–218 (1975).

14

Liver Regeneration

Icterus: Gr., *ikteros*, a yellowish-green oriole

The largest organ in the body is the one that regenerates best. It does not regenerate in the epimorphic sense, for the lobes that may have been surgically removed are not replaced. It is the part left behind that, in a burst of mitotic activity, restores the original mass of the liver. Its regeneration, therefore, is achieved by compensatory hyperplasia, the magnitude of which is so striking as to have found its way into Greek mythology.

As the story goes, fire was originally the exclusive property of the gods until Prometheus gave it to mortal men on earth. Angered by his act, the gods decreed that he should be punished. Chained to a rock, an eagle came each day and plucked out his liver. Each night, however, a new one grew in its place. One wonders if some unsung Greek surgeon might not have performed the world's first partial hepatectomy and discovered the phenomenon of liver regeneration. If so, the data were never published, except in the not altogether inaccurate accounts of Greek mythology.

Few organs and tissues of the body have been the object of so much research as has the liver. The apparent homogeneity of its cells makes it a model system in which to analyze a multitude of biochemical pathways, and the liver's vital necessity has emphasized the importance of such research. However, it is the remarkable capacity of the liver to grow in compensation for reductions in its mass that has provided much of the impetus for the innumerable investigations of this unsolved problem. Notwithstanding the magnitude of these efforts, the unparalleled complexities of hepatic physiology leave unexplained the cause of liver regeneration (8, 36).

First and foremost, the liver is an exocrine gland that secretes bile into the small intestine. Its cells must therefore be capable of producing all of the components in this complex fluid, as well as recycling some of them in the enterohepatic circulation following their absorption from the small intestine. The liver's proximity to the digestive tract also places it in a strategic location for processing the end products of digestion. All of the blood from the gut and its associated organs drains into the liver via the hepatic portal vein. The copious blood flow through the liver greatly facilitates the many hepatic functions that depend upon exchange between plasma and hepatic parenchyma. The storage of glycogen in the liver bears witness to its responsiveness to insulin and glucagon. Its prodigious capacity as a site for lipid storage makes the liver a comercially valuable source of oils, especially in fishes. The liver is also a reservoir for vitamin A; so much so that its toxic concentration in the polar bear liver makes this the one organ in these animals which Eskimos wisely refrain from eating.

The enzymatic armamentarium of the liver permits the decarboxylation and deamination of amino acids absorbed from the gut as well as the synthesis of urea. Not unrelated to these processes is the detoxification of noxious substances. Whatever may be the nature of the poison, whether narcotic or carcinogen, the liver seems capable of synthesizing the right enzymes specific for the detoxification of each new substrate. Hence, the liver is unsurpassed in its capacity for enzyme induction and indispensable as the body's first line of defense against ingested poisons, even to the point of self-immolation.

In addition to its functions as an endocrine organ and its job in processing a variety of substrates, the liver is the source of many of the plasma proteins. These include albumin, some of the α- and β-globulins, fibrinogen, prothrombin, angiotensinogen, and heparin, to name but a few. It may also be an extrarenal source of erythropoietin, a function possibly correlated with the hemopoietic functions of the prenatal liver. Clearly, no other cell can rival the versatile specializations of the hepatocyte, yet the differentiated state does not preclude its potential for cell division.

The liver is an expanding organ. During maturation it grows by cell proliferation, extending its lobules by the organization of new hepatic cords. Although such growth ceases in the adult animal, the tissues of the liver never lose their capacity for renewed mitotic activity. In the infant rat, the characteristically high rate of proliferation begins to decline after about 10 days of age (22). Binucleate cells and polyploid nuclei appear at about the time of weaning, possibly in response to the ingestion of solid food. Nuclei become increasingly polyploid with age (1), and multinucleate cells are not uncommon (Fig. 112). In fetal and neonatal rat livers the pattern of proliferation is evenly distributed throughout the parenchyma, but with the organization of the adult configuration of secretory acini during the second week after birth mitotic activity becomes predominantly localized in the central zone of the lobules. Prenatally, the liver as a whole grows faster than the rest of the

would otherwise have proceeded into the S phase. When treatment is discontinued, all such cells commence synthesis of DNA at once, as can be demonstrated autoradiographically following injections of ^{3}H-thymidine (Fig. 115).

HUMORAL GROWTH REGULATORS

Man's quest for the cause of liver growth and regeneration has focused on the hypothetical roles of stimulators versus inhibitors. One school of thought contends that growth is an active process triggered by external stimulators without which enlargement ceases. The alternative hypothesis is that growth is an intrinsic attribute of the liver expressed in inverse proportion to the concentration of circulating inhibitors. Whatever may be the true explanation, it also remains to be determined whether such growth regulating factors are produced in the liver itself or elsewhere in the body, and whether the liver is subject to the actions of a single substance or to many agencies operating additively. Finally, it is important to learn if such growth regulating factors as may exist are solely involved in controlling liver mass *per se* or do so in conjunction with their role in mediating physiological activities in the liver. Clearly, the problem promises to be a complex one not likely to yield to oversimplified solutions.

The importance of growth regulators in hepatic homeostasis cannot be exaggerated. Unlike many other organs in the body, the enlargement of the liver is not limited by a fixed number of functional units incapable of being multiplied beyond certain stages of development. It is an organ of indeterminate size capable of potentially unlimited growth, in part because its functional capacities can keep pace with its structural enlargement. Any organ capable of unrestricted growth runs the risk of becoming too large. In the case of the liver, it is of the utmost importance that there be some mechanism by which liver growth can be turned off when the organ reaches the right size. Hence, it is logical to explore the possibility that the liver might slow down its own growth by negative feedback mediated by self-inhibitors.

Various investigators over the years have tried to determine if liver growth is affected by homogenates or extracts of hepatic origin. The bewildering profusion of results from such experiments, ranging from inhibition to stimulation, makes the interpretation of such effects something of a guessing game. Equally confusing have been the many attempts to analyze the action of serum factors in regulating liver growth. Plasma can be removed or supplemented. It can also be transfused from intact animals to partially hepatectomized ones, or vice versa. Such variables as the timing of its withdrawal or injection with respect to partial hepatectomy, and the dose and route of administration are seldom the same from one experiment to another. The net result has been a spectrum of findings, both positive and negative, which defies interpretation. Yet the possible existence of humoral factors responsible for hepatic growth regulation is as difficult to deny as it is to prove.

Attempts to demonstrate once and for all the existence of blood-borne growth regulators have focused on the technique of parabiosis. This approach involves the grafting together of rats or mice in such a way as to allow exchange of blood through the capillaries which link the two partners. Partial hepatectomy of one parabiotic partner would be expected to result in a proliferative reaction in the intact liver of the other animal, as well as in the remaining liver of the operated one. The results are not always as anticipated. Although positive results have been obtained more often than not, the lack of consistency has prompted the development of techniques for the more direct and copious crossflow of blood between the two partners. This has been achieved by the reciprocal union of the carotid artery and jugular vein between two rats, an arrangement that cannot be maintained indefinitely but long enough to measure DNA synthesis in the livers. The positive outcome of such experiments prove that when one partner is partially hepatectomized the liver of the intact one can respond. Although these results can be explained in terms of either inhibitors or stimulators, they confirm the existence of humoral factors in regulating liver growth (15, 26, 30, 38).

The role of circulating factors in governing the growth of the liver is particularly relevant to the many experiments which have been carried out over the years involving the experimental manipulation of the hepatic blood supply. The liver receives its arterial blood via the hepatic artery and venous blood from the hepatic portal vein. Branches of these vessels carry the arterial venous blood to the peripheries of the lobules where it is delivered to the hepatic sinusoids. Here the oxygenated and unoxygenated blood intermingles during its centripetal passage across the width of the lobules to the central veins which combine eventually to form the hepatic vein through which the blood is channeled to the inferior vena cava. Ligation of the hepatic artery depriving the liver of oxygenated blood has been reported to result in infarction of the liver in some cases (53), but in others there is no apparent effect even to the extent of permitting delayed regeneration after partial hepatectomy (7). The effects of portal vein ligation are more consistent, regularly causing atrophy of the liver but not necessarily preventing its regeneration after partial hepatectomy (5, 53). If aortic blood is shunted to the portal vein to replace venous blood with arterial, the atrophy that otherwise occurs is prevented and the regenerative response preserved (34).

These results have been interpreted to mean that the atrophy of livers deprived of portal inflow is due more to the hemodynamic disturbance *per se* than to the lack of factors in the portal blood upon which the integrity of the liver depends. Another approach to the problem is to create a portacaval shunt (Eck fistula) whereby the portal blood is diverted to the inferior vena cava thus bypassing the liver. Again, this results in hepatic atrophy and reduced liver regeneration following partial hepatectomy (23). If the hepatic portal circulation is replaced by blood from the inferior vena cava, as when the portal vein and vena cava are cross-anastomosed (portacaval transposition), liver weight and regeneration have also been reported to decline. The overall results seem to indicate that maintenance and growth of the liver do in fact depend upon something in the portal blood delivered from the abdominal viscera to the liver.

THE GROWTH OF LIVER GRAFTS

The importance of the portal blood is demonstrated in transplanted livers. Early attempts to graft hepatic tissue involved the autotransplantation of liver fragments. This can be achieved by attaching the end of one lobe to some nearby organ, allowing adhesions and vascularization to occur and then separating the graft from the rest of the liver. Such transplants are unavoidably abnormal since it is impossible to reestablish a normal outflow of bile. Nevertheless, the cells usually survive, although the fragment tends gradually to lose weight. Atrophy of such grafts, however, is greater in subcutaneous sites than when attached to the intestines or spleen, the difference presumably being attributable to the presence or absence of a splanchnic blood supply. In either case, partial hepatectomy of the main liver elicits a proliferative response in the graft, retarding the normal rate of atrophy (25, 39).

With the development of microvascular surgical techniques, it has been possible in recent years to transplant entire organs to other sites in the abdominal cavity. There are many ways to connect the blood supply to such grafts, but the maintenance of auxiliary livers vis-à-vis that of the host's own organ depends primarily upon how the portal circulation is distributed. Without portal blood flow, auxiliary liver grafts tend to atrophy (35, 40). If the portal blood is diverted from the host's liver to the transplant, then the former atrophies and the latter does not. However, the atrophy of the auxiliary liver graft that occurs in the absence of portal inflow can be at least partly prevented by resection of the host's liver or by inducing its atrophy by a portacaval shunt. Partial hepatectomy of the host's liver nevertheless elicits vicarious hypertrophy in the graft, while the presence of the latter tends to reduce the regenerative response in the host remnant. However important the portal circulation may be in maintaining the normal size of the liver, hepatic tissue can survive in its absence and is not without some capacity for at least limited growth.

Attempts to pinpoint the source, if not the nature, of factors in the portal blood responsible for promoting liver growth have involved a variety of deletion experiments. Resection or isolation of various parts of the small intestine, for example, has been shown to reduce the extent of regeneration following partial hepatectomy (16). If the digestive and absorptive functions of the intestines are bypassed by intravenous feeding, hyperplasia still takes place after partial hepatectomy (29). The effects of the various gastroenteric hormones on the liver have not been sufficiently explored except to show that administration of pentagastrin has no apparent effect on RNA or protein synthesis in the liver.

Aside from the digestive tract itself, the pancreas is a possible source of factors affecting liver growth. In order to explore the possible roles of hormone-rich blood from the pancreas versus nutrient-rich blood from the intestines, it has been possible to divert splanchnic venous blood from the pancreas to the right lobes of the liver while irrigating the left lobes with blood draining from the intestines (41). Under these circumstances, the right side of the liver (receiving hormones) exhibits greater DNA synthesis than the left side (receiving nutrients). Pancreatectomy or the administration of alloxan to destroy the β cells of the islets, reduces the hormone-induced growth on the right side, suggesting that insulin may have a hepatotrophic effect (42). More direct studies have shown that insulin reduces atrophy of the liver following portacaval shunt and stimulates DNA synthesis *in vivo* and *in vitro* (24, 32, 43). Glucagon tends to have opposite effects. Others have shown, however, that liver regeneration in eviscerated rats is not restored by the administration of insulin or glucagon alone, but that combined treatment by both hormones promotes hepatocyte proliferation in such animals (10). Curiously, intact livers are not stimulated by such interventions. In all, the visceral role in regulating liver growth is a promising field of investigation, but confusion may outweigh clarification for some time to come.

DETOXIFICATION OF DRUGS

Perhaps the most rewarding approach to the problem of how liver growth is controlled has been the pharmacological one. An important attribute of liver physiology is the enzymatic processing of various compounds, ranging from the end products of digestion to narcotics. Studies have shown that a variety of poisons elicits a reaction on the part of the liver that has much in common with regeneration. Although many different compounds have been explored, the reaction of the liver to phenobarbital is typical. When confronted with this narcotic, the liver is induced to produce enzymes (e.g., demethylase) capable of metabolizing the drug (2). Indeed, it is this response that explains the development of tolerance to phenobarbital, the narcotic effects of which are progressively reduced upon daily administration. Concomitant with the induction of enzymes appropriate to the degradation of this compound, the liver undergoes an increase in weight accompanied by a rise in total protein, proliferation of smooth endoplasmic reticulum, and increased DNA synthesis and mitotic activity (4). The extent to which these responses occur is modest in comparison with the regenerative reaction to partial hepatectomy. Nevertheless, the parallels are unmistakable.

Comparable studies have been carried out on the reactions of the liver to administration of carcinogens (e.g., 3-methylcholanthrene, dibenzanthracene). Again, the response by the liver involves an increase in weight, protein, RNA, and cell size (3). In this case, DNA synthesis and cellular proliferation are less prominent than following phenobarbital, but demethylase and reductase activities are elevated.

In view of these interesting results, it is particularly important to note that combined treatments yield additive effects. Not only is the response to the simultaneous administration of phenobarbital and 3-methylcholanthrene greater than that observed following either drug alone, but these compounds also potentiate the regenerative response following partial hepatectomy. The implication of these findings is that a variety of stimuli can elicit a response in the liver which is both functional and hypertrophic (33). The possibility arises, therefore, that liver growth and regeneration may be a reaction to increased functional demands, and that the exaggerated regenerative response to partial hepatectomy may represent the sum of numerous physiological deficiencies, each one inducing a modest growth response which, when added together, results in the remarkable adaptations for which the liver is famous.

REFERENCES

1. Alfert, M., and Geschwind, I.I. The development of polysomaty in rat liver. *Exp. Cell Res.* **15**, 230–232 (1958).
2. Argyris, T.S. Additive effects of phenobarbital and high protein diet on liver growth in immature male rats. *Dev. Biol.* **25**, 293–309 (1971).

3. Argyris, T.S., and Heinemann, R. Ribosome accumulation in 3-methylcholanthrene-induced liver growth in adult male rats. *Exp. Mol. Pathol.* **22**, 335–341 (1975).

4. Augenlicht, L.H., and Argyris, T.S. Stimulation of immature male rat liver by phenobarbital and 3-methylcholanthrene. *Exp. Mol. Pathol.* **22**, 1–10 (1975).

5. Becker, F.F. Restoration of liver mass following partial hepatectomy: "Surgical hepartrophy." *Am. J. Pathol.* **43**, 497–510 (1963).

6. Becker, F.F. Structural and functional correlation in regenerating liver. *In* "Biochemistry of Cell Division" (R. Baserga, ed.), pp. 113–118. Thomas, Springfield, Illinois, 1969.

7. Bengmark, S., Hafstrom, L.O., and Loughridge, B. Studies of the influence of hepatic artery ligation on liver regeneration in partially hepatectomized rats. *Acta Hepato-Splenol.* **16**, 349–355 (1969).

8. Bucher, N.L.R., and Malt, R.A. "Regeneration of Liver and Kidney." Little, Brown, Boston, Massachusetts, 1971.

9. Bucher, N.L.R., and Swaffield, M.N. Rate of incorporation of (6-¹⁴C)orotic acid into uridine 5′-triphosphate and cytidine 5′-triphosphate and nuclear ribonucleic acid in regenerating rat liver. *Biochim. Biophys. Acta* **108**, 551–567 (1966).

10. Bucher, N.L.R., and Weir, G.C. Insulin, glucagon, liver regeneration and DNA synthesis. *Metab. Clin. Exp.* **25**, 1423–1425 (1976).

11. Church, R.B., and McCarthy, B.J. Ribonucleic acid synthesis in regenerating and embryonic liver. I. The synthesis of new species of RNA during regeneration of mouse liver after partial hepatectomy. *J. Mol. Biol.* **23**, 459–475 (1967).

12. Dettmer, C.M., Kramer, S., Driscoll, D.H., and Aponte, G.E. A comparison of the chronic effects of irradiation upon the normal, damaged, and regenerating rat liver. *Radiology* **91**, 990, 993–997 (1968).

13. Fabrikant, J.I. Cell proliferation in the regenerating liver and the effect of prior continuous irradiation. *Radiat. Res.* **32**, 804–826 (1967).

14. Fabrikant, J.I. Size of proliferating pools in regenerating liver. *Exp. Cell Res.* **55**, 277–279 (1969).

15. Fisher, B., Szuch, P., Levine, M., and Fisher, E.R. A portal blood factor as the humoral agent in liver regeneration. *Science* **171**, 575–577 (1971).

16. Fisher, B., Szuch, P., Levine, M., Saffer, E., and Fisher, E.R. The intestine as a source of a portal blood factor responsible for liver regeneration. *Surg. Gynecol. Obstet.* **137**, 210–214 (1973).

17. Gerhard, H., Schultze, B., and Maurer, W. Die Proliferation des Gallengangsepithels bei der Regeneration der CCl₄-Leber der Maus. *Virchows Arch. B* **17**, 213–227 (1975).

18. Hays, D.M., Matsushima, Y., Tedo, I., and Tsunoda, A. Liver regeneration: Influence of interval postextirpation on *in vitro* growth of cells from the remnant liver. *Proc. Soc. Exp. Biol. Med.* **138**, 658–660 (1971).

19. Iatropoulos, M. Cytoarchitecture of rat liver during compensatory growth. *Anat. Rec.* **169**, 509–514 (1971).

20. Ingle, D.J., and Baker, B.L. Histology and regenerative capacity of liver following multiple partial hepatectomies. *Proc. Soc. Exp. Biol. Med.* **95**, 813–815 (1957).

21. Kennedy, G.C., Pearce, W.M., and Parrott, D.M.V Liver growth in the lactating rat. *J. Endocrinol.* **17**, 158–160 (1958).

22. LeBouton, A.V. DNA synthesis and cell proliferation in the simple liver acinus of 10 to 20-day-old rats: Evidence for cell fusion. *Anat. Rec.* **184**, 679–688 (1976).

23. Lee, S., Broelsch, C.E., Flamant, Y.M., Chandler, J.G., Charters, A.C., and Orloff, M.J. Liver regeneration after portacaval transportation in rats. *Surgery* **77**, 144–149 (1975).

24. Leffert, H.L. Growth control of differentiated fetal rat hepatocytes in primary monolayer

culture. 7. Hormonal control of DNA synthesis and its possible significance to the problem of liver regeneration. *J. Cell Biol.* **62**, 792–801 (1974).

25. Leong, G.F., Grisham, J.W., Hole, B.V., and Albright, M.L. Effect of partial hepatectomy on DNA synthesis and mitosis in heterotopic partial autografts of rat liver. *Cancer Res.* **24**, 1496–1501 (1964).
26. Levi, J.U., and Zeppa, R. Source of the humoral factor that initiates hepatic regeneration. *Ann. Surg.* **174**, 364–370 (1971).
27. MacDonald, R.A., and Pechet, G. Liver cell regeneration due to biliary obstruction. *Arch. Pathol.* **72**, 133–141 (1961).
28. MacDonald, R.A., Rogers, A.E., and Pechet, G. Regeneration of the liver. Relation of regenerative response to size of partial hepatectomy. *Lab. Invest.* **11**, 544–548 (1962).
29. Max, M.H., Price, J.B., Jr., Takeshige, K., and Voorhees, A.B., Jr. The role of factors of portal origin in modifying hepatic regeneration. *J. Surg. Res.* **12**, 120–123 (1972).
30. Moolten, L., and Bucher, N.L.R. Regeneration of rat liver: Transfer of humoral agent by cross circulation. *Science* **158**, 272–273 (1967).
31. Munro, H.N., and Downie, E.D. Relationship of liver composition to intensity of protein metabolism in different mammals. *Nature (London)* **203**, 603–604 (1964).
32. Ozawa, K., Yamaoka, Y., Nanbu, H., and Honjo, I. Insulin as the primary factor governing changes in mitochondrial metabolism leading to liver regeneration and atrophy. *Am. J. Surg.* **127**, 669–675 (1974).
33. Popper, H. Implications of portal hepatotrophic factors in hepatology. *Gastroenterology* **66**, 1227–1233 (1974).
34. Price, J.B., Jr., Takeshige, K., Max, M.H., and Voorhees, A.B., Jr. Glucagon as the portal factor modifying hepatic regeneration. *Surgery* **72**, 74–82 (1972).
35. Price, J.B., Jr., Voorhees, A.B., Jr., and Britton, R.C. The role of portal blood in regeneration and function of completely revascularized partial hepatic autografts. *Surgery* **62**, 195–203 (1967).
36. Rabes, H.M. Kinetics of hepatocellular proliferation after partial resection of the liver. *Prog. Liver Dis.* **5**, 83–99 (1976).
37. Rabes, H.M., Iseler, G., and Tuczek, H.V. Induced cell-cycle synchrony of hepatocytes after partial hepatectomy. *Naturwissenschaften* **62**, 142 (1975).
38. Sakai, A. Humoral factor triggering DNA synthesis after partial hepatectomy in the rat. *Nature (London)* **228**, 1186–1187 (1970).
39. Sigel, B., Baldia, L.B., Menduke, H., and Feigl, P. Independence of hyperplastic and hypertrophic responses in liver regeneration. *Surg. Gynecol. Obstet.* **125**, 95–100 (1967).
40. Starzl, T.E., Marchioro, T.L., and Porter, K.A. Progress in homotransplantation of the liver. *Adv. Surg.* **2**, 295–370 (1966).
41. Starzl, T.E., Porter, K.A., Kashiwagi, N., and Putnam, C.W. Portal hepatotrophic factors, diabetes mellitus and acute liver atrophy, hypertrophy and regeneration. *Surg. Gynecol. Obstet.* **141**, 843–858 (1975).
42. Starzl, T.E., Porter, K.A., and Putnam, C.W. Insulin, glucagon, and the control of hepatic structure, function, and capacity for regeneration. *Metab. Clin. Exp.* **25**, 1429–1434 (1976).
43. Starzl, T.E., Watanabe, K., Porter, K.A., and Putnam, C.W. Effects of insulin, glucagon, and insulin/glucagon infusions on liver morphology and cell division after complete portacaval shunt in dogs. *Lancet* **1**, 821–825 (1976).
44. Spolter, P.D., and Harper, A.E. Effect of leucine-isoleucine and valine antagonism and comparison with the effect of ethionine on rat liver regeneration. *Arch. Biochem. Biophys.* **100**, 369–377 (1963).
45. Stenger, R.J., and Confer, D.B. Hepatocellular ultrasturcture during liver regeneration after subtotal hepatectomy. *Exp. Mol. Pathol.* **5**, 455–474 (1966).

46. Stenram, U., and Willén, R. Effect of actinomycin D on ultrastructure and radioauto-graphic ribonucleic acid and protein labeling in rat liver after partial hepatectomy. *Cancer Res.* **26,** 765–772 (1966).
47. Stirling, G.A., Bourne, L.D., and Marsh, T. Effect of protein deprivation and a reduced diet on the regenerating rat liver. *Br. J. Exp. Pathol.* **56,** 502–509 (1975).
48. Thomson, J.F., Straube, R.L., and Smith, D.E. The effect of hibernation on liver regenera-tion in the ground squirrel (*Citellus tridecemlineatus*). *Comp. Biochem. Physiol.* **5,** 297–305 (1962).
49. Tilak, T.B.G., and Krishnamurthi, D. Liver regeneration following partial hepatectomy. Effects of aflatoxin B$_1$. *Arch. Pathol.* **96,** 18–20 (1973).
50. Tsukada, K., and Lieberman, I. Metabolism of nucleolar ribonucleic acid after partial hepatectomy. *J. Biol. Chem.* **239,** 1564–1568 (1964).
51. Tuczek, H.V., and Rabes, H. Verlust der Proliferationsfähigkeit der Hepatozyten nach subtotaler Hepatektomie. *Experientia* **27,** 526 (1971).
52. Van Vroonhoven, T.J., Malamud, D., and Malt, R.A. Reversal of azathioprine-induced inhibition of hyperplasia in regenerating liver. *Transplantation* **14,** 603–607 (1972).
53. Weinbren, K., and Tarsh, E. The mitotic response in the rat liver after different regenera-tive stimuli. *Br. J. Exp. Pathol.* **45,** 475–480 (1964).
54. Widmann, J.-J., and Fahimi, H.D. Proliferation of mononuclear phagocytes (Kupffer cells) and endothelial cells in regenerating rat liver. A light and electron microscopic cytochemi-cal study. *Am. J. Pathol.* **80,** 349–366 (1975).

15

Functional Demand in the Digestive Tract

Bowel: L., *botulus*, sausage

The gut is a one-way street. Its mucosal landscape varies with the changing composition of its contents, and successive segments, each designed to play its own role in processing the "disassembly line" of digestion and absorption, are responsible for the conversion of food to feces. There is mounting evidence that the contents of the gut may themselves be responsible for shaping its histology.

THE GROWTH OF GASTRIC GLANDS

The functional units of the stomach are the gastric glands, the walls of which are lined with numerous chief cells which secrete pepsin, and fewer but larger acidophilic parietal cells from which HC1 is produced. The low pH of the stomach creates optimal conditions for the activity of its enzymes but puts the stomach wall in jeopardy. The mucus-secreting cells lining the necks of the gastric glands afford protection against this intraluminal acidity. Nevertheless, the regulation of gastric secretion is of the utmost importance in preserving the normal function and structure of the stomach.

The secretory activity and the histology of the gastric glands are sensitive to a variety of physiological factors which can be demonstrated experimentally. One of the long-standing approaches to gastric physiology has been the utilization of the Heidenhain pouch (Fig. 116). This is created by the surgical subdivision of the stomach, one part of which is sealed along its cut edges and attached to the abdominal wall by a gastric fistula. This arrangement is

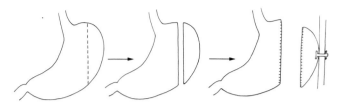

Fig. 116.   Steps in the creation of a Heidenhain pouch. The isolated portion of the stomach, deprived of its innervation, is sealed along its cut margins. A fistula is then established through the abdominal wall.

convenient for monitoring gastric secretions under different physiological conditions without food getting in the way. It is an abnormal situation, however, in that the parasympathetic (vagus) nerves to the pouch are interrupted. Cholinergic innervation is important in gastric physiology, for vagotomy tends to decrease the normal proliferation of parietal cells resulting in a reduction in their population (11, 35). HCl secretion is correspondingly reduced (55).

Humoral factors are of even greater importance, for the secretory activity of the digestive tract is coordinated by the interplay between a variety of hormones. In the case of the stomach, physiological activity is stimulated by gastrin, a hormone secreted by the antrum which constitutes the posterior end of the stomach. Accordingly, antrectomy results in a decrease in proliferative activity among the epithelial cells of the gastric mucosa leading to atrophy of the fundus (37). The administration of gastrin has the opposite effects. It increases the mucosal height, promoting mitotic activity, RNA synthesis, and amino acid incorporation into proteins (13, 48, 57). The population of parietal cells rises as does their secretion of acid. The chief cells remain unaffected. A similar response occurs in the Zollinger–Ellison syndrome, a condition usually attributable to the presence of a pancreatic adenoma, the cells of which secrete excessive amounts of gastrin (45). The hypersecretion of acid in the stomachs of such patients is accompanied by a six- to eightfold increase in the parietal cell population (44). Not surprisingly, ulcers are commonly associated with this disease.

Other experimental interventions resulting in elevated gastrin levels have similar effects. For example, if the intestine is resected or bypassed, parietal cell hyperplasia leads to gastric hypersecretion (55). This effect is more pronounced when it is the proximal, rather than the distal, segment of the small intestine which is removed (46). It is believed to be due to the absence of secretin, normally produced in the small intestine, which otherwise inhibits gastrin secretion (4, 7, 49). This may explain the similar effects of transplanting the antrum to the colon (32). Following antrocolic transposition, parietal cell hyperplasia is again accompanied by acid hypersecretion pre-

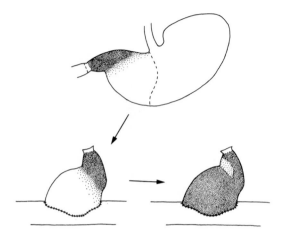

Fig. 117.   Antrocolic transposition in the rat. The distribution and density of gastrin-secreting cells is indicated by stippling. Following transplantation of the antrum from the stomach to the colon, the number and distribution of gastrin cells increase markedly owing to the unavailability of secretin which otherwise suppresses gastrin secretion (32).

sumably because in its new location the secretion of gastrin by the antrum is no longer subject to inhibition by secretin. In such grafts there is a marked increase in the numbers and distribution of the cells that produce gastrin (Fig. 117). These effects emphasize how important the anatomical relations between tissues are in maintaining normal physiological controls.

THE SIZE OF THE STOMACH

Aside from those factors that regulate secretion from gastric cells, it is necessary to control the overall size of the stomach. Like other parts of the gut, the stomach grows to keep pace with the maturation of the body. Whether it is this enlargement that determines how much an animal will eat or the other way around is not an easy question to answer. It is relatively simple, however, to study the growth of the stomach as a function of the amount of food consumed. Although the rate of cellular proliferation in the gastric mucosa declines and the overall size of the stomach shrinks in starved or hibernating animals, most studies have focused on the opposite effects of hyperphagia. There are several different ways to induce overeating, the effects depending upon the experimental methods used. Perhaps the simplest approach is to feed an animal a diet bulked out with inert ingredients (e.g., talc or kaolin). This forces an animal to consume excessive

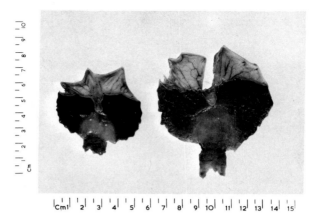

Fig. 118. Partial obstruction of the pylorus causes marked enlargement of the rat stomach (right) compared with controls (left) by 11 weeks after operation, despite equivalent age and body weights (*12*). Reproduced with permission. © 1969 The Williams & Wilkins Co., Baltimore.

quantities of such rations in order to meet its nutritional needs. Despite overeating, however, the stomachs of animals held on this regimen do not hypertrophy (*16, 28*). In marked contrast are the effects of experimentally induced hyperphagia in which increased nutrients are consumed in the absence of inert ingredients. One way to achieve this is by inducing alloxan diabetes in rats. Such animals consume 2–3 times as much food as normal, and their stomachs are enlarged accordingly (*28*). Hyperphagia associated with pregnancy and lactation also stimulates gastric growth (*14*). Still another strategy is to allow rats to eat only every few days, their food being withheld at other times. Under these conditions of intermittent starvation, animals consume several times more food than normal on their feeding days, resulting in increased gastric capacity together with hypertrophy of the mucosa and muscularis (*17*). These findings lead to the conclusion that simple stretching of the stomach is not necessarily sufficient to promote its growth, but that increased nutrition and absorption are important factors in triggering gastric hypertrophy.

An argument in favor of both kinds of stimuli, stretching and nutrition, derives from experiments in which gastric outflow is obstructed. If a ligature

Fig. 119. Wound healing in the rat stomach 2, 5, 10, 18, and 36 days (top to bottom) after electric cauterization of the mucosa. Mucous neck cells play a major role in the regeneration of missing gastric glands (*27*).

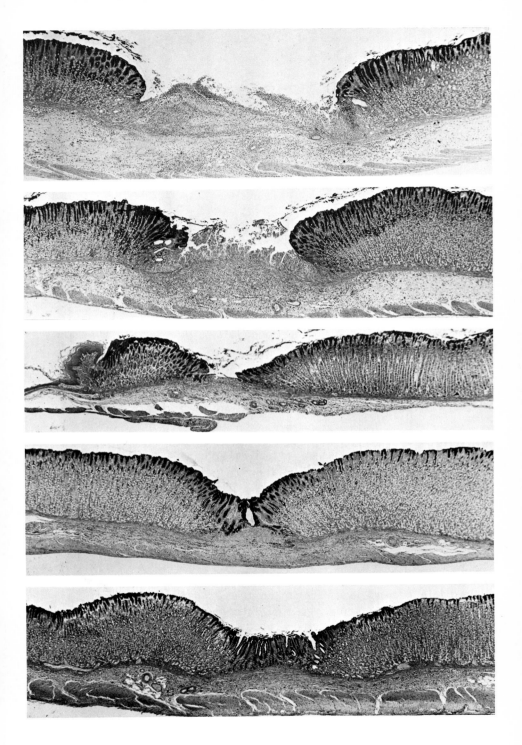

is placed around the pylorus to reduce, but not abolish, the emptying of the stomach, considerable hypertrophy ensues (Fig. 118) (12, 29). The gastric mucosal epithelial cells become hyperplastic, the height of the mucosa increases, and the total number of gastric glands is augmented. Whether this marked expansion and thickening of the stomach wall is to be attributed to the experimental dilation of the stomach or to its greater nutritional contents is a problem yet to be resolved. Of particular significance in these results, however, is the demonstration that even the adult stomach retains its capacity to make new gastric glands, something normally confined to immature stages of development.

A similar phenomenon is seen in the healing of wounds in the stomach. In the intact organ, it is in the mucous neck cells of the gastric glands that mitotic activity occurs, suggesting that they may represent the stem cells from which the chief and parietal cells lining the body of the glands are derived. The mucous neck cells could also be the principal source of repair following injury. Wounds may be inflicted in various ways. The gastric mucosa can be mechanically removed (27), it can be destroyed by exposure to alcohol or acid instilled into the lumen (24), or ulcers can be induced by the injection of histamine (58). Whatever may be the nature of the insult, the stomach normally heals its lesions with remarkable alacrity (Fig. 119). The denuded mucosa may be resurfaced with epithelium in a matter of hours, and it is from such cells that new gastric glands are derived by invagination from the surface followed by differentiation of chief and parietal cells. The demonstration of such remarkable capacities for repair in experimental lesions stands in contrast to the stubborn failure of clinical ulcers to do likewise, a situation not likely to be rectified until more is learned about the nexus between the function and the growth of the stomach.

INTESTINAL VILLI

The mucosa of the small intestine, at least in warm-blooded vertebrates, is dominated by the presence of villi, and of the crypts of Lieberkühn interspersed between them. In lower vertebrates, renewal of the mucosal epidermis is achieved by the proliferation of cells in epithelial "nests." In the metamorphosing tadpole, the shortening gut develops longitudinal folds in the troughs of which are produced cells destined to be lost at the crests after a 16-day transit time (38). In birds and mammals the site of epithelial renewal lies in the walls of the crypts. Here there is an active proliferation of cells, the descendants migrating upward onto the villous walls, eventually to be sloughed off their tips (1). It is this incessant escalation of mucosal epithelial cells which is responsible for the renewal of the villous epithelium every few days (5).

Th total number of villi in the small intestine varies with the size of the animal. Counts range from 130×10^3 in the mouse to 1×10^6 in the cat, 10×10^6 in man, and 50×10^6 in the horse (26). In immature animals the villus population is less than in the adult, indicating that their numbers are augmented throughout embryonic and postnatal life until the final population is attained at maturity.

A similar situation is encountered with respect to the neurons in the myenteric plexus. Using appropriate staining techniques it is possible to count the number of nerve cells in whole mounts of representative segments of the small intestine (21). In the newborn rat there are over $64 \times 10^3/cm^2$, or 420×10^3 in the entire intestine. In adults, the concentration of neurons decreases to $9.4 \times 10^3/cm^2$, the total number in the intestine being 1.85×10^6. These figures represent a sevenfold decrease in the concentration of

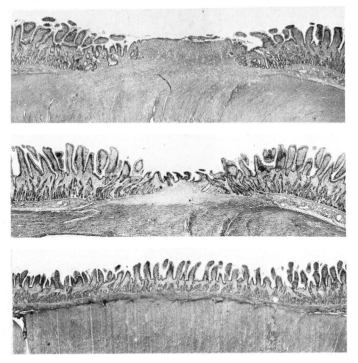

Fig. 120.   Cross sections through the small intestine of the cat 14, 42, and 100 days (top to bottom) after surgical excision of 1 cm² of mucosa and submucosa. Healing is achieved by the regeneration of the missing layers, including restoration of crypts and villi (39).

neurons, but because the total area increases 28 times, there is about a fourfold increase in the total number of nerve cells from birth to maturity. In view of the paucity of mitoses among these cells, it must be assumed that the increase in their numbers during maturation is by the differentiation of new ones, the source of which remains obscure.

Inasmuch as the villi are the most conspicuous features of the small intestinal mucosa, it is their adaptations to change which have dominated intestinal experimentation. Perhaps the most remarkable attribute of the intestinal mucosa is its capacity for regeneration following the infliction of wounds (Fig. 120). New epithelium migrates over the denuded area in a day or so, accompanied by high rates of proliferation. Crypts of Lieberkühn begin to appear several days later, followed closely by the incipient differentiation of villi. Regeneration may be completed within several months (39).

While the mammalian intestine is capable of remarkable mucosal regeneration, the intestines of amphibians are famous for even more extraordinary feats of regrowth. Here it is possible to cut the intestines into segments, stuff them back into the abdominal cavity with little regard for their orientation, and within several weeks the parts unite and reestablish continuity of their lumens (43). Proliferation in the smooth muscle and serosa of transected intestines suggests that these tissues give rise to the blastema from which regeneration occurs (25).

COMPENSATORY GROWTH OF THE INTESTINE

Although the adaptations of the intestine to changes in its workload are less spectacular than its regeneration, they are no less interesting. Indeed, it is such changes as these which may reveal what factors might be responsible for establishing and maintaining the normal histology of the gut. As in the case of the stomach, it is important to determine whether the growth of the intestines occurs in response to increasing food consumption by a maturing animal, or if the enlarging gut simply makes it possible to eat more. Attempts to resolve this enigma have prompted experiments similar to those on the stomach, namely, increasing the bulk versus the nutritional content of the diet.

When an animal's food is bulked out by the addition of indigestible ingredients (e.g., agar, talc, kaolin) the extra amounts consumed cause only a slight increase in the weight of the small intestine, a result comparable to that previously noted in the stomach. This may be due to the liquid nature of its contents, but more likely to the unaltered demands for digestion and reabsorption (16,28). The large intestine, however, undergoes considerable hypertrophy, a reaction presumably associated with the more solid nature of its contents and the increased fecal output necessitated by the bulk diet.

When hyperphagia involves an actual increase in nutriments, the small intestines react by undergoing considerable hypertrophy, a reaction that is much the same whether it is induced by lactation (9), forced feeding via stomach tube (20), withholding food three days out of four to compel intermittent gorging (17), alloxan diabetes (28), or hypothalamic lesions. Whatever method is used to increase the food intake, the small intestines are stimulated to enlarge in length as well as circumference, but the numbers of villi and crypts of Lieberkühn remain unchanged. Thus, although the quantity of material consumed is not without some effect on the small intestines, it is the nutritional content of what is eaten that elicits the lion's share of intestinal hypertrophy.

While it is possible to heighten the amount eaten without elevating nutritional intake, it is not feasible to diminish consumption without decreasing nutrition and thereby reducing the need for digestion and reabsorption. Nevertheless, decreased food consumption results in intestinal atrophy. While starvation for up to five days does not affect the number of villi in the small intestine, there is a slight decrease in the number of crypts and a marked decline in mitotic activity and rate of cell migration up the villi (10, 56). In view of the necessarily short-term nature of fasting experiments, the development of intravenous feeding techniques for rats has permitted more long-range studies of how the intestines react to relative disuse. Under these conditions the mass of both the small and large intestines decreases, an atrophy characterized especially by thinning of the mucosal layer (8, 34). IV feeding also reduces the extent of compensatory hypertrophy in intestinal remnants following resection or bypass of proximal segments (19, 33).

Adaptive changes in the intestine following its partial resection are part of the short gut syndrome (36). Which portion of the intestines may have been removed is very important, for proximal resection elicits reactions in the distal remnant different from those seen in the proximal segment following distal resection (6, 15, 42, 50, 54). If the proximal (jejunal) half of the small intestine is removed, considerable enlargement in the distal (ileal) segment ensues. In contrast, resection of the distal half of the small intestines elicits little or no hypertrophy upstream (Fig. 121). When intestinal hypertrophy occurs under these circumstances, it is achieved solely by increasing the thickness and circumference of the gut, not its length. It involves all layers of the intestine, but is especially prominent in the muscularis and mucosa (Fig. 122). Although the number of villi remains unchanged, they increase considerably in height. This is accompanied by increases in the number of epithelial cells lining their surfaces and enhanced proliferative activity in the

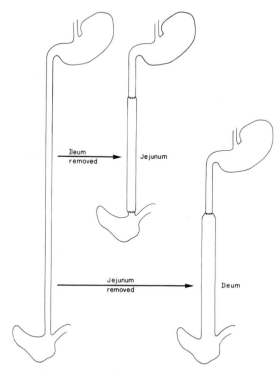

Fig. 121. The magnitude of intestinal hypertrophy differs after proximal versus distal resection of the small intestine. Removal of the ileum causes relatively little jejunal hypertrophy while jejunectomy results in considerable compensatory hypertrophy of the ileum.

crypts. Such changes can double the absorptive surface of the remaining distal segment of the small intestine.

These morphological compensations are accompanied by physiological changes. Blood flow is increased in the distal remnant, but not in the proximal one (53). Catecholamines are decreased distally (due to denervation) but not proximally (52). An animal deprived of its distal intestine increases its food intake, while no such change occurs following proximal resection (59). These responses may be correlated with the physiological activity of the proximal versus distal remnants.

It normally takes about four hours for the contents of a rat's stomach to traverse the small intestine and reach the cecum. In animals deprived of the distal segment, there is a shortening of transit time from stomach to cecum, indicating that the jejunum does not alter its motility in response to ileal resection. Under the opposite circumstances, transit time is not seriously affected in the ileal remnant after jejunectomy (Fig. 123) (6, 41). This may be

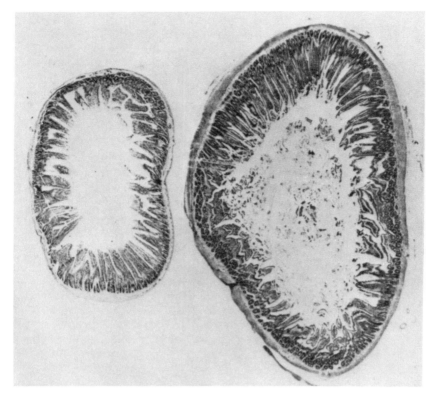

Fig. 122.   Cross sections of control rat ileum (left) and one that has hypertrophied (right) 5 weeks after partial resection of the jejunum (36).

taken to indicate that ileal motility decreases following proximal resection, as a result of which the contents of the intestines are exposed to the mucosal surface for longer periods of time. This slowdown in the transportation of material through the lumen allows more time for digestion and absorption to occur. Hence, there is increased absorption in ileal remnants after proximal resections compared with jejunal remnants after distal resections, which may account for why more food is consumed in the latter case.

   The foregoing evidence points to the fact that growth and maintenance of the intestine are responsive to the amounts eaten and the resultant demands for digestion and absorption. There is reason to believe, however, that other intraluminal influences may play important roles in regulating the structure and function of the intestines. This is based on experiments involving the transposition of the jejunum and ileum. Such operations are made possible by the fact that successive intestinal segments are each supplied by their own

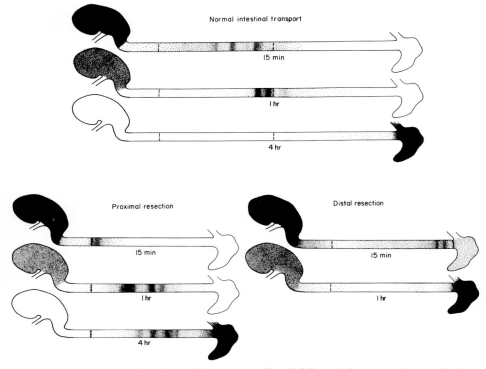

Fig. 123. Transport along the small intestine is affected differently by proximal versus distal resection (41). It normally takes about 4 hours for a marker (dense strippling) introduced into the stomach of a rat to reach the cecum, transport being faster in the jejunum than the ileum (above). If the jejunum is resected, ileal motility is diminished so that it still takes about 4 hours for transit from stomach to cecum (lower left). Distal resection, however, does not alter jejunal motility, resulting in the transport of the stomach contents to the cecum in only 1 hour (lower right).

blood vessels which fan out in the mesenteries. Consequently, it is convenient to carry the blood supply along with a segment to be transplanted, the major surgical task remainining only to anastomose the sectioned ends of the gut (31).

In a classic series of experiments, Altmann and Leblond reversed the sequence of jejunum and ileum in the small intestines (3). Normally, the ileal villi are smaller than those of the jejunum, but following ileojejunal transposition they enlarge in much the same way as following jejunal resection. These results suggest that hypertrophy of the distal remnant following proximal resection may be attributable to the position of the segment along the intestine, rather than compensation for the lose of the jejunum *per se*. Indeed, this could be interpreted to mean that the part of the small intestine to which the food is first exposed is stimulated to grow larger than that which is farther downstream.

Additional experiments have suggested an alternative possibility, namely, that following jejunal resection or ileojejunal transposition, the hypertrophy of the ileum may be stimulated by its proximity to the duodenum. When a segment of duodenum containing the bile duct and pancreatic duct is transplanted to the ileum, the ileal villi distal to the graft enlarge. When only the bile duct or pancreatic duct is included in the grafted segment, the effect is found to be promoted largely by the latter, indicating that pancreatic juice may contain a factor responsible for villus growth (2). It is conceivable, therefore, that the apparent effect of food consumption on intestinal growth and function may be mediated via pancreatic secretions—which are themselves affected by intraluminal nutrition (see Chap. 13).

Still another way to probe the factors responsible for maintaining intestinal integrity is by excluding a segment from the normal continuity of the gut. If the two ends in such an isolated segment are ligated, or anastomosed to form a circle, the contents cannot be eliminated and the isolated segment dilates and eventually becomes perforated unless bacterial action is controlled by antibiotics (47). A more nearly normal arrangement is to provide an exit for the contents. This can be achieved by making a fistula in the abdominal wall (Thiry–Vella loop) through which the contents of the segment can be voided (Fig. 124). Alternatively, the isolated segment can be drained into the gut. In a nonfunctional, self-emptying segment, the proximal end is tied off and the distal one left in continuity with the rest of the gut. The posterior ileum is usually used for this, preserving its continuity with the cecum while anastomosing the more anterior part of the ileum elsewhere to the cecum. It is also possible to create a self-emptying loop of intestine by performing a side-to-side anastomosis bypassing the intervening length of intestine. Depending upon the size and patency of the anastomosis, the loop may or may not be completely bypassed.

Disuse atrophy consistently occurs in isolated nonfunctional segments of the intestine (22, 23, 30). This may be characterized by decreases in circumference as well as length, although in young growing animals some elongation of the isolated segment may accompany normal body growth. The villi decrease in height, their configurations sometimes changing from leaflike to finger-shaped. Mitotic activity declines in the crypts. These effects, however, are reversible, for if an isolated segment of small intestine is perfused with nutrients (methionine, glucose) the lengths of the crypts and villi increase (40). Mechanical stimulation alone likewise counteracts the atrophy of isolated intestinal segments. When a series of short plastic tubes strung on a thread is introduced into a Thiry–Vella loop its atrophy is prevented (18).

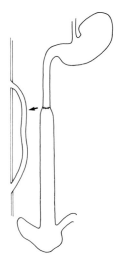

Fig. 124. Isolation of the jejunum (left) from continuity with the rest of the intestine (right) has two effects. The isolated segment (Thiry–Vella loop) atrophies, while the *in situ* ileum hypertrophies.

Atrophy is also reversed when a bypassed loop is reunited with the intestine to restore its function (30).

The fact that hypertrophy of an intestinal remnant is just as great following partial resection as it is after isolation of a comparable length of intestine proves that this kind of compensatory growth is triggered by functional demands rather than alterations in tissue mass *per se* (29, 23). There is compelling evidence that the physiological factors to which the dimensions of the intestine adapt operate within the lumen, partly in terms of nutrients consumed but perhaps also involving the secretion of digestive juices. It has been suggested that humoral factors may also influence intestinal growth (50, 51). If this were the case, however, one would expect isolated segments of ileum to hypertrophy, or at least not to atrophy following jejunal resection. Since not all experiments along these lines have yielded negative results, the possibility that intraluminal influences may not be the only factors affecting the growth of the gut cannot be categorically ruled out.

REFERENCES

1. Al-Dewachi, H.S., Wright, N.A., Appleton, D.R., and Watson, A.J. Cell population kinetics in the mouse jejunal crypt. *Virchows Arch. B* **18**, 225–242 (1975).
2. Altmann, G.G. Influence of bile and pancreatic secretions on the size of the intestinal villi in the rat. *Am. J. Anat.* **132**, 167–178 (1971).

3. Altmann, G.G., and Leblond, C.P. Factors influencing villus size in the small intestine of adult rats as revealed by transposition of intestinal segments. *Am. J. Anat.* **127**, 15–36 (1970).

4. Berstad, A., Petersen, H., Roland, M., and Liavåg, I. Effect of secretin on pentagastrin-stimulated gastric acid and pepsin secretion after vagotomy in man. *Scand. J. Gastroenterol.* **8**, 119–122 (1973).

5. Bhartiya, H.C., and Srivastava, P.N. Cell migration in the intestine of Indian desert gerbil (*Meriones hurrianae* Jerdon) and its relationship with the radiosensitivity of the animal. *Experientia* **30**, 1397 (1974).

6. Booth, C.C., Evans, K.T., Menzies, T., and Street, D.F. Intestinal hypertrophy following partial resection of the small bowel in the rat. *Br. J. Surg.* **46**, 403–410 (1959).

7. Bradley, E.L., III and Galambos, J.T. Failure of a secretin feedback loop in the Zollinger-Ellison syndrome? The "gastric factor" revisited. *Am. J. Dig. Dis.* **17**, 939–944 (1972).

8. Cameron, I.L., Pavlat, W.A., and Urban, E. Adaptive responses to total intravenous feeding. *J. Surg. Res.* **17**, 45–52 (1974).

9. Campbell, R.M., and Fell, B.F. Gastro-intestinal hypertrophy in the lactating rat and its relation to food intake. *J. Physiol. (London)* **171**, 90–97 (1964).

10. Clarke, R.M. The effect of growth and of fasting on the number of villi and crypts in the small intestine of the albino rat. *J. Anat.* **112**, 27–33 (1972).

11. Crean, G.P., Gunn, A.A., and Rumsey, R.D.E. The effects of vagotomy on the gastric mucosa of the rat. *Scand. J. Gastroenterol.* **4**, 675–680 (1969).

12. Crean, G.P., Hogg, D.F., and Rumsey, R.D.E. Hyperplasia of the gastric mucosa produced by duodenal obstruction. *Gastroenterology* **56**, 193–199 (1969).

13. Crean, G.P., Marshall, M.W., and Rumsey, R.D.E. Parietal cell hyperplasia induced by the administration of Pentagastrin (ICI 50, 123) to rats. *Gastroenterology* **57**, 147–155 (1969).

14. Crean, G.P., and Rumsey, R.D.E. Hyperplasia of the gastric mucosa during pregnancy and lactation in the rat. *J. Physiol. (London)* **215**, 181–197 (1971).

15. Dowling, R.H. Small bowel resection and bypass—recent developments and effects. *In* "Modern Trends in Gastro-enterology" (W.I. Card and B. Creamer, eds.), Vol. 4, pp. 73–104. Appleton-Century-Crofts, New York, 1970.

16. Dowling, R.H., Riecken, E.O., Laws, J.W., and Booth, C.C. The intestinal response to high bulk feeding in the rat. *Clin. Sci.* **32**, 1–9 (1967).

17. Fabry, P., and Kujolova, V. Enhanced growth of the small intestine in rats as a result of adaptation to intermittent starvation. *Acta Anat.* **43**, 264–271 (1960).

18. Fenyo, G. Role of mechanical stimulation in maintaining small intestinal mass in Thiry-Vella loops in the rat. *Eur. Surg. Res.* **8**, 419–427 (1976).

19. Fenyo, G., Hallberg, D., Soda, M., and Roos, K.A. Morphological changes in the small intestine following jejuno-ileal shunt in parenterally fed rats. *Scand. J. Gastroenterol.* **11**, 635–640 (1976).

20. Forrester, J.M. The number of villi in rat's jejunum and ileum: Effect of normal growth, partial enterectomy and tube feeding. *J. Anat.* **111**, 283–291 (1972).

21. Gabella, G. Neuron size and number in the myenteric plexus of the newborn and adult rat. *J. Anat.* **109**, 81–95 (1971).

22. Gleeson, M.H., Cullen, J., and Dowling, R.H. Intestinal structure and function after small bowel by-pass in the rat. *Clin. Sci.* **43**, 731–742 (1972).

23. Gleeson, M.H., Dowling, R.H., and Peters, T.J. Biochemical changes in intestinal mucosa after experimental small bowel by-pass in the rat. *Clin. Sci.* **43**, 743–757 (1972).

24. Grant, R. Rate of replacement of the surface epithelial cells of the gastric mucosa. *Anat. Rec.* **91**, 175–185 (1945).

25. Grubb, R.B. An autoradiographic study of the origin of intestinal blastemal cells in the newt. *Notophthalmus viridescens. Dev. Biol.* **47**, 185–195 (1975).

26. Hilton, W.A. The morphology and development of intestinal folds and villi in vertebrates. *Am. J. Anat.* **1**, 459–505 (1902).

27. Hunt, T.E. Regeneration of the gastric mucosa in the rat. *Anat. Rec.* **131**, 193–211 (1958).

28. Jervis, E.L., and Levin, R.J. Anatomic adaptation of the alimentary tract of the rat to the hyperphagia of chronic alloxan-diabetes. *Nature (London)* **210**, 391–393 (1966).

29. Kaye, M.D. The effect of partial pyloric obstruction on gastric secretion and stomach size in the rat. *Am. J. Dig. Dis.* **16**, 217–226 (1971).

30. Keren, D.F., Elliott, H.L., Brown, G.D., and Yardley, J.H. Atrophy of villi with hypertrophy and hyperplasia of Paneth cells in isolated (Thiry-Vella) ileal loops in rabbits. Light-microscopic studies. *Gastroenterology* **68**, 83–93 (1975).

31. Lambert, R. "Surgery of the Digestive System in the Rat." Thomas, Springfield, Illinois, 1965.

32. Lehy, T., Voillemot, N., Dubrasquet, M., and Dufougeray, F. Gastrin cell hyperplasia in rats with chronic antral stimulation. *Gastroenterology* **68**, 71–82 (1975).

33. Levine, G.M., Deren, J.J., and Yezdimir, E. Small-bowel resection. Oral intake is the stimulus for hyperplasia. *Am. J. Dig. Dis.* **21**, 542–546 (1976).

34. Levine, G.M., Steiger, E., and Deren, J.J. The importance of oral intake in maintenance of rat small intestinal mass and disaccharidase activity. *Clin. Res.* **21**, 828 (1973).

35. Ley, R., Willems, C., and Vansteenkiste, Y. Influence of vagotomy on parietal cell kinetics in the rat gastric mucosa. *Gastroenterology* **65**, 764–772 (1973).

36. Loran, M.R., Althausen, T.L., and Irvine, E. Effects of "minimal" resection of the small intestine on absorption of vitamin A in the rat. *Gastroenterology* **31**, 717–726 (1956).

37. Martin, F., MacLeod, I.B., and Sircus, W. Effect of antrectomy on the fundic mucosa of the rat. *Gastroenterology* **59**, 437–444 (1970).

38. McAvoy, J.W., and Dixon, K.E. Cell proliferation and renewal in the small intestinal epithelium of adult *Xenopus laevis. J. Exp. Zool.* **202**, 129–138 (1977).

39. McMinn, R.M.H., and Mitchell, J.E. The formation of villi following artificial lesions of the mucosa in the small intestine of the cat. *J. Anat.* **88**, 99–107 (1954).

40. Menge, H., Müller, K., Lorenz-Meyer, H., and Riecken, E.O. Untersuchungen zum Einfluss von Methionin- und Methionin-Glucose-Lösungen auf Morphologie und Funktion der Schleimbaut selbstentleerender jeunaler Blindschlingen der Ratte. *Virchows Arch. B* **18**, 135–143 (1975).

41. Nygaard, K. Resection of the small intestine in rats. IV. Adaptation of gastro-intestinal motility. *Acta Chir. Scand.* **133**, 407–416 (1967).

42. Nylander, G., and Olerud, S. Intestinal adaptation following extensive resection in the rat. *Acta Chir. Scand.* **123**, 51–56 (1962).

43. O'Steen, W.K. Regeneration and repair of the intestine in *Rana clamitans* larvae. *J. Exp. Zool.* **141**, 449–476 (1959).

44. Polacek, M.A., and Ellison, E.H. A comparative study of parietal cell mass and distribution in normal stomachs, in stomachs with duodenal ulcer, and in stomachs of patients with pancreatic adenoma. *Surg. Forum* **14**, 313–315 (1963).

45. Royston, C.M.S., Brew, D.S.J., Garnham, J.R., Stagg, B.H., and Polak, J. The Zollinger-Ellison syndrome due to an infiltrating tumour of the stomach. *Gut* **13**, 638–642 (1972).

46. Santillana, M., Wise, L., Schuck, M., and Ballinger, W.F. Changes in gastric acid secretion following resection or exclusion of different segments of the small intestine. *Surgery* **65**, 777–782 (1969).

47. Stafford, E.S., Schnaufer, L., and Cone, D.F. The isolated circular intestinal loop. *Bull. Johns Hopkins Hosp.* **104**, 260–261 (1959).
48. Stanley, M.D., Coalson, R.E., Grossman, M.I., and Johnson, L.R. Influence of secretin and pentagastrin on acid secretion and parietal cell number in rats. *Gastroenterology* **63**, 264–269 (1972).
49. Straus, E., Gerson, C.D., and Yalow, R.S. Hypersecretion of gastrin associated with the short bowel syndrome. *Gastroenterology* **66**, 175–180 (1974).
50. Tilson, M.D. Compensatory hypertrophy of the gut. Testing of the tissue mass, intraluminal nutrition and functional demand hypothesis. *Arch. Surg.* **104**, 69–72 (1972).
51. Tilson, M.D., and Wright, H.K. Adaptational changes in the ileum following jejunectomy. *In* "Regulation of Organ and Tissue Growth" (R.J. Goss, ed.), pp. 257–270. Academic Press, New York, 1972.
52. Touloukian, R.J., Aghajanian, G.K., and Roth, R.H. Adrenergic denervation of the hypertrophied gut remnant. *Ann. Surg.* **176**, 633–637 (1972).
53. Touloukian, R.J., and Spencer, R.P. Ileal blood flow preceding compensatory intestinal hypertrophy. *Ann. Surg.* **175**, 320–325 (1972).
54. Weser, E., and Hernandez, M.H. Studies of small bowel adaptation after intestinal resection in the rat. *Gastroenterology* **60**, 69–75 (1971).
55. Wickbom, G., Landor, J.H., Bushkin, F.L., and McGuigan, J.E. Changes in canine gastric acid output and serum gastrin levels following massive small intestinal resection. *Gastroenterology* **69**, 448–452 (1975).
56. Wiebecke, B., Heybowitz, R., Lohrs, U., and Eder, M. The effect of starvation on the proliferation kinetics of the mucosa of the small and large bowel of the mouse. *Virchows Arch. B.* **4**, 164–175 (1969).
57. Willems, G., and Lehy, T. Radioautographic and quantitative studies on parietal and peptic cell kinetics in the mouse. A selective effect of gastrin on parietal cell proliferation. *Gastroenterology* **69**, 416–426 (1975).
58. Williams, A.W. Acute histamine erosions in the stomach and duodenum of guinea-pigs. *J. Pathol. Bacteriol.* **63**, 465–469 (1951).
59. Young, E.A., and Weser, E. Nutritional adaptation after small bowel resection in rats. *J. Nutr.* **104**, 994–1001 (1974).

16

Compensatory Pulmonary Hypertrophy

Trachea: Gr., *trachus*, rough

Nowhere in the body is the separation between the inside and the outside so tenuous as it is in the respiratory organs. Designed to provide the maximum surface area in the smallest possible volume, the average human lung if flattened out would cover about one-quarter of a tennis court. The 300 million alveoli in the adult lung represent a tenfold increase in their numbers since birth (Fig. 125), and although this may be several times as much respiratory surface as is needed for survival, the margin of safety can be considerably reduced in old age (*21, 29, 49*). Indeed, the complement of alveoli may decline to one-fifth of normal adult levels in cases of emphysema (*27*). The problem here is not that the lung is incapable of compensating for pulmonary deficiencies, but that it does so by enlarging its alveoli instead of multiplying them.

THE NUMBER OF ALVEOLI

Alveoli are the functional units of the lung (Fig. 126). Lined with a thin veneer of cytoplasm belonging to simple squamous epithelial cells, the dimensions of the alveoli are as elusive as is the size of a balloon. Most estimates of their diameters approximate 100 μm or more in adult mammals. Although slower-moving animals tend to have larger alveoli than do more active ones, the alveolar dimensions from bat to whale are remarkably uniform (*48*).

284

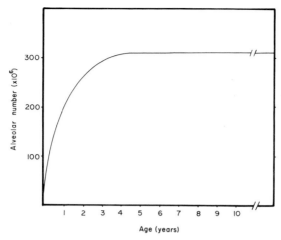

Fig. 125.   Increase in the alveolar population of human lungs as a function of age. Although estimates vary considerably, the rate at which alveoli are formed decelerates with age, approaching adult levels during the first decade (*21, 29, 49*).

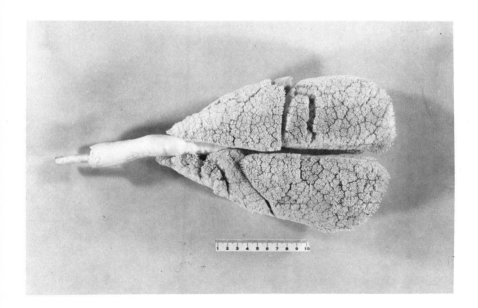

Fig. 126.   Cast of a dog lung made by injecting silicone rubber intratracheally, then digesting away the lung tissue (*39*).

Mammalian alveoli trace their origins to the surface compartmentalizations found in the simple lungs of amphibians. These lungs are little more than air-filled sacs floating free in the abdominal cavity and inflated by positive pressure. Lungs are superfluous in many aquatic amphibians which can sustain their respiratory needs by cutaneous respiration. In terrestrial forms, such as toads, the heightened importance of the lungs is reflected in the more highly developed air sacs honeycombed with vascularized partitions. In reptiles the lungs are no longer unattached to the surrounding tissues and air for the first time is pulled into them by expansion of the rib cage. Reptiles have evolved a more elaborate respiratory surface in their lungs than is found in amphibians, although the lungs are still basically open sacs. It was by thickening the respiratory surface of such primitive lungs that those of warm-blooded vertebrates evolved into organs that are about 90% alveoli and 10% airways, instead of the other way around.

Ontogeny of the lung is characterized by repeated dichotomous branchings of the original outgrowths from the embryonic pharynx. In man, there are as many as 28 generations of pulmonary bifurcations leading to the respiratory bronchioles and alveoli at their ends (50). The number of alveoli continues to increase after birth, albeit at a decelerating rate. In man, alveoli may be added throughout the first decade of life until the adult complement is attained, after which further growth is achieved only by their enlargement (21). Alveoli as such are absent altogether in neonatal rats, which are born in a very premature condition (14, 49). Gas exchange occurs in the still undifferentiated airways until alveoli begin to develop by the fourth postnatal day.

The alveolar interface is lined with a simple epithelium made up of squamous type 1 cells and the more cuboidal type 2 cells (Fig. 127). In the newborn rat lung, type 2 cells predominate for the first two weeks after which the type 1 cell population increases (32). Autoradiographic evidence shows that only type 2 cells synthesize DNA, suggesting that these may represent the stem cell pool from which type 1 cells differentiate. Such cells contain osmiophilic lamellar bodies in their cytoplasm believed to be associated with the production of surfactant. Without surfactant, the alveoli cannot overcome the surface tension forces which would normally oppose their expansion.

The alveolar epithelial cells could not be in a more vulnerable situation, exposed as they are to whatever noxious gases may be inhaled. Not the least of these is oxygen itself, which in the pure state destroys the type 1 cells of the alveolar lining, the loss of which is followed by increased proliferation of type 2 cells to replace them (26, 30). Ozone (O_3) is also toxic, causing the loss of ciliated epithelial cells from the terminal bronchioles and destruction of type 1 cells in the alveoli (46). Again, the latter are replaced by proliferation of type 2 cells (23). Breathing nitrous oxide (NO_2) results in similar reactions

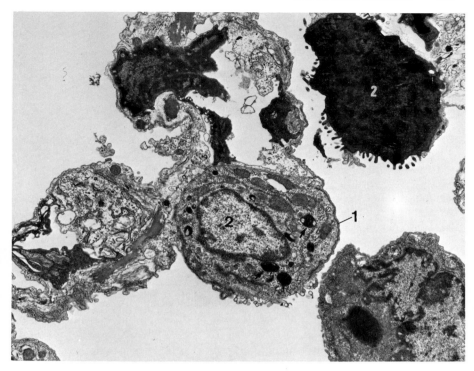

Fig. 127.   Electron photomicrograph of type 1 and 2 cells in the rat lung. The type 1 cell makes up the squamous lining of the alveolus, while the type 2 cells are the more spherical ones with microvilli on their surfaces and containing osmiophilic lamellar bodies (arrows). (Courtesy of Dr. William H.J. Douglas, W. Alton Jones Cell Science Center, Lake Placid, N.Y.)

(22). Not all deleterious substances need be inhaled, for if a trypsin solution is instilled into mouse lungs it triggers a proliferative response in type 2 (not type 1) cells (2). Finally, if urethane is administered in the drinking water or given by intraperitoneal injection it stimulates alveolar epithelial hyperplasia which eventually leads to the production of numerous type 2 cell adenomas in the lungs (31). It would appear that the sensitivity of alveolar epithelial cells to mitogenic stimulation may account for their carcinogenic susceptibilities.

At some point during the development of the mammalian lung, an inequity is established between the left and right lungs which is reflected not only in the smaller size of the left one but in the different number of lobes that are formed on the two sides. The smaller size of the left lung is of course correlated with the displacement of the heart to that side, although there is no way to tell whether this relationship is causal or coincidental. Curiously, an even greater inequity between the two lungs is encountered in snakes and certain amphibians with elongate bodies. Here the right lung is considerably larger than the left one, and in the more highly evolved snakes the left lung is represented by nothing more than a rudimentary diverticulum off the anterior end of the functional lung (Fig. 128). In the common garter snake, the rudimentary left lung is less than 2 mm in diameter, while the right one extends for virtually half the length of the entire body. Most of this length is represented by the posterior air sac, a smooth-walled extension of the lung which lacks respiratory surfaces (37). Only about 20% of the entire length of the lung contains alveoli in its walls, although in some species of snakes the trachea may be expanded into respiratory surfaces too. As in other cases of asymmetry in animals, it remains to be determined what factors are responsible for the overdevelopment of one organ presumably at the expense of its opposite partner. Experiments on snakes have shown that when most of the functional right lung has been tied off the left rudimentary one is capable of only very limited compensatory growth, hardly enough to make up for the functional deficiency. Such is not the case in higher forms.

Compensatory growth of the remaining mammalian lung after unilateral pneumonectomy may restore most or all of the entire mass of the original organs, filling the chest cavity and pushing the heart and mediastinum to the right (9, 13, 16, 38, 47). The mechanism by which this is accomplished depends on the age of the animal, particularly with reference to the adaptive growth of the functional units. In adult mammals, the number of alveoli in the remaining lung is not augmented, their physiological compensation being achieved by enlargement of preexisting ones (7, 10, 35). A similar response is sometimes observed in young growing animals still in the process of developing their adult complements of alveoli (43). On the other hand,

Fig. 128.   Rudimentary left lung (L) of a garter snake as seen in ventral dissection next to the functional right lung (R). The heart is visible anteriorly to the right.

young mammals have not infrequently been found capable of multiplying their numbers of alveoli per lung to counts in excess of the normal adult complement following resection of the opposite lung, a phenomenon that has been observed in rats (28, 38), rabbits (34), cats (7), dogs (35), and humans (28).

RESPONSES TO PHYSIOLOGICAL ALTERATIONS

Whatever may be the mechanism by which compensatory pulmonary hypertrophy occurs, the physiological control of this prodigious enlargement is not easily explained. It is not unreasonable to attribute compensatory pulmonary hypertrophy to increased respiratory demands, but respiration is itself a complicated process. It involves such responses as changes in tidal volume, alterations in rates of ventilation, and changes affecting the rate and pressure of pulmonary blood flow. These parameters are responsive to altered demands for oxygen by the body, which are in turn affected by hemoglobin levels and cardiac output.

The basic approach to the problem of how the growth of the lung is controlled is to remove one lung and study the compensatory growth of the other (Fig. 129). Unilateral pneumonectomy, however, does many things to the respiratory system, any one of which might trigger compensation in the remaining lung. First and foremost, it creates a vacancy in the chest cavity which, by providing room for expansion, might bring about growth of the other lung (16). However oversimplified such a mechanistic interpretation might be, it remains a difficult explanation to disprove. Although filling the vacated pleural cavity with inert material may reduce the extent of compensatory hypertrophy, it is not prevented altogether (8, 17, 44, 47). Unfortunately, it is not possible to increase the space available for lung expansion without resecting part of the pulmonary mass. The opposite situation, however, is found in cases of scoliosis in which the space available for lung growth is limited. Under these circumstances, lungs remain underdeveloped (19). A similar situation obtains in clinical cases of visceral herniation through the diaphragms of newborn infants (33). The lung on the affected side remains small and immature, suggesting that its normal development must in part be contingent upon the availability of space for growth.

Still another approach to the spatial aspects of the problem is to collapse one lung (atelectasis) by filling the space it would normally occupy (Fig. 129). Although such procedures are known to elicit contralateral hypertrophy, the extent of compensatory growth is considerably less than it is following lung resection (8, 17, 44, 47). Is this reduction in compensatory growth due to the presence, albeit collapsed, of the other lung, or is hypertrophy impeded by the lack of room to grow? This dilemma can be resolved by collapsing the

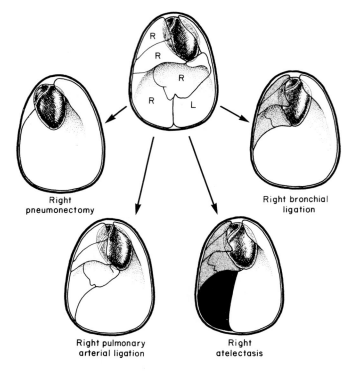

Fig. 129.   Relative degrees of hypertrophy of the left rat lung (L) in compensation for various experimental interventions on the right lung (R) within the nearly inflexible confines of the thoracic cavity (heavy lines) (47). Right pneumonectomy or ligation of the right bronchus results in maximal hypertrophy of the left lung with displacement of the heart. Ligation of the right pulmonary artery promotes moderate left hypertrophy. Collapse of the right lung (heavily stippled) by intrathoracic injection of silastic (black) permits only slight enlargement of the left lung.

lung without obliterating its space. If one lung is exteriorized through the chest wall, but allowed to persist in a collapsed state between the ribs and the overlying skin, there should be no impediment to the growth of the other lung if it were stimulated to compensate for the still present but nonfunctional opposite lung. Experiments have proved that compensatory growth still occurs under these conditions.

Two basic processes are arrested when a lung is removed. One is inhalation, the other pulmonary circulation. To assess the impact of these changes on the growth responses of the opposite lung it is necessary to eliminate each one independently by surgical interventions. As might be expected, ligation of the bronchus on one side brings about enlargement of the other lung to approximately the same extent as after unilateral pneumonectomy (Fig. 129) (47). These findings are significant in that they prove that compensatory pulmonary hypertrophy can occur even when no pulmonary tissue has been removed, the incapacitated lung remaining in the chest cavity. It is also significant that following bronchial occlusion the blood flow from the affected lung is diverted to the opposite side (1). One wonders if this reaction, which accompanies unilateral pneumonectomy too, might play some role in promoting the growth of the hyperemic lung.

Hyperemia is also induced by ligating one pulmonary artery, an operation designed to stimulate pneumonectomy without otherwise disrupting the activities of the operated lung. Ventilation continues, and the lung survives by virtue of its collateral bronchial circulation. Theoretically, there should be little respiratory function in terms of oxygen exchange because the pulmonary circulation has been eliminated. In practice, however, the possibility cannot be ruled out that the bronchial circulation might participate in respiratory gas exchange, particularly if vascular anastomoses are established with the pulmonary vessels (see Chap. 6). Enlargement of the bronchial artery following pulmonary arterial ligation testifies to this probability. Nevertheless, ligation of one pulmonary artery results in increased blood flow and elevated pulmonary arterial pressure in the opposite lung. There is also a rise in oxygen uptake and as much as 50% increase in weight (12, 47). The lack of complete compensatory hypertrophy may be attributable to the development of vascular anastomoses with collateral blood vessels in the operated lung.

Ultimately, the problem of how growth of the lung is controlled comes down to a balance between the quantity of oxygen available in the atmosphere and the amount needed by the body. Decrease the former or increase the latter, and there should be a growth response if the condition of the lung is at all responsive to functional demands. This is exactly what happens in waltzing mice (25). Owing to lesions in their brains, these animals spend virtually all of their waking hours turning in circles. Such excessive feats of exercise increase the oxygen consumption almost twofold. This heightened respiratory activity is accompanied by an augmentation of the number and size of the alveoli, increasing the pulmonary surface area by more than 50%.

Comparable responses have been even more thoroughly documented in studies of animals exposed to high altitude. This has profound effects on the circulation to the lung, increasing blood flow, elevating the pulmonary arterial pressure, and ultimately bringing about an enlargement of the arteries themselves (36, 51). The immediate physiological response of animals held under hypoxic conditions is an increase in their ventilation. In due course, the lung increases in size, a response that in younger animals may be accompanied by the production of extra alveoli (3, 15, 18). When unilateral pneumonectomy is combined with hypoxia, the extent of compensatory pulmonary hypertrophy is enhanced (9).

Theoretically, one might expect the opposite reaction in animals exposed to excess oxygen. However, there are limits to the degree of hyperoxia that can be tolerated by an animal. For example, exposure to nearly pure oxygen destroys the alveolar epithelium in a few days and may eventually be fatal (4, 6, 30). These and other deleterious consequences of oxygen poisoning are preceded by sharp declines in the rates of DNA and protein synthesis in the lungs (6, 24). Exposure to more tolerable levels of hyperoxia have been shown to counteract compensatory pulmonary hypertrophy after unilateral pneumonectomy (9).

Physiological experiments in which one lung is rendered hypoxic while the other is allowed to breathe normal air have shed light on the factors that might be responsible for regulating pulmonary growth. Although such experimental conditions cannot be maintained long enough to determine if the untreated lung will compensate for contralateral hypoxia, they have been shown to bring about hyperventilation in treated animals (20). Moreover, the pattern of blood flow resembles that already found to be induced by other surgical interventions, namely, reduced flow and pulmonary blood pressure in the hypoxic lung and increased blood flow in the opposite one (5). Indeed, contralateral hyperemia would appear to be the most consistent physiological response to any treatment that interferes with the functional efficiency of the other lung. Inasmuch as compensatory hypertrophy is logically regarded as a long-range adaptation to short-range physiological adjustments, the diversion of blood from the affected lung to the other one ranks as a leading candidate for the role of growth stimulator.

In view of the diversity of physiological influences that have been shown to affect pulmonary growth, it is logical to seek a common denominator through which these various factors might operate. Some experimental interventions (e.g., unilateral pneumonectomy, atelectasis, bronchial occlusion, pulmonary arterial ligation) effectively incapacitate one lung and overburden the other. Exercise and hypoxia exert comparable physiological demands on the respiratory system by creating an imbalance between the body's need for oxygen and the lungs' capacity to deliver it. Metabolic activity would seem to be the key to the control of respiratory function and pulmonary growth. Not surprisingly, it has been discovered that growth hormone stimulates disproportionate growth of the lungs. In acromegaly, the lungs are enlarged (11). On the other hand, hypophysectomy reduces the extent of compensatory growth after unilateral pneumonectomy in rats, while the transplantation of pituitary tumors that secrete excess growth hormone promotes lung growth (10).

It is not known whether growth hormone acts directly on the tissues of the lung, or indirectly by enhancing the metabolic rate of the animal. If the latter possibility is true, it follows that other metabolic stimulants should also promote lung growth—as is indeed the case with exercise and hypoxia. Recent studies have also shown that if hamsters are exposed to lowered temperatures, their lungs (and hearts) enlarge (40). The administration of triiodothyronine likewise stimulates pulmonary growth (40). These findings suggest that the responsiveness of lung growth to so many seemingly different experimental conditions is to be explained by the heightened metabolic activity to which they all contribute.

Notwithstanding the abundance of evidence testifying to the adaptation of lung size to respiratory demands, there remains the possibility that lung growth may be self-regulating. It is conceivable that the amount of pulmonary tissue might be maintained by a negative feedback mediated via inhibitory factors produced by the lung itself (45). If so, any reduction in pulmonary mass might be expected to bring about a decline in the concentration of such inhibitors resulting in a concomitant growth response until the original mass of lungs has been restored. Although such a hypothesis is inconsistent with the fact that compensatory growth occurs after incapacitation of one lung without actually reducing its mass, it is nevertheless possible that self-inhibitors might operate in addition to, not instead of, functional demands. The problem is to separate structure from function. The prenatal lung is an excellent system in which to test this hypothesis.

GROWTH OF THE NONFUNCTIONAL LUNG

Whatever other functions it may perform, the fetal mammalian lung does not engage in respiratory activities. Nevertheless, the orderly pattern of its morphogenesis and increase in size suggests the operation of important growth controls, controls that are probably exempt from the various respiratory influences known to affect the growth of the postnatal lung. Two crucial experiments suggest themselves. One is to learn if compensatory hypertrophy occurs after unilateral pneumonectomy in the fetus. The other is to see if fetal lungs compensate for maternal deficiencies. The latter approach is the easier.

When one lung is removed from a pregnant rat, there ought to be no respiratory compensation on the part of her fetal lungs as would be expected in the remaining maternal lung. Hence, if the fetal lungs enlarge beyond their normal proportions, it would be compelling evidence in favor of the existence of growth regulating factors independent of respiratory physiology.

Experiments along these lines have yielded positive results, although they remain to be confirmed (*42*). Not only does the remaining maternal lung enlarge, but those of her fetuses have been reported to undergo a vicarious compensation presumably owing to the transfer of some message across the placenta. The nature of this communication, whether it is simply a transient oxygen starvation in the pneumonectomized mother or a decrease in circulating growth inhibitors, is for the future to determine.

A more direct approach to the problem involves unilateral pneumonectomy of the fetus. There are occasional cases of human infants born with one lung collapsed due to diaphragmatic herniation. This causes the lung on the affected side to be hypoplastic, but is not known to result in compensatory enlargement of the opposite lung (*33*). The relative unresponsiveness of the remaining fetal lung has been confirmed in experiments on rats from which the left lung was removed on the twentieth day of gestation. The relative weights of the right lungs at term two days later were unaffected, although mitotic activity was higher than in littermate controls (*41*).

Studies on fetal lungs, difficult as they may be, are of the utmost importance in contributing to an understanding of the factors responsible for controlling pulmonary growth. While such compensatory growth as may occur before birth cannot be attributed to respiratory influences, it is conceivable that other physiological activities (e.g., regulation of amniotic fluid) might not be without their effects on the control of prenatal pulmonary development. It will be interesting to follow research along these lines as technical advances in intrauterine experimentation make it possible to attack increasingly crucial problems of how the growth of the lung is governed.

REFERENCES

1. Arborelius, M., Jr. Influence of unilateral hypoventilation on distribution of pulmonary blood flow in man. *J. Appl. Physiol.* **26**, 101–104 (1969).
2. Aronson, J.F., Johns, L.W., and Pietra, G.G. Initiation of lung cell proliferation by trypsin. *Lab. Invest.* **34**, 529–536 (1976).
3. Bartlett, D., Jr., and Remmers, J.E. Effects of high altitude exposure on the lungs of young rats. *Respir. Physiol.* **13**, 116–125 (1971).
4. Bonikos, D.S., Bensch, K.G., Ludwin, S.K., and Northway, W.H., Jr. Oxygen toxicity in the newborn. The effect of prolonged 100 per cent O_2 exposure on the lungs of newborn mice. *Lab. Invest.* **32**, 619–635 (1975).
5. Borst, H.G., Whittenberger, J.L., Berglund, E., and McGregor, M. Effects of unilateral hypoxia and hypercapnia on pulmonary blood flow distribution in the dog. *Am. J. Physiol.* **191**, 446–452 (1957).
6. Bowden, D.H., and Adamson, I.Y.R. Reparative changes following pulmonary cell injury: Ultrastructural, cytodynamic and surfactant studies in mice after oxygen exposure. *Arch. Pathol.* **92**, 279–283 (1971).

7. Bremer, J.L. The fate of the remaining lung tissue after lobectomy or pneumonectomy. *J. Thorac. Surg.* **6**, 336–343 (1937).
8. Brody, J.S. Factors responsible for post-pneumonectomy lung growth. *Am. Rev. Respir. Dis.* **109**, 732 (1974).
9. Brody, J.S. Time course of and stimuli to compensatory growth of lung after pneumonectomy. *J. Clin. Invest.* **56**, 897–904 (1975).
10. Brody, J.S., and Buhain, W.J. Hormonal influence on post-pneumonectomy lung growth in the rat. *Respir. Physiol.* **19**, 344–355 (1973).
11. Brody, J.S., Fisher, A.B., Gocmen, A., and DuBois, A.B. Acromegalic pneumonomegaly: Lung growth in the adult. *J. Clin. Invest.* **49**, 1051–1060 (1970).
12. Brofman, B.L., Charms, B.L., Kohn, P.M., Elder, J., Newman, R., and Rizika, M. Unilateral pulmonary artery occlusion in man. Control studies. *J. Thorac. Surg.* **34**, 206–227 (1957).
13. Buhain, W.J., and Brody, J.S. Compensatory growth of the lung following pneumonectomy. *J. Appl. Physiol.* **35**, 898–902 (1973).
14. Burri, P.H. The postnatal growth of the rat lung. III. Morphology. *Anat. Rec.* **180**, 77–98 (1974).
15. Burri, P.H., and Weibel, E.R. Morphometric estimation of pulmonary diffusion capacity. II. Effect of P_{O_2} on the growing lung. Adaption of the growing rat lung to hypoxia and hyperoxia. *Respir. Physiol.* **11**, 247–264 (1971).
16. Cohn, R. Factors affecting the postnatal growth of the lung. *Anat. Rec.* **75**, 195–206 (1939).
17. Cowan, M.J., and Crystal, R.G. Lung growth after unilateral pneumonectomy: Quantitation of collagen synthesis and content. *Am. Rev. Respir. Dis.* **111**, 267–276 (1975).
18. Cunningham, E.L., Brody, J.S., and Jain, B.P. Lung growth induced by hypoxia. *J. Appl. Physiol.* **37**, 362–366 (1974).
19. Davies, G.M., and Reid, L. Effect of scoliosis on growth of alveoli and pulmonary arteries and on right ventricle. *Arch. Dis. Child.* **46**, 623–632 (1971).
20. Dirken, M.N.J., and Heemstra, H. The adaptation of the lung circulation to the ventilation. *Q. J. Exp. Physiol.* **34**, 213–226 (1948).
21. Dunnill, M.S. Postnatal growth of the lung. *Thorax* **17**, 329–333 (1962).
22. Evans, M.J., Cabral, L.J., Stephens, R.J., and Freeman, G. Transformation of alveolar type 2 cells to type 1 cells following exposure to NO_2. *Exp. Mol. Pathol.* **22**, 142–150 (1975).
23. Evans, M.J., Johnson, L.V., Stephens, R.J., and Freeman, G. Cell renewal in the lungs of rats exposed to low levels of ozone. *Exp. Mol. Pathol.* **24**, 70–83 (1976).
24. Gacad, G., and Massaro, D. Hyperoxia: Influence on lung mechanics and protein synthesis. *J. Clin. Invest.* **52**, 559–565 (1973).
25. Geelhaar, A., and Weibel, E.R. Morphometric estimation of pulmonary diffusion capacity. III. The effect of increased oxygen consumption in Japanese waltzing mice. *Respir. Physiol.* **11**, 354–366 (1971).
26. Hackney, J.D., Evans, M.J., and Christie, B.R. Effects of 60 and 80% oxygen on cell division in lung alveoli of squirrel monkeys. *Aviat. Space Environ. Med.* **46**, 791–794 (1975).
27. Hieronymi, G. Über den durch das Alter bedingten Formwandel menschlicher Lungen. *Ergeb. Allg. Pathol. Pathol. Anat.* **41**, 1–62 (1961).
28. Hilber, H. Embryonale Wachstumspotenzen der jugendlichen Lunge in Dienste der funktionellen Anpassung. *Klin. Wochschr.* **25**, 244–246 (1947).
29. Hislop, A., and Reid, L. Lung development in relation to gas exchange capacity. *Bull. Physio-Pathol. Respir.* **9**, 1317–1343 (1973).
30. Kapanci, Y., Weibel, E.R., Kaplan, H.P., and Robinson, F.R. Pathogenesis and reversibil-

ity of the pulmonary lesions of oxygen toxicity in monkeys. II. Ultrastructural and morphometric studies. *Lab. Invest.* **20**, 101–118 (1969).

31. Kauffman, S.L. Kinetics of alveolar epithelial hyperplasia in lungs of mice exposed to urethane. I. Quantitative analysis of cell populations. *Lab. Invest.* **30**, 170–175 (1974).
32. Kauffman, S.L., Burri, P., and Weibel, E.R. The postnatal growth of the rat lung. II. Autoradiography. *Anat. Rec.* **180**, 63–76 (1974).
33. Kitagawa, M., Hislop, A., Boyden, E.A., and Reid, L. Lung hypoplasia in congenital diaphragmatic hernia. A quantitative study of airway, artery and alveolar development. *Br. J. Surg.* **58**, 342–346 (1971).
34. Langston, C., Sachdeva, P., Cowan, M.J., Haines, J., Crystal, R.G., and Thurlbeck, W.M. Alveolar multiplication in the contralateral lung after unilateral pneumonectomy in the rabbit. *Am. Rev. Respir. Dis.* **115**, 7–13 (1977).
35. Longacre, J.J., and Johansmann, R. An experimental study of the fate of the remaining lung following total pneumonectomy. *J. Thorac. Surg.* **10**, 131–149 (1940).
36. Malik, A.B., and Kidd, B.S. Adrenergic blockade and the pulmonary vascular response to hypoxia. *Respir. Physiol.* **19**, 96–106 (1973).
37. McDonald, H.S. Respiratory functions of the ophidian air sac. *Herpetologica* **15**, 193–198 (1959).
38. Nattie, E.E., Wiley, C.W., and Bartlett, D., Jr. Adaptive growth of the lung following pneumonectomy in rats. *J. Appl. Physiol.* **37**, 491–495 (1974).
39. Phalen, R.F., Yeh, H-C., Raabe, O.G., and Velasquez, D.J. Casting the lungs *in-situ*. *Anat. Rec.* **177**, 255–263 (1973).
40. Richards, M.E.D. Personal communication (1977).
41. Romanova, L.K., Leikina, E.M., Antipova, K.K., and Sokolova, T.N. The role of function in the restoration of damaged viscera. *Sov. J. Dev. Biol.* **1–2**, 384–390 (1970–1971). (Engl. transl.)
42. Romanova, L.K., and Zhikhavera, I.A. Humoral regulation of regeneration in the lungs, kidneys, and liver. *Biull. Eksp. Biol. Med.* **73**, 84–87 (1972). (Engl. transl.)
43. Šerý, Zd., Keprt, E., and Obručník, M. Morphometric analysis of late adaptation of the residual lung following pneumonectomy in young and adult rabbits. *J. Thorac. Cardiovasc. Surg.* **57**, 549–557 (1969).
44. Simnett, J.D. Stimulation of cell division following unilateral collapse of the lung. *Anat. Rec.* **180**, 681–686 (1974).
45. Simnett, J.D., and Fisher, J.M. Description of growth phenomena and the formulation of growth control models. *Nat. Cancer Inst. Monogr.* **38**, 29–36 (1973).
46. Stephens, R.J., Sloan, M.F., Evans, M.J., and Freeman, G. Early response of lung to low levels of ozone. *Am. J. Pathol.* **74**, 31–58 (1974).
47. Tartter, P.I., and Goss, R.J. Compensatory pulmonary hypertrophy after incapacitation of one lung in the rat. *J. Thorac. Cardiovasc. Surg.* **66**, 147–152 (1973).
48. Tenney, S.M., and Remmers, J.E. Comparative quantitative morphology of the mammalian lung: Diffusing area. *Nature (London)* **197**, 54–56 (1963).
49. Thurlbeck, W.M. Postnatal growth and development of the lung. *Am. Rev. Respir. Dis.* **111**, 803–844 (1975).
50. Weibel, E.R., and Gomez, D.M. Architecture of the human lung. *Science* **137**, 577–585 (1962).
51. Will, D.H., Alexander, A.F., Reeves, J.T., and Grover, R.F. High altitude induced pulmonary hypertension in normal cattle. *Circ. Res.* **10**, 172–177 (1962).

17

The Physiology of Renal Growth

Nephrite: a stone worn on the side as a remedy for kidney disease

The kidney is organized around its blood supply. It is from the plasma that material is filtered or secreted into the intraluminal fluid of the nephron, and into which materials are reabsorbed. The kidneys of most vertebrates are provided with both arterial and venous blood, the latter reaching the kidney via renal portal veins. The purpose of the portal veins is not clear, for the kidney seems unaffected even if they are tied off (31). Mammals are unique in that theirs are the only vertebrate kidneys lacking renal portal veins, a condition that also testifies to the relative unimportance of these vessels for renal function.

The concept of similitude refers to changes in shape as a function of size, of which the mammalian kidney is an interesting example. The weight of the kidney is proportional to the surface area of the body. That is, except in infancy it tends to grow at a rate only about two-thirds that of the overall increase in body mass. This negative allometry is also found in phylogenetic comparisons, smaller species having relatively larger kidneys than bigger ones.

The shape of the kidney also changes with animal size (Fig. 130). In small mammals the kidney possesses a single renal papilla through which the collecting ducts convey the urine to the renal pelvis. In larger species, such as man, there may be a dozen or so papillae, suggesting that during phylogenetic enlargement of the kidney there is a limit beyond which the renal papilla cannot grow without jeopardizing its functional efficiency. In still larger creatures, such as the cow or the elephant, the entire kidney is partially subdivided into fused lobes, each of which has a single papilla. This trend reaches its extreme in the case of whales whose kidneys resemble a bunch of grapes, each one of which is a renculus, or small kidney with a

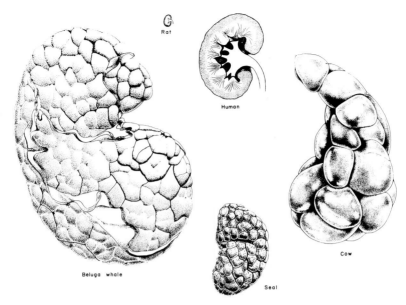

Fig. 130.   Examples of mammalian kidneys drawn to same scale to illustrate how renal morphology is affected by organ size. There is a tendency for the structure of the kidney to become subdivided with increasing size. From rat to human, the number of renal papillae is augmented. The bovine kidney is divided into lobes with partially fused cortices. The kidneys of marine mammals are composed of numerous renculi, each with a single papilla.

single papilla. These renculi number in the hundreds in smaller whales such as the beluga, or in the thousands in larger species. Although the number of renculi is not known to increase beyond early developmental stages, their diameters may double up to 3 cm during maturation (60).

The rationale for such anatomical changes during the evolution from small to large mammals may reside in the maximum length to which a nephron can grow without sacrificing its functional efficiency. While the length of the proximal tubule measures only a few millimeters in small rodents, it may reach 2–3 cm in larger ungulates (85). Extrapolation to whales would presumably extend nephron length severalfold beyond these limits if the shape of the kidney were to be preserved as its size increased. The hydrodynamic problems of propelling glomerular filtrates along such narrow tubules for such great lengths may have militated against the continued elongation of nephrons in the evolution of our largest mammals. Accordingly, the problem was solved by subdividing each kidney into smaller ones of more manageable size, a step that permitted the renal mass to increase by developing more numerous nephrons of shorter lengths rather than longer but fewer ones. This cannot be the entire explanation, however, for even smaller marine mammals, such as seals and dugongs, also possess compound kidneys made up of many renculi. The significance of this adaptation among aquatic mammals remains to be fathomed.

THE NUMBER OF NEPHRONS

The ontogeny of the kidney reflects the changing physiological demands of the maturing organism. In metamorphosing cyclostomes and amphibians, for example, there may be considerable degeneration of larval renal tissue and its replacement with adult kidneys (61). In those forms which switch from ammonia to urea excretion, there is a metamorphosis in the enzymes too. The fetal kidneys of mammals form their first nephrons long before birth, nephrons that acquire sufficient functional competence to contribute dilute urine to the amniotic fluid (6). Although such kidneys are not vitally essential at this stage of development owing to the excretory role of the placenta, their congenital absence results in oligohydramnios and hypoplastic development of the lungs, an interesting concatenation of pathologies known as Potter's syndrome.

Nephrons differentiate in the embryo and fetus from nephrogenic mesenchyme under the inductive stimulus of the ureteric bud (21). The nephrogenic zone at the periphery of the renal cortex (25) may virtually disappear before birth in organisms such as man, but persists for 40 days or so in the rat (7) and for up to 100 days in the quokka (5). Hence, nephrogenesis in the mammalian kidney may occupy a variable span of the developmental process, depending on the species. There eventually develops a species-specific number of glomeruli per kidney that despite wide variations from one individual to another would appear to be genetically fixed. Thus, there are about 12×10^3 glomeruli per mouse kidney, 35×10^3 in the rat, around 2×10^5 in the rabbit and cat, 1×10^6 in the cow, and 7.5×10^6 in the elephant (75). These figures are most conveniently determined by maceration of kidneys in concentrated HCl and enumeration of the glomeruli in aliquots of the resulting suspension in a counting chamber. Application of such techniques to immature animals has shown that in the rat at birth there are about 10×10^3 glomeruli, a number that doubles during the first ten days of life and reaches the adult population of 35×10^3 between 40 and 50 days after birth (7). In the human there are some 200 nephrons per kidney at 8 weeks of gestation, over 5×10^5 at 24 weeks, and 8×10^5 at 40 weeks (62). Thus, the adult complement of 10^6 nephrons per kidney is attained almost entirely before birth in man, subsequent growth being achieved by the elongation and thickening of nephron tubules and the doubling of glomerular diameters. Once the adult population of nephrons is

reached early in development, there is no further possibility for nephrogenesis even after the depletion of renal mass or heightened functional
loads. Indeed, there is a gradual attrition of nephrons as an individual ages,
sometimes accounting for reductions to one-half to two-thirds of the normal
young adult complement. In man, this may average as many as two nephrons
lost per hour over a 50-year period!

Whatever may be the factors determining the fixed number of gomeruli
per kidney in birds and mammals, it would appear not to operate in cold-
blooded vertebrates, many of which have indeterminate body sizes. In the
cyclostome, *Petromyzon*, for example, glomeruli continue to be added posteriorly in the growing animal (*61*). In teleost fishes there is also a linear
increase in glomerular number with body size (*23*), a phenomenon matched
in amphibians and reptiles of different sizes. It would appear, therefore, that
the kidneys of these creatures continue to augment their populations of
functional units to keep pace with the overall growth of the body which is
itself not so predetermined as in warm-blooded vertebrates. Presumably,

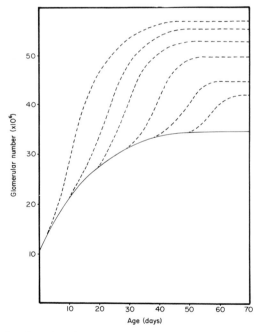

Fig. 131.   Effects of unilateral nephrectomy on glomerular numbers in immature rats (*7, 12*).
Solid curve represents the normal postnatal rise in glomerular population per kidney. Broken
curves indicate the production of supplemental glomeruli following removal of the other kidney
at different ages from birth to 50 days.

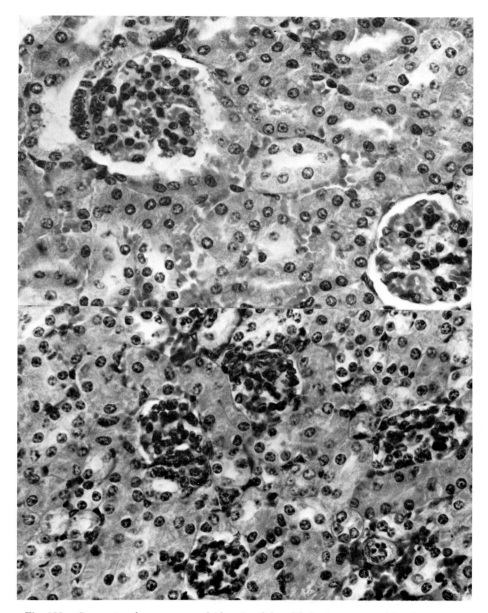

Fig. 132. Comparison between normal (above) and dwarf (below) mouse renal histology (92). Although the cell size is the same in the two strains, despite a two- to threefold difference in body and kidney weights, the glomerular size is greatly reduced and their numbers slightly increased in the dwarfs.

even in the aglomerular kidneys of such fishes as the goose fish, sculpin, and sea horse, the number of nephrons may increase with growth.

This mode of development, whereby the capacity for nephrogenesis is not lost early in development, suggests the persistence of cells capable of differentiating into nephrons throughout life. Yet there still remains the problem of how the rate of nephrogenesis is regulated. Is it adjusted to the overall rate of body growth *per se*, or does it represent an adaptation to increased physiological demands? The kidneys of marine teleosts have smaller and fewer glomeruli than those in fresh water, and when fishes raised in salt water are compared with the same kinds raised in fresh water, the latter develop more nephrons in their kidneys (23). Indeed, there is reason to suspect that following partial ablation of renal tissue in adult lower vertebrates the remaining kidneys compensate by supplementing their numbers of glomeruli, something that is not seen in the adult kidneys of higher vertebrates (31). Even in mammals, however, the number of nephrons is not necessarily immutable. Malnutrition during early developmental stages results in the development of fewer nephrons than in well-nourished mice (95). Further, unilateral nephrectomy of infant rats accelerates nephrogenesis, leading to the development of more nephrons than are normally produced per kidney (7). The earlier such animals are operated on, the greater the effect, with a 63% increase in the complement of gomeruli in the remaining kidney following unilateral nephrectomy of neonatal rats (Fig. 131) (12). These findings clearly implicate the importance of physiological adaptation in establishing the adult population of neprhons per kidney.

Other factors related to body size and growth are not without their effects. For example, in a strain of dwarf mice, the relative weight of the kidneys is nearly normal, as is the number of glomeruli, but the glomerular diameters are reduced about 40% (Fig. 132) (92). On the other hand, comparisons between dogs of different breeds over a sixfold range in body weight reveal that although the glomerular diameters are greater in the larger breeds, the number of glomeruli is not correlated with body weight (22).

COMPENSATORY RENAL HYPERTROPHY

The growth of the kidney can be analyzed by altering its function, the complexities of which provide ample opportunity for numerous experimental innovations. They also complicate the interpretation of results. The simplest approach is to remove one kidney and study compensation in the remaining organ. It has been known since the early nineteenth century that animals can survive on only one kidney. Not until 1869, following the advent of anesthetics and sterile techniques, was the first successful human nephrectomy carried out by the German surgeon, Gustav Simon. Since then, innumerable experiments on unilaterally nephrectomized animals, as well as donors of kidneys for organ transplants, have yielded valuable information concerning the physiological and morphological adaptations by the remaining kidney to reductions in renal mass (10, 58).

Studies on the congenital absence or hypoplasia of one kidney in man and other animals have confirmed the hypertrophy of solitary kidneys in such cases (54). The same result is found after the surgical resection of one kidney. The degree of hypertrophy is inversely related to the age of the animal, although the capacity for compensatory growth persists into old age (71). It is significant, however, that compensatory renal hypertrophy seldom exceeds 50% of the original weight. This does not necessarily represent the limits of renal hypertrophy, for if more than half the renal mass is excised the residual portions are capable of even greater compensatory growth. Indeed, it is possible to resect up to five-sixths of the total renal mass, a surgical tour de force not incompatible with survival if carried out in a two-step operation to allow provisional hypertrophy to occur (57). Conversely, ablation of less than one kidney results in correspondingly reduced degrees of compensation (27). Nevertheless, the failure of remaining kidneys to make up completely for the loss of renal tissue may be attributed to the fact that not all parts of the remaining kidney participate equally in the compensatory response. Indeed, most of the growth is to be found in the cortex, particularly in the proximal convoluted tubules, while relatively little enlargement takes place in the medulla (68). It is perhaps significant, therefore, that the greatest response occurs in those parts of the kidney that are most metabolically active.

In keeping with the normal loss of nephrogenic capacities during development, it is not surprising that compensatory hypertrophy of adult kidneys should take place without the formation of new nephrons. As mentioned above, in immature animals nephrectomized while the population of neprhons is still expanding, compensatory growth is accompanied by the production of supernumerary nephrons (7, 12). Lower vertebrates, even as adults, seem capable of compensatory nephrogenesis throughout life (23). The abridgement of this potential in adult birds and mammals may be partly responsible for the less than complete compensation for reductions in renal mass noted above. Instead, growth is achieved by the enlargement of preexisting nephrons (44). Glomerular diameters increase and the widths and lengths of the convoluted tubules enlarge. This is achieved by a mixture of cellular hyperplasia and hypertrophy. In young growing animals, proliferation is the dominant feature of compensatory renal growth. In older ones, cell hypertrophy predominates (41). In either case, the burst of mitosis that occurs is relatively short-lived. The earliest divisions may occur by the end of the first day, reaching a peak during the second day to magnitudes approximately 5–6 times the normal preoperative levels. Thereafter, there is a rapid decline in the incidence of mitosis, returning to nearly normal levels within several days.

The hyperplastic response is correlated with the extent of nephrectomy. When all of one kidney and half of the other are removed, the maximum magnitude of proliferation is greater than after lesser resections, although the onset and peak occur at the same time. After the loss of only half of one kidney, the mitotic activity drops to normal in four days or so. Following removal of 75% of the renal mass, hyperplasia may persist for a week or more (27).

In such cases mitotic activity is expressed as a percent of the total number of cells counted. Hence, the greater hyperplasia found after more radical renal resections represents an increase in the density of mitotic figures. However, if the lesser amounts of kidney remaining in such instances is taken into account, it is found that the absolute number of dividing cells at a given time is approximately the same following different extents of nephrectomy. That is, the greater density of proliferating cells in smaller amounts of residual kidney is roughly equivalent to the lesser density of mitoses in larger quantities of remaining tissue (27).

The antecedent of mitosis is DNA synthesis. Autoradiographic studies following the injection of ^{3}H-thymidine reveal a rise in DNA synthesis by 18–24 hours after unilateral nephrectomy, preceding by some hours the onset of mitosis (40). The maximum DNA synthesis is found during the second day after operation (Fig. 133), eventually leading to as much as a 25% increase in total DNA. It is noteworthy that the administration of hydroxyurea inhibits DNA synthesis in the compensating kidney but does not prevent renal growth (38). Since this drug does not directly interfere with RNA or protein synthesis, it would appear that the enlargement of the kidney may be achieved solely by cell hypertrophy if necessary.

What DNA synthesis is to mitosis, RNA synthesis is to hypertrophy. First detectable as early as one hour after unilateral nephrectomy, the synthesis of RNA is a dominant feature of compensatory renal hypertrophy (40, 73). The sustained and elevated rates of RNA synthesis, accompanied by increases in polyribosomes, reaches a peak several days postoperatively (Fig. 133), increasing the total RNA content by one-third. The RNA/DNA ratio rises more modestly. The importance of RNA synthesis in compensatory hypertrophy is demonstrated by the effects of actinomycin D. Not only does this drug inhibit the production of DNA-dependent RNA, but it also depresses kidney growth after unilateral nephrectomy (50).

Protein synthesis parallels RNA synthesis in the kidney (90), rising during the first hours after operation to maximum levels on the second day (Fig. 133). The hypertrophic kidney may attain protein contents one-third to

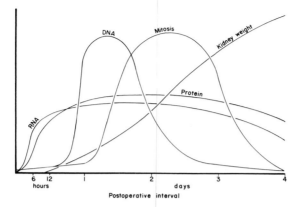

Fig. 133.   Relationships between various growth parameters in the remaining kidney following unilateral nephrectomy. Scale of ordinate is arbitrary.

one-half above normal. Although much of this protein is structural, as required for the production of new cells and organelles, there is a considerable increase in renal enzymes. Some of these, such as RNase and DNA polymerase, play important roles in increasing the amount of renal tissue. Others, involved in the enhancement of renal function, exhibit heightened activities commensurate with the increased work load carried by the remaining kidney (44). Particularly impressive is the severalfold increase in Na-K-ATPase activity after unilateral nephrectomy, a response associated with the active transport of electrolytes across tubule walls (76).

PHYSIOLOGICAL ADAPTATIONS

In seeking the cause for compensatory hypertrophy, it is logical to explore the functional compensation which accompanies it. There is a marked and immediate increase in blood flow through the remaining kidney after unilateral nephrectomy (42, 47). It is not inconceivable that this prompt hyperemia, which may eventually lead to a 40% increase in the caliber of the renal artery, plays an important role in intiating the functional, if not morphological, compensation of the remaining kidney.

Early investigators were impressed with the fact that after the loss of one kidney the amount of urine produced remained approximately normal, and even rose in some cases (20). This testified to the ability of the remaining kidney to do the work of both organs, work it was capable of performing due to the increased glomerular filtration rate (GFR), a response in turn made possible by the postoperative increase in renal blood flow (42, 43, 65). Not all physiological parameters remain unchanged, however. The extra urine produced by the remaining kidney, for example, is more dilute than normal because the degree of diuresis exceeds the decrease in Na$^+$ reabsorption in the tubules. Soon after the loss of one kidney, therefore, the nephrons in the opposite one not only put out more urine but are faced with the need to increase their rates of reabsorption. When the growth response eventually catches up with the functional overload, the physiological activities of the tubular cells can return to normal. Whether physiological compensation precedes or succeeds the growth response depends in part upon the methods of analysis. Those who have noted prompt increases in physiological activity look upon this as the stimulus for growth. Others, not finding increased function until after the hypertrophic response is under way, tend to attribute the latter to factors other than increased work load. Like the riddle of the chicken versus the egg, the answer may be that both come first.

ENDOCRINE INFLUENCES

In searching for the cause of renal growth, whether natural or compensatory, it is logical to consider various physiological factors or conditions that normally prevail in the body. Chief among these are certain hormones, the renal effects of which have been extensively explored. The pituitary has been found to play an important role in kidney growth. Hypophysectomy results in a decrease in kidney weight and reduces compensatory growth after unilateral nephrectomy, effects partially reversed by the administration of growth hormone (53, 64). It is worth noting, however, that this is not the principal cause of renal growth, for compensatory hypertrophy still occurs after hypophysectomy if one takes into account the decrease in absolute renal size in the absence of the pituitary.

The thyroid gland has similar effects (70). Following thyroidectomy there is a decrease in the protein content and weight of the kidneys, as well as the extent of compensatory renal hypertrophy after unilateral nephrectomy. Hypothyroidism caused by the administration of goitrogens such as thiouracil exert similar effects while the administration of thyroxine stimulates both normal and compensatory renal growth. Whether the influence of the thyroid is a direct effect on the tissues of the kidney or is mediated indirectly via heightened metabolic rates is not yet known.

The kidney is also subject to gonadal influences. Castration of males results in the reduction of normal kidney weights and decreased compensatory hypertrophy (3). The administration of testosterone, on the other hand, promotes hyperplasia and hypertrophy of tubule cells, reversing the effects of castration. If testosterone is given after unilateral nephrectomy, there is an additive effect on subsequent enlargement of the kidney. The size of the kidney is decreased under the influence of estrogen, presumably accounting for the normally smaller sizes of female kidneys.

The influence of mineralocorticoids on electrolyte balance is mediated by the kidneys. The decline in blood pressure and GFR and the abolition of compensatory renal hypertrophy following bilateral adrenalectomy are prevented in animals simultaneously given 1% NaCl to drink or following replacement therapy with deoxycorticosterone (28). The adrenal cortex, in particular the zona glomerulosa, exerts no small influence on the growth of the kidneys and must be listed as one of the major factors in regulating renal growth (56). This is in contrast to other endocrine glands which have modulating influences on the kidney but do not appear to secrete hormones that are primary stimuli of kidney growth.

RENAL REACTIONS TO THE COLD

The interesting effects of cold exposure and hibernation on the kidneys may not be unrelated to endocrine influences. If hamsters are held at 4°–5°C the mitotic activity of their proximal tubules rises after two days and their kidneys may increase in weight as much as 40% in two weeks, an effect that is additive to the compensatory hypertrophy following unilateral nephrectomy (69, 72). This curious response is believed to be related to the animal's preparation for hibernation which may last several months. [Even nonhibernators such as the rat, however, develop enlarged kidneys in the cold (34).] While the physiological activity of cold-exposed kidneys is increased, it is almost abolished once the animal actually goes into hibernation. Even then, a trickle of dilute urine production may persist in such animals as the bat, ground squirrel, and marmot (94), but most excretion in hibernation occurs during the periodic arousals every week or two when the body temperature rises and the bladder is filled and voided before hibernation resumes.

In the case of the marmot, the world's largest hibernating mammal, it is not known if the kidneys enlarge in the cold prior to the onset of hibernation. Curiously, these animals have been shown capable of prolonged survival following bilateral nephrectomy (8). If both kidneys are removed in the winter, despite being maintained at room temperature, marmots may survive for 3–5 weeks. In the summer, this operation is fatal in six days. Their survival after bilateral adrenalectomy follows a similar course. What the nature of the physiological adaptations are that account for this remarkable phenomenon is not known, except that it is profoundly affected by the seasons and may therefore have to do with such annual changes as are associated with the ability to hibernate.

The bear is not a true hibernator because its body temperature does not drop to that of the environment in winter. However, those inhabiting temperate zones tend to sequester themselves for several months in the winter without food or water. It is during this time that the cubs are born and suckled. In other mammals food consumption may double during lactation, but in bears fat reserves probably make important contributions to milk production. In the absence of food and water, defecation and excretion do not occur during the winter months. Although various parameters of renal physiology are decreased, such urine as may be produced is not dilute (9). It is not known if the kidneys of dormant bears undergo atrophy in the winter concomitant with their reduced functions, nor if indeed such bears could long survive the loss of both kidneys the way marmots can.

HUMORAL GROWTH CONTROL

A mammal is least dependent on its kidneys before birth when nitrogenous wastes are eliminated via the placenta into the maternal circulation. Thus, the congenital absence of both kidneys is not necessarily fatal until soon after birth. Unilateral nephrectomy of fetal rats elicits hypertrophy of the opposite kidney before parturition, but not to the extent seen postnatally (29). The moderate compensation may be correlated with the reduced function of fetal kidneys and the dilute urine they produce compared with postnatal ones. Removal of one or both of the maternal kidneys does not stimulate compensatory hypertrophy in the kidneys of her fetuses (26). It would be difficult to imagine how a fetal response to maternal renal insufficiency could have evolved. To whatever extent the kidneys of the fetuses might compensate, their increased excretory efficiency would simply be passed back to the mother. The lack of a transplacental reaction suggests that if humoral factors are involved in regulating renal growth they either do not cross to the fetal circulation or the fetal kidneys are insensitive to them.

There can be little doubt that humoral transfer between the two kidneys is responsible for the reaction of one to the absence of the other. Parabiotic experiments have confirmed the nature of this communication (Fig. 134). When both kidneys are removed from one partner, the survival of the other is contingent upon the confluence of their peritoneal cavities in addition to the usual capillary connections. Under these circumstances, both kidneys of the intact partner hypertrophy (19). If three of the four kidneys are removed, the single remaining one hypertrophies to an even greater degree. Unilat-

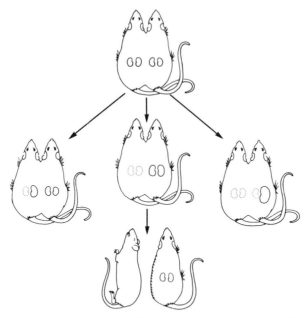

Fig. 134. Effects of varying degress of kidney loss on compensatory renal hypertrophy in parabiotic rats. Removal of one of the four kidneys results in preferential enlargement of the remaining kidney in the operated rat (left). Bilateral nephrectomy of one rat stimulates hypertrophy of both of the partners' kidneys, an effect that is reversible following separation of the parabionts (middle). If three out of four kidneys are resected, extreme hypertrophy occurs in the remaining one (right).

eral nephrectomy of one partner brings about compensatory growth in the three remaining kidneys, but more so in the ipsilateral one (48). These studies confirm the existence of blood-borne factors responsible for regulating kidney growth, but do not disclose whether such factors are stimulators or inhibitors.

The most direct experimental approach to this problem is to inject serum from unilaterally nephrectomized animals into normal recipients to see if renal growth might be stimulated. The results of such experiments are difficult to interpret, although about twice as many authors have reported positive results as have found no effect (67, 84). When serum from bilaterally nephrectomized donors is used, there is again a difference of opinion, one group of investigators finding no effect, the other reporting increased DNA synthesis. There are too many variables for experiments along these lines to be repeated with the same techniques from one laboratory to another, including the timing of donor serum collections and the dose and timing of injections into the recipient. Despite these disadvantages, there would seem to be evidence for the existence of renotrophic factors in the serum of animals deprived of one and perhaps both kidneys. On the other hand, the injection of a variety of renal preparations into partially nephrectomized recipients has been found more often than not to inhibit compensatory growth (17, 51). The possibility exists, therefore, that organ-specific growth inhibitors may be present in the kidney while stimulators of growth exist in the serum. If this interpretation is accepted in spite of the wide variation in results of such experiments, the next task is to determine the nature of such growth regulators. Do they exert their effects on growth in conjunction with possible influences on renal physiology, or are they concerned solely with size determination irrespective of functional adaptations?

EXCRETION VERSUS REABSORPTION

One of the earliest hypotheses to account for compensatory growth following contralateral nephrectomy was that the remaining kidney increases its growth rate to accommodate the need to excrete as much urea as was formerly handled by both kidneys. To test the possible role of urea excretion in promoting kidney growth, animals have been subjected to diets enriched with urea, sometimes in heroic doses. Although this treatment markedly increases the concentration of urea in the urine, it does not promote renal hypertrophy (4). Alternatively, it is possible to recirculate urine by severing one ureter or anastomosing it to the vena cava or intestine. The net result of these treatments is to overburden the kidneys' excretory role, and although the GFR increases as the result of such treatments there is no growth response (74, 93). In the case of ureteroperitoneostomy, it is conceivable that whatever growth responses might have occurred could be precluded by the chronic irritation of the peritoneal cavity by urine. Peritoneal irritation with talc following unilateral nephrectomy actually prevents compensatory hypertrophy of the remaining kidney (74). Nevertheless, it must still be concluded that renal growth is not stimulated by the demands for the excretion of nitrogenous waste products, an interpretation consistent with the fact that the elimination of urea by glomerular filtration requires little or no metabolic work on the part of the kidney (39). It is the pumping of the heart that creates the blood pressure to drive glomerular filtration.

Urea is produced by the deamination of amino acids. Consequently, a high protein diet would be expected to increase urea production and excretion. Numerous studies have shown that such diets promote renal hypertrophy and enhance compensatory growth following unilateral nephrectomy (37). Conversely, low protein diets reduce the size of the kidneys. Such results, inconsistent as they are with the ineffectiveness of direct administration of urea, may possibly be explained in terms of the deamination of amino acids which is necessary on a high protein diet. Although deamination goes on in the liver (an organ also stimulated to enlarge on high protein diets) it is also carried out in the kidney itself. Furthermore, high protein diets are known to be nephrotoxic, eventually causing glomerulosclerosis (49). Hence, the renal hypertrophy caused by protein rich diets could be more pathological than compensatory.

In view of the important role played by the kidney in regulating electrolyte balance, it is logical to explore its reactions to the administration of various salts. One of the most effective salts in promoting renal growth is NH_4Cl. Whether administered in the drinking water or mixed in the diet with water available *ad lib*, NH_4Cl stimulates considerable renal hypertrophy after 4–7 days, an effect that is additive to that which follows unilateral nephrectomy (37, 66). It is achieved primarily by cellular hypertrophy with little DNA replication, although the rate of protein synthesis rises. This enlargement of the kidneys is associated with the metabolic acidosis brought about by NH_4Cl. There is increased excretion of NH_4^+ and glutamine, with increased glutaminase activity.

NaCl also promotes renal growth, but in a different way (83). It is less effective if given in the diet with water *ad lib* than when mixed in the drinking water. In the latter situation, the experimental animal is more effectively compelled to become overloaded with sodium since it cannot flush out the excess salt with water. Given in the drinking water, NaCl becomes effective in doses that equal or exceed isotonicity. Above approximately 1% NaCl, the promotion of renal growth increases until the salt concentration in the drinking water becomes so high as to depress consumption. When given to unilaterally nephrectomized animals, it increases the extent of compensatory renal hypertrophy (88). Physiologically, NaCl promotes diuresis due to increased GFR. Excess Na^+ is therefore eliminated in the urine, although after unilateral nephrectomy NaCl may be retained, at least until the remaining kidney has hypertrophied. Increased salt also tends to elevate the blood pressure, especially when combined with deoxycorticosterone. The hypertensive effect of salt promotes cardiac hypertrophy, and may eventually lead to hypertensive vascular disease, including glomerulosclerosis (49).

The influence of salt may also be observed in marine environments. Fishes living in salt water, which is hypertonic to the body fluids, tend to lose water to the environment due to the osmotic differential. It is to their advantage, therefore, to minimize water loss by having smaller, and sometimes fewer, glomeruli. Thus, marine telosts characteristically have small glomeruli compared with their fresh water relatives. Indeed, fishes in the Black Sea where the salinity is only half that of the open ocean, have larger and more numerous glomeruli than do those inhabiting the Mediterranean. If salmon fry are raised in fresh water they develop more glomeruli in their kidneys than do those in salt water (23). Conversely, guppies reared in seawater develop 40% fewer glomeruli than in their fresh water controls. The extreme adaptation is that of aglomerular fishes in which such glomeruli as may develop early in life usually degenerate (30). It is no coincidence that all aglomerular fishes are marine.

The apparent inconsistency between the reduction of kidneys in marine fishes compared with their hypertrophy in salt-loaded mammals may be an expression of how phylogenetic adaptations differ from physiological ones. In salt water fishes the problem is not so much that of getting rid of excess salt as it is of conserving water, which is achieved primarily by the reduction in glomerular filtration. Mammals faced with superfluous salt depend solely upon their glomeruli for its elimination. Marine birds and reptiles have evolved salt glands to do the job by modifying their lacrimal or salivary glands for the elimination of excess NaCl. Not unexpectedly, these salt glands undergo considerable enlargement when birds are given salt water to

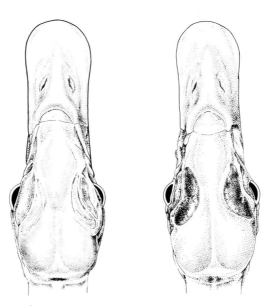

Fig. 135. Comparison between the salt glands of a control (left) and a salt-loaded duck (right). If ducklings are given salt water to drink their salt glands, located dorsally in the orbits, become hypertrophied (24). (Courtesy of Dr. Richard A. Ellis, Brown University.)

drink (Fig. 135) (24). Nevertheless, the kidneys of marine birds are typically larger than are those of terrestrial ones.

Sodium and potassium tend to be in reciprocal balance in the body. The retention of one is accompanied by the excretion of the other. Thus, the increased proliferation of renal cells and the enlargement of the kidney following K^+ deficiency could be attributed as much to the attending retention of sodium as to the depletion of potassium *per se*. Nevertheless, the development of lesions, particularly in the distal tubules and collecting ducts in K^+-deficient animals together with the impaired concentrating capacity of such kidneys, suggests a specific role that potassium may play in the normal integrity of renal structure and function (55). The proliferation of renal epithelial cells and the hypertrophy of the kidney may therefore be a response to epithelial damage in the absence of potassium.

HYDRONEPHROSIS AND RENAL COUNTERBALANCE

One school of thought has held that compensatory hypertrophy of the kidney is triggered by reduction in renal mass *per se* rather than functional overload following partial nephrectomy. This hypothesis can be tested by the functional incapacitation of one kidney without actually removing it, an experiment that should result in no contralateral renal growth if functional demand is not the stimulus for compensatory hypertrophy. It will be remembered that if one ureter is severed, neither kidney hypertrophies (74, 93). In this case, however, the ipsilateral kidney continues to function as usual since it has no way of reacting to a downstream operation. A more effective method of interfering with the function of one kidney is to ligate its ureter. This leads to hydronephrosis owing to continued glomerular filtration for a number of days despite the buildup of hydrostatic pressure in the kidney sufficient to cause its extensive distension. The internal pressure of hydronephrosis also obstructs the flow of blood through the kidney. It is the resulting ischemia which may be largely responsible for the eventual destruction of most of the renal parenchyma. Long before this happens, however, the opposite intact kidney undergoes the typical sequence of events leading to hypertrophy, a reaction in compensation for the physiological deficiency of the opposite hydronephrotic kidney (18, 63, 93).

It was once suspected that compensatory renal hypertrophy following unilateral hydronephrosis might be related to the marked rise in blood pressure following ureteral obstruction. This is brought about by the reduced blood flow through the hydronephrotic kidney resulting in the secretion of excess renin from its juxtaglomerular cells (see Chap. 19). Renin is an enzyme that catalyzes the conversion of angiotensinogen to angiotensin in the blood. This in turn stimulates the secretion of mineralocorticoids from the zona glomerulosa, promoting the retention of sodium and the development of hypertension. This sequence of events is probably not directly related to the initiation of renal hypertrophy since unilateral nephrectomy causes neither a detectable increase in serum Na^+ levels nor a rise in blood pressure.

There is a growth response in the hydronephrotic kidney also. Its tubular epithelial cells proliferate extensively after the ureter is ligated, presumably owing to the deleterious effects of tubular distension. Although prolonged hydronephrosis results in the ultimate destruction of the kidney, the damage caused by its limited duration is reversible following removal of the obstruction (91). The extent of recovery, however, is affected by the response of the opposite intact kidney which in the meantime will have undergone compensatory hypertrophy to take over the functions of its hydronephrotic partner (Fig. 136). If this has happened, there is only minimal recovery of the hydronephrotic kidney. On the other hand, removal of the intact hypertrophic kidney at the time the ureteral obstruction is alleviated shifts the burden of renal function to the hydronephrotic kidney. In some such cases, when the duration of hydronephrosis has been too long, the kidney does not recover in time to prevent death. In animals managing to survive, however,

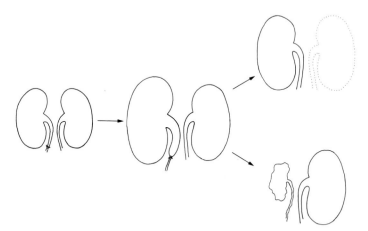

Fig. 136. Influence of intact kidney on recovery of a contralateral hydronephrotic organ. Ligation of one ureter (left) causes ipsilateral hydronephrosis and compensatory hypertrophy of the opposite kidney (middle). Resection of the latter concomitant with removal of the ureteral obstruction permits recovery of the formerly hydronephrotic kidney provided the damage was not irreversible (right, above). Otherwise the injured kidney is no longer needed and continues to degenerate in accordance with the principle of "renal counterbalance" (right, below).

the formerly hydronephrotic kidney recovers more promptly and to a greater extent than in animals not forced to rely on damaged organs (33, 35). Compulsory physiological activity, therefore, appears to promote the reversal of hydronephrosis. The failure of this to occur when the contralateral hypertrophic kidney has not been removed illustrates an important principle of renal physiology which Hinman referred to as "renal counterbalance" (32).

Normally both kidneys share the excretory needs of the body equally. However, when one kidney is made to assume more than its normal load, the functional inequity tends to persist even after the original cause of the imbalance has been corrected. Thus, once hypertrophy has been established it is not reversed so long as the physiological load persists even if another kidney is added to the system. Only when the magnitude of the original functional demand is diminished will such kidneys revert to their normal dimensions. This is illustrated in the case of bilateral nephrectomy in one partner of a parabiotic pair of rats. The kidneys of the intact partner hypertrophy in the process of taking over the excretory needs of the nephrectomized individual (19). When such rats are subsequently separated, the hypertrophic kidneys lose the weight they gained because the excess physiological demand has been withdrawn.

Glomerular filtration depends on a sufficiently high blood pressure without which the kidneys cease to function. Perhaps this is why the juxtaglomerular apparatus (JGA) is situated next to the glomerulus where its baroreceptors can monitor the blood pressure in the afferent arteriole. When one renal artery is partially constricted the JGA hypertrophies in the process of producing renin which promotes hypertension. Meanwhile, the reduced blood flow through the kidney decreases or arrests glomerular filtration, as a result of which the tubules of the nephrons collapse and the kidney atrophies (78). Histologically, such kidneys appear to be composed of solid cords of cells resembling certain endocrine glands, although the term "endocrine kidney" is a physiological misnomer.

In rats, it is possible to reduce the blood flow through the left kidney by placing a ligature around the aorta in the short distance separating the more anterior right renal artery from the more posterior left one. When this operation is performed on weanling rats the constriction tightens as the aorta enlarges during subsequent growth. The left nonfunctioning kidney involutes completely, and in the absence of its juxtaglomerular cells hypertension fails to develop. In adult rats subjected to the same operation, the left (downstream) kidney ceases to produce urine and undergoes considerable atrophy (Fig. 137). The opposite right (upstream) kidney hypertrophies in compensation for its nonfunctioning partner (52, 59). Even if the ligature is then removed the right and left kidneys remain hypertrophic and atrophic, respectively. However, if the right kidney is removed (without relieving the

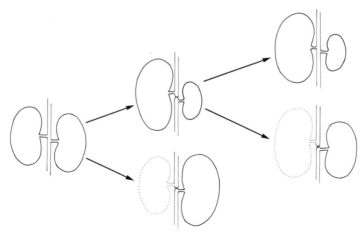

Fig. 137.   Role of blood flow in the control of kidney size in rats (52, 59). Partial ligation of the aorta between the right and left renal arteries (ventral view) causes the right kidney to enlarge and the left one to atrophy owing to insufficient blood pressure for glomerular filtration (above, middle). Right nephrectomy concomitant with aortic constriction results in hypertrophy instead of atrophy of the downstream kidney (below, middle). Removal of the ligature following establishment of right hypertrophy and left atrophy is followed by persistence of the renal inequities (above, right). Removal of the right kidney once left renal atrophy has occurred reverses the latter despite continued presence of the ligature (below, right).

aortic ligation) the atrophic left kidney then enlarges to normal dimensions. Removal of the ligature allows hypertrophy of the formerly atrophic left kidney to above normal dimensions. If the right kidney is resected at the time the ligation is applied to the aorta, the left kidney undergoes hypertrophy instead of atrophy. While such results are in part to be explained in terms of alterations in blood flow, other factors come into play after unilateral nephrectomy which override the vascular disturbance.

REPAIR OF THE DAMAGED KIDNEY

If the kidney is completely tied off for a short period, the resulting ischemia does considerable damage. Complete cessation of blood flow for up to three hours, however, is not incompatible with recovery despite the considerable necrosis of tubular epthelial cells which may have occurred (16). Four hours of ischemia usually results in irreversible damage (46). As in case of temporary hydronephrosis, the recovery of ischemic kidneys is hastened by removal of the opposite intact kidney (although survival may sometimes be contingent upon retention of the functioning kidney until the dam-

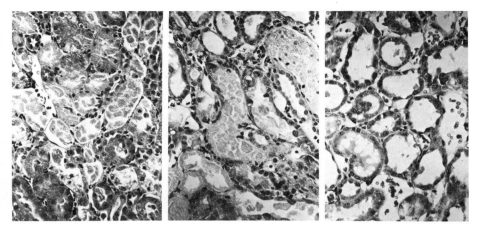

Fig. 138. Regeneration of renal tubules following intravenous administration of mercuric chloride (*15*). One day afterward the proximal tubules exhibit extensive necrosis (left). By 3 days, regeneration of the proximal tubular epithelium has begun by proliferation of surviving cells (middle). After 5 days the damaged tubules have become lined with new cuboidal epithelium (right). Reproduced with permission. © 1972 U.S.–Canadian Division of the International Academy of Pathology.

aged one is at least partly repaired) (*11*). Indeed, experiments have shown that if the intact kidney is removed following temporary ischemia, the damaged kidney may undergo greater compensatory hypertrophy than is observed in residual kidneys not previously deprived of their blood flow (*77*). It would seem that the wound healing response following ischemia may be additive to compensatory growth after unilateral nephrectomy.

Like most other organs of the body, the tissues of the kidney are capable of repairing injuries. Direct mechanical trauma to the kidney results in a localized proliferative response. Decapsulation stimulates tubular epithelial proliferation in the periphery of the cortex (*13*). In addition, a wide variety of chemical agents damages the kidney, particularly compounds containing heavy metals such as cadmium, mercury, lead, chromium, and uranium (Fig. 138) (*14, 15, 36*). Injection of glycerin, chloroform, alloxan, or folic acid also damages the kidney (*86, 87*). Such treatments usually cause injury or necrosis in the tubules leading to the desquamation of their cells into the lumens. Surviving epithelial cells then proliferate prodigiously, often at rates many times higher than the maximum proliferation observed following unilateral nephrectomy. Although the kidney is more sensitive than most organs to a wide variety of noxious compounds, which in sufficient doses often inflict permanent damage, the epithelial cells of the nephron are endowed with an impressive capacity for histological regeneration.

THE GROWTH OF GRAFTED KIDNEYS

The ultimate size to which the mammalian kidney is destined to grow is in part determined early in life when the number of nephrons becomes fixed. Subsequent development is achieved by enlargement of these nephrons to adult dimensions, but can go beyond this in cases of compensatory hypertrophy. The mechanisms by which the normal adult kidney hypertrophies are extensions of the modes of growth that prevailed during ontogeny, namely, hyperplasia and hypertrophy of the cells. In this sense, it is logical to regard compensatory renal hypertrophy simply as a supplement to normal development and presumably subject to the same regulatory mechanisms. If

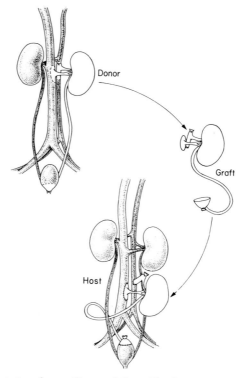

Fig. 139. Transplantation of an auxiliary rat kidney. The donor organ is removed with a length of attached dorsal aorta and a cuff of inferior vena cava. The urethra and contralateral ureter are tied off, and the proximal portion of the bladder is resected with the kidney. While a length of host aorta and vena cava is clamped posterior to the kidneys, the donor aorta and vena caval cuff are anastomosed by microvascular surgery to the corresponding vessels of the host. The dome of the host bladder is replaced with the lower part of the donor bladder. Such grafts may resume function despite an hour's ischemia.

this interpretation is correct, then whatever stimulates compensatory growth must also be responsible for the normal enlargement of the kidneys during maturation. To test this possibility, techniques have become available in recent years to transplant kidneys in rats using microvascular surgery (Fig. 139). Originally applied to the transplantation of adult kidneys, these methods have now been adapted to infant ones making it possible to graft kidneys between immature and adult animals.

When an adult kidney is grafted to an adult host still in possession of both of its own kidneys the result is a 50% superabundance of renal mass (Fig. 140). One might expect all three kidneys to undergo sufficient atrophy to reduce the total amount of kidney tissue to the equivalent of two organs. Such is not the case, however. The host's kidneys retain their original sizes while the transplant loses weight (45). If one of the host's kidneys is removed, the grafted kidney still loses weight, but not so much; the host's remaining kidney hypertrophies (though only about half as much as after

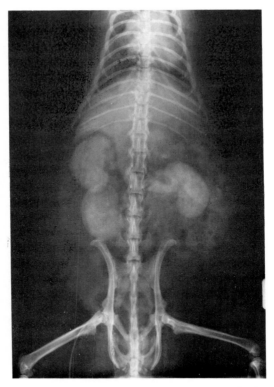

Fig. 140.   Intravenous pyelogram of a rat with three kidneys, 6 weeks after transplantation (82).

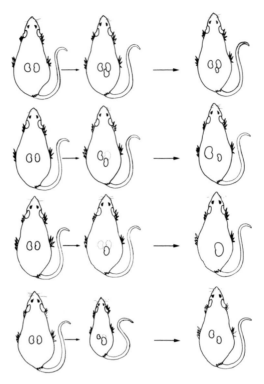

Fig. 141.

unilateral nephrectomy). The differential in response between the graft and host kidneys would appear to be an expression of renal counterbalance such that the kidneys of the host, already at advantage, preserve their dominance over the transplant. Only when both host kidneys are removed does the grafted kidney undergo hypertrophy, confirming that its atrophy in the former cases was due to the lack of stimulation, not its inability to grow (Fig. 141).

The transplantation of infant kidneys is particularly interesting, especially in view of the clinical need for a more convenient supply of donors for patients in need of kidney transplants. The possibility of utilizing the kidneys from the many babies that die soon after birth from unrelated causes has raised the question of whether or not an adult can survive on such small and immature kidneys. Experiments on dogs in which puppy kidneys have been transplanted to bilaterally nephrectomized adults have proved that they are adequate to the task, a conclusion subsequently confirmed in man (1, 2, 79). In these cases the infant kidney not only takes over the renal functions of its

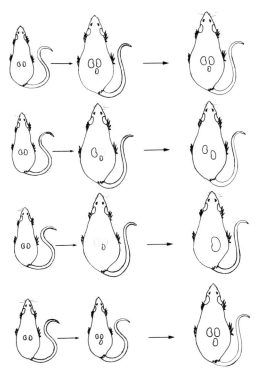

Fig. 141. Effects of auxiliary kidney grafts on renal growth in rats. Opposite page, top to bottom, a third kidney tends to atrophy, but loses less weight if one of the host's own kidneys has been removed and its opposite one hypertrophies. If both host kidneys are missing, the grafted kidney now hypertrophies. Transplantation of an adult kidney to a unilaterally nephrectomized infant host results in both kidneys being of normal size when the recipient grows up. Above, infant kidneys grafted to adult hosts attain sizes that are atrophic, normal, or hypertrophic depending on whether the host has both, one, or none of its own kidneys, respectively. An infant kidney transplanted to one of its littermates grows at the same rate as do the host's organs, resulting in an adult animal with three normal-sized kidneys.

host, but undergoes enlargement in the postoperative months preceding its eventual immunological rejection (2, 79).

Experiments on inbred strains of rats have made it possible, in the absence of rejection, to follow the growth course of infant kidneys grafted to adults. When the adult host retains both of its own kidneys, the extra infant graft exhibits relative atrophy, i.e., it continues to grow, but fails to attain normal adult dimensions (80). If one of the host's kidneys has been removed, the grafted kidney grows to normal adult dimensions (81). When both host kidneys have been resected, the infant kidney grows beyond the normal adult size in compensation for the missing host kidneys (81). When an extra infant kidney is grafted into an otherwise intact infant host, all three kidneys continue their enlargement during maturation of the host, eventually attaining normal adult dimensions in all three cases (80). Finally, if an adult kidney is grafted to a unilaterally nephrectomized infant host, the recipient's remaining kidney continues to grow to normal adult size while the transplanted adult kidney remains the same (Fig. 141) (80).

The significance of these findings cannot be exaggerated. They show that a third kidney in an adult animal undergoes compensatory atrophy, but that an extra infant kidney graft continues to grow even in the presence of the normal complement of renal mass in the host. The conclusion is inescapable that infant kidneys possess an intrinsic potential for growth irrespective of the amount of the other renal tissue present and presumably in the absence of a full work load. This "obligatory growth" is apparently subject to different regulatory mechanisms from those responsible for compensatory renal hypertrophy in adult kidneys. Such a conclusion is strengthened by studies in which both donor and host are unilaterally nephrectomized and allowed to undergo compensatory hypertrophy of their remaining kidneys. When the hypertrophic kidney from one is grafted to the other, restoring the host's missing kidney but creating a superabundance of renal mass, both hypertrophic kidneys revert to normal size (82). This proves that compensatory hypertrophy is reversible, and apparently more readily so than is the obligatory growth by which the normal adult dimensions of the kidney are attained. Even the latter is reversible when a kidney is deprived of functional stimulation, but the extent to which it atrophies is limited. It is this limit which may mark the point of transition from obligatory growth in the maturing kidney to the onset of compensatory growth dependent upon physiological demands.

In view of the many physiological influences which have been shown to affect kidney growth, one might assume that compensatory renal hypertrophy, if not the normal maturation of the kidneys, occurs in response to increased functional demands. A note of caution is in order, however. Kidneys are sometimes subject to the development of Wilms tumor, a renal adenocarcinoma that would appear not to share in the work load of the normal kidneys. Yet when these tumors are transplanted subcutaneously into growing rats, the hosts' kidneys later average 10% below normal dimensions. If such rats are unilaterally nephrectomized, the magnitude of compensatory hypertrophy in the remaining kidneys is only half that in controls (89). The implications of these results are as important for our interpretation of organ growth regulation as they are for our understanding of cancer.

REFERENCES

1. Andersen, O.S., Jonasson, O., and Merkel, F.K. En bloc transplantation of pediatric kidneys into adult patients. *Arch. Surg.* **108**, 35–37 (1974).
2. Baden, J.P., Wolf, G.M., and Sellers, R.D. The growth and development of allo-transplanted neonatal canine kidneys. *J. Surg. Res.* **14**, 213–220 (1973).
3. Basinger, G.T., and Gittes, R.F. Effect of testosterone propionate on compensatory renal hypertrophy. *Endocrinology* **94**, 599–601 (1974).
4. Baxter, J.H., and Cotzias, G.C. Effects of proteinuria on the kidney. Proteinuria, renal enlargement, and renal injury consequent on protracted parenteral administration of protein solutions in rats. *J. Exp. Med.* **89**, 643–688 (1949).
5. Bentley, P.J., and Shield, J.W. Metabolism and kidney function in the pouch young of the macropod marsupial *Setonyx brachyurus*. *J. Physiol. (London)* **164**, 127–137 (1962).
6. Berton, J.-P. Développement normal de l'arbre vasculaire métanéphrique chez le foetus de lapin et modifications apparaissant après hydronéphrose unilatérale expérimentale. *Bull. Assoc. Anat.* **141**, 546–567 (1968).
7. Bonvalet, J.P., Champion, M., Wanstok, F., and Berjal, G. Compensatory renal hypertrophy in young rats: Increase in the number of neprhons. *Kid. Int.* **1**, 391–396 (1972).
8. Britton, S.W., and Silvette, H. Survival of marmots after nephrectomy and adrenalectomy. *Science* **85**, 262–263 (1937).
9. Brown, D.C., Mulhausen, R.O., Andrew, D.J., and Seal, U.S. Renal function in anesthetized dormant and active bears. *Am. J. Phsyiol.* **220**, 293–298 (1971).
10. Bucher, N.L.R., and Malt, R.A. "Regeneration of Liver and Kidney." Little, Brown, Boston, Massachusetts, 1971.
11. Campbell, J.C., Edwards, E.C., Grindlay, J.H., Maher, F.T., and Thompson, G.J. Temporary renal ischemia with immediate and delayed contralateral nephrectomy. *J. Urol.* **85**, 875–878 (1961).
12. Canter, C.E., and Goss, R.J. Induction of extra nephrons in unilaterally nephrectomized immature rats. *Proc. Soc. Exp. Biol. Med.* **148**, 294–296 (1975).
13. Choie, D.D., and Richter, G.W. Stimulation of nuclear DNA-synthesis in rat kidney by renal decapsulation. *Beitr. Pathol.* **148**, 86–93 (1973).
14. Choie, D.D., and Richter, G.W. Cell proliferation in mouse kidney induced by lead. I. Synthesis of deoxyribonucleic acid. *Lab. Invest.* **30**, 647–651 (1974).
15. Cuppage, F.E., Chiga, M., and Tate, A. Cell cycle studies in the regenerating rat nephron following injury with mercuric chloride. *Lab. Invest.* **26**, 122–126 (1972).
16. Cuppage, F.E., Neagoy, D.R., and Tate, A. Repair of the nephron following temporary occlusion of the renal pedicle. *Lab. Invest.* **17**, 660–674 (1968).
17. Dicker, S.E., and Morris, C.A. Investigation of a substance of renal origin which inhibits the growth of renal cortex explant *in vitro*. *J. Embryol. Exp. Morphol.* **31**, 655–665 (1974).
18. Dicker, S.E., and Shirley, D.G. Compensatory hypertrophy of the contralateral kidney after unilateral ureteral ligation. *J. Physiol. (London)* **220**, 199–210 (1972).
19. Dijkhuis, C.M., Van Urk, H., Malamud, D., and Malt, R.A. Rapid reversal of compensatory renal hypertrophy after withdrawal of the stimulus. *Surgery* **78**, 476–480 (1975).
20. Emmanouel, D.S., Lindheimer, M.D., and Katz, A.I. Urinary concentration and dilution after unilateral nephrectomy in the rat. *Clin. Sci. Mol. Med.* **49**, 563–572 (1975).
21. Erickson, R.A. Inductive interactions in the development of the mouse metanephros. *J. Exp. Zool.* **169**, 33–42 (1968).
22. Finco, D.R., and Duncan, J.R. Relationship of glomerular number and diameter to body size of the dog. *Am. J. Vet. Res.* **33**, 2447–2450 (1972).

23. Ford, P. Studies on the development of the kidney of the Pacific pink salmon (*Onchorynchus gorbuscha* (Walbaum)). II Variation in glomerular count of the kidney of the Pacific pink salmon. *Can. J. Zool.* **36**, 45–47 (1951).
24. Goertemiller, C.C., Jr., and Ellis, R.A. Specificity of sodium chloride in the stimulation of growth in the salt glands of ducklings. *Z. Mikroskop.-Anat. Forsch.* **74**, 296–302 (1966).
25. Goncharevskaya, O.A., and Dlouhá, H. The development of various generations of nephrons during postnatal ontogenesis in the rat. *Anat. Rec.* **182**, 367–375 (1975).
26. Goss, R.J. Effects of maternal nephrectomy on foetal kidneys. *Nature (London)* **198**, 1108–1109 (1963).
27. Goss, R.J. Kinetics of compensatory growth. *Q. Rev. Biol.* **40**, 123–146 (1965).
28. Goss, R.J. Renal and adrenal relationships in compensatory hyperplasia. *Proc. Soc. Exp. Biol. Med.* **118**, 342–346 (1965).
29. Goss, R.J., and Walker, M.J. Compensatory renal hypertrophy in fetal rats. *J. Urol.* **106**, 360–362 (1971).
30. Grafflin, A.L. Glomerular degeneration in the kidney of the daddy sculpin (*Myoxocephalus scorpius*). *Anat. Rec.* **57**, 59–73 (1933).
31. Hammarstrom, C.J. Personal communication (1976).
32. Hinman, F. Renal counterbalance. An experimental and clinical study with reference to the significance of disuse atrophy. *J. Urol.* **9**, 289–314 (1923).
33. Hinman, F. The condition of renal counterbalance and the theory of renal atrophy of disuse. *J. Urol.* **49**, 392–400 (1943).
34. Holečková, E., and Baudyšová, M. Stimulation of DNA synthesis in rat liver and kidney by cold acclimation. *Physiol. Bohemoslov.* **24**, 311–313 (1975).
35. Huguenin, M.E., Thiel, G.T., Brunner, F.P., Torhorst, J., Flückiger, E.W., and Wirz, H. Régénération du rein après levée d'une occlusion de l'urètere chez le rat. *Kid. Int.* **5**, 221–232 (1974).
36. Itokawa, Y., Abe, T., Tabei, R., and Tanaka, S. Renal and skeletal lesions in experimental cadmium poisoning. *Arch. Environ. Health* **28**, 149–154 (1974).
37. Janicki, R.H., and Argyris, T.S. Kidney growth and adaptation of phosphate-dependent glutaminase in the mouse. *Am. J. Physiol.* **217**, 1389–1395 (1969).
38. Janicki, R., and Lingis, J. Unabated renal hypertrophy in uninephrectomized rats treated with hydroxyurea. *Am. J. Physiol.* **219**, 1188–1191 (1970).
39. Johnson, H.A. Cytoplasmic response to overwork. *In* "Compensatory Renal Hypertrophy" (W.W. Nowinski and R.J. Goss, eds.), pp. 9–27. Academic Press, New York, 1969.
40. Johnson, H.A., and Vera Roman, J.M. Compensatory renal enlargement. Hypertrophy versus hyperplasia. *Am. J. Pathol.* **49**, 1–13 (1966).
41. Karp, R., Brasel, J., and Winick, M. Compensatory kidney growth after uninephrectomy in adult and infant rats. *Am. J. Dis. Child.* **121**, 186–188 (1971).
42. Kaufman, J.M., Siegel, N.J., and Hayslett, J.P. Functional and hemodynamic adaptation to progressive renal ablation. *Circ. Res.* **36**, 286–293 (1975).
43. Kaufman, J.M., Siegel, N., Lytton, B., and Hayslett, J.P. Compensatory renal adaptation after progressive renal ablation. *Invest. Urol.* **13**, 441–444 (1976).
44. Kazimierczak, J., Chavaz, P., Krstić, R., and Bucher, O. Morphometric and enzyme histochemical behaviour of the kidney of young rats before and after unilateral nephrectomy. *Histochemistry* **46**, 107–120 (1976).
45. Klein, T.W., and Gittes, R.F. Three-kidney rat: Renal isografts and renal counterbalance. *J. Urol.* **109**, 19–27 (1973).
46. Koletsky, S. Effects of temporary interruption of renal circulation in rats. *Arch. Pathol.* **58**, 592–603 (1954).

47. Krohn, A.G., Peng, B.B.K., Antell, H.I., Stein, S., and Waterhouse, K. Compensatory renal hypertrophy: Role of immediate vascular changes in its production. *J. Urol.* **103**, 564–568 (1970).
48. Kurnick, N.B., and Lindsay, P.A. Compensatory renal hypertrophy in parabiotic mice. *Lab. Invest.* **19**, 45–48 (1968).
49. Lalich, J.J., Burkholder, P.M., and Paik, W.C.W. Protein overload nephropathy in rats with unilateral nephrectomy. *Arch. Pathol.* **99**, 72–79 (1975).
50. Lotspeich, W.D. Metabolic aspects of acid-base change. *Science* **155**, 1066–1075 (1967).
51. Lucas, J., Buzelin, F. Soulillou, J.-P. Fontenaille, C., Ginet, J., and Guenel, J. Hypertrophie compensatrice du rein. I. Recherche sur l'existence d'une substance rénale inhibitrice. *Pathol. Biol.* **24**, 383–390 (1976).
52. Masson, G.M.C., and Hirano, J. Prevention and reversal of renal atrophy. *In* "Compensatory Renal Hypertrophy" (W.W. Nowinski and R.J. Goss, eds.), pp. 235–249. Academic Press, New York, 1969.
53. McCreight, C.E., and Reiter, R.J. Effects of hormones on the renal response to unilateral nephrectomy in hypophysectomized rats. *J. Exp. Zool.* **166**, 65–70 (1967).
54. Meares, E.M., Jr., and Gross, D.M. Hypertension owing to unilateral renal hypoplasia. *J. Urol.* **108**, 197–201 (1972).
55. Milne, M.D., Muehrcke, R.C., and Heard, B.E. Potassium deficiency and the kidney. *Brit. Med. Bull.* **13**, 15–18 (1957).
56. Moraski, R. Renal hyperplasia in the intact rat. *Proc. Soc. Exp. Biol. Med.* **121**, 838–840 (1966).
57. Morrison, A.B. Experimentally induced chronic renal insufficiency in the rat. *Lab. Invest.* **11**, 321–332 (1962).
58. Nowinski, W.W., and R.J. Goss (eds.) (1969) "Compensatory Renal Hypertrophy." Academic Press, New York.
59. Omae, T., and Masson, G.M.C. Reversibility of renal atrophy caused by unilateral reduction of renal blood supply. *J. Clin. Invest.* **39**, 21–27 (1960).
60. Ommanney, F.D. The urino-genital system of the fin whale (*Balaenoptera physalus*). *Discovery Rep.* **5**, 363–466 (1932).
61. Ooi, E.C., and Youson, J.H. Growth of the opisthonephric kidney during larval life in the anadromous sea lamprey, *Petromyzon marinus* L. *Can. J. Zool.* **54**, 1449–1458 (1976).
62. Osathanondh, V., and Potter, E.L. Development of human kidney as shown by microdissection. IV. Development of tubular portions of nephrons. *Arch. Pathol.* **82**, 391–402 (1966).
63. Paulson, D.F., and Fraley, E.E. Compensatory renal growth after unilateral ureteral obstruction. *Kid. Int.* **4**, 22–27 (1973).
64. Poffenbarger, P.L., and Prince, M.J. The role of serum nonsuppressible insulin-like activity (NSILA) in compensatory renal growth. *Growth* **40**, 83–97 (1976).
65. Potter, D.E., Leumann, E.P., Sakai, T., and Holliday, M.A. Early responses of glomerular filtration rate to unilateral nephrectomy. *Kid. Int.* **5**, 131–136 (1974).
66. Preuss, H.G., and Goldin, H. Ammoniagenesis in growing nephrons of uninephrectomized rats. *Lab. Invest.* **31**, 454–457 (1974).
67. Preuss, H.G., and Goldin, H. Renotropic system in rats. *J. Clin. Invest.* **57**, 94–101 (1976).
68. Reiter, R.J. Cellular proliferation and deoxyribonucleic acid synthesis in compensating kidneys of mice and the effect of food and water restriction. *Lab. Invest.* **14**, 1636–1643 (1965).
69. Reiter, R.J. Early response of the hamster kidney to cold exposure and unilateral nephrectomy. *Comp. Biochem. Physiol.* **25**, 493–500 (1968).

70. Reiter, R.J. Endocrines and compensatory renal enlargement. *In* "Compensatory Renal Hypertrophy" (W.W. Nowinski and R.J. Goss, eds.), pp. 183–204. Academic Press, New York, 1969.

71. Reiter, R.J., McCreight, C.E., and Sulkin, N.M. Age differences in cellular proliferation in rat kidneys. *J. Gerontol.* **19**, 485–489 (1964).

72. Rosen, J.K. Renal physiology of cold exposed and uninephrectomized golden hamsters. *Comp. Biochem. Physiol.* **44**, 1277–1287 (1973).

73. Ross, J.S., Malamud, D., Caulfield, J.B., and Malt, R.A. Differential labeling with orotic acid and uridine in compensatory renal hypertrophy. *Am. J. Physiol.* **229**, 952–954 (1975).

74. Royce, P.C. Inhibition of renal growth following unilateral nephrectomy in the rat. *Proc. Soc. Exp. Biol. Med.* **113**, 1046–1049 (1963).

75. Rytand, D.A. The number and size of mammalian gomeruli as related to kidney and to body weight, with methods for their enumeration and measurement. *Am. J. Anat.* **62**, 507–520 (1938).

76. Schmidt, U., and Dubach, U.C. Induction of Na K ATPase in the proximal and distal convolution of the rat nephron after uninephrectomy. *Plfluegers Arch.* **346**, 39–48 (1973).

77. Segaul, R.M., Lytton, B., and Schiff, M., Jr. Functional aspects of compensatory renal growth after ischemic injury. *Invest. Urol.* **10**, 235–238 (1972).

78. Selye, H. Transformation of kidney into an exclusively endocrine organ. *Nature (London)* **158**, 131 (1946).

79. Silber, S.J. Renal transplantation between adults and children: Differences in renal growth. *J. Am. Med. Assoc.* **228**, 1143–1145 (1974).

80. Silber, S.J. Compensatory and obligatory renal growth in babies and adults. *Austral. N.Z. J. Surg.* **44**, 421–423 (1974).

81. Silber, S.J. Growth of baby kidneys transplanted into adults. *Arch. Surg.* **111**, 75–77 (1976).

82. Silber, S., and Malvin, R.L. Compensatory and obligatory renal growth in rats. *Am. J. Physiol.* **226**, 114–117 (1974).

83. Solomon, S., Romero, C., and Moore, L. The effect of age and salt intake on growth and renal development of rats. *Arch. Int. Physiol. Biochim.* **80**, 871–882 (1972).

84. Soulillou, J.-P., Lucas, J., Fontenaille, C., Guenel, J., and Ginet, J. L'hypertrophie compensatrice renale. II. Recherche d'une facteur serique stimulant. *Pathol. Biol.* **24**, 391–396 (1976).

85. Sperber, I. Studies on the mammalian kidney. *Zool. Bidrag Från Uppsala* **22**, 249–432 (1944).

86. Suzuki, T., and Mostofi, F.K. Electron microscopic studies of acute tubular necrosis. Vascular changes in the rat kidney after subcutaneous injection of glycerin. *Lab. Invest.* **23**, 29–38 (1970).

87. Threlfall, G. Cell proliferation in the rat kidney induced by folic acid. *Cell. Tissue Kinet.* **1**, 383–392 (1968).

88. Tingle, L.E., and Cameron, I.L. Cell proliferation response in several tissues following combined unilateral nephrectomy and high salt diet in mice. *Texas Rep. Biol. Med.* **31**, 537–549 (1973).

89. Tomashefsky, P., Lattimer, J.K., Priestly, J., Jr., Furth, J., Vakili, B.F., and Tannenbaum, M. An experimental Wilms' tumor suitable for therapeutic and biologic studies. II. The inhibition of renal compensatory hypertrophy by a transplantable tumor. *Invest. Urol.* **11**, 141–144 (1973).

90. Tomashefsky, P., and Tannenbaum, M. Macromolecular metabolism in renal compensatory hypertrophy. II. Protein turnover. *Lab. Invest.* **23**, 190–195 (1970).

91. Vaughan, E.D., Jr., and Gillenwater, J.Y. Recovery following complete chronic unilateral

ureteral occlusion: Functional, radiographic and pathologic alterations. *J. Urol.* **106**, 27–35 (1971).

92. Wegelius, O., Pasternack, A., and Friman, C. The influence of pituitary dwarfism on the development of the kidney in mice. *Acta Pathol. Microbiol. Scand.* **78**, 238–240 (1970).

93. Weinman, E.J., Renquist, K., Stroup, R., Kashgarian, M., and Hayslett, J.P. Increased tubular reabsorption of sodium in compensatory renal growth. *Am. J. Physiol.* **224**, 565–571 (1973).

94. Zatzman, M.L., and South, F.E. Concentration of urine by the hibernating marmot. *Am. J. Physiol.* **228**, 1336–1340 (1975).

95. Zeman, F.J. Effects of maternal protein restriction on the kidney of the newborn young of rats. *J. Nutr.* **94**, 111–116 (1968).

18

Pressure on the Bladder

Pudendal nerve: L., *pudere*, to be ashamed

In casting about for the most physiologically simple organ of the body in which to study growth regulation in its least complicated expression, nothing would seem to be more suitable than the urinary bladder. Functioning solely in the storage of urine, it might be anticipated that the urinary bladder grows or atrophies only in response to the mechanical influences to which it is subjected. Although there is some truth to this assumption, the simplicity of bladder function is deceptive.

The bladders of fishes and amphibians are more than storage organs. They participate actively in the process of concentrating the urine, being capable of reabsorbing water and certain solutes from the urine. Some higher forms lack bladders altogether. Most birds and egg laying mammals do not possess bladders, suggesting a correlation with oviparity. Such birds as the ostrich, however, do possess bladders, which may account for the more liquid nature of their excretory products. In other avian species the urine, with its uric acid crystals in suspension, is conveyed via the ureters directly to the cloaca where water and electrolytes are reabsorbed prior to elimination. Experiments have shown that if the ureters are made to bypass the cloaca, birds drink extra water in accommodation for the lack of cloacal and rectal reabsorption (8).

The mammalian bladder is not known to alter the composition of its urine. Its function is therefore exclusively mechanical, a function to which its histology is well adapted. The transitional epithelium, or urothelium, which lines the bladder and its tributaries consists of cells capable of considerable rearrangement, an attribute necessary for accommodation to the prodigious distension of which the bladder is capable. The urothelium is a renewing tissue, its cells capable of slow turnover in association with DNA synthesis and mitosis. Even those cells in the surface layer are occasionally found in the process of synthesizing DNA (27). In the newborn mouse all of the urothelial cells are diploid, but in the adult bladder a large percentage of them are tetraploid and some are octoploid (10).

REPAIR OF THE UROTHELIUM

Exposed as they are to the noxious constituents of urine, the cells of the urothelium are vulnerable to chemical injury. This is particularly true of carcinogens. For example, 20-methylcholanthrene implanted in the bladder in a pellet of paraffin induces carcinomas in the urothelium of the mouse (5). Systemic factors also affect bladder epithelium, but whether the exposure is vascular or via excretory products in the urine, remains to be determined. If 4-ethylsulfonylnaphthalene-1-sulfonamide is fed in the diet, it induces

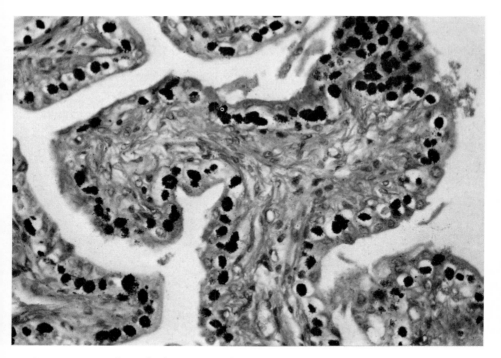

Fig. 142. Autoradiograph of a rat urinary bladder 22 hours after intraperitoneal injection of isoproterenol and 1 hour after ^{3}H-thymidine. The number of labeled nuclei has increased a hundredfold over control levels (42).

hyperplasia in the urothelium from renal pelvis to bladder, even among polyploid cells on the surface (26). Carcinogenesis may be a long-term result (28). The intraperitoneal injection of isoproterenol into rats elicits a hundred-fold increase in DNA synthesis of the urothelial cells during the first day (Fig. 142), an effect indicative of an apparent association between the bladder and its innervation (42). The urothelium is also sensitive to vitamin A deficiency which promotes cornification reminiscent of comparable responses in epidermal tissues (12, 17, 18).

Wound healing in the bladder can be studied by surgically stripping away the sheet of urothelium, a procedure facilitated by the submucosal injection of saline (16). Healing can also be initiated by the instillation of formalin into the bladder to cause necrosis of the urothelium (33). Such injuries elicit a burst in DNA synthesis in peripheral cells, including those migrating over the wound (32). In large denuded areas, islands of epithelium may be found in isolation from the centripetally migrating sheet. Whether these represent the progeny of residual epithelial cells adhering to the denuded area or the seeding of the wound by cellular flotsam in the urine is a problem awaiting solution (13).

REGENERATION OF BLADDER AND URETERS

Both the bladder and its ureters are capable of extensive regeneration following their subtotal resection, an operation not infrequently necessitated by the carcinogenic susceptibility of these tissues. Following subtotal cystectomy, the bladder is regenerated primarily by the expansion of what remains. Excision of a length of ureter or of part of its circumference, however, requires a repair process that depends in large measure upon the outgrowth of new tissues. In either case, it is often clinically desirable to patch a lesion with grafts of other tissues in a stopgap measure to reconstruct the bladder or ureters for the conduction of urine pending regeneration. A variety of organs has served as the source of such tissues, including skin, abdominal wall, fallopian tube, ileum, serosa, peritoneum, artery, and dura (21, 23, 37). Such grafts frequently become covered with urothelium, particularly when their own epithelium is not viable. However, plastic reconstructions never turn into bladder or ureter, and are replaced by such tissues only to limited extents. It is better to encourage self-regeneration if functional competence is to be optimally restored.

To this end, attempts have been made to insert molds over which urothelium and submucosal tissues can migrate in repairing defects. This has been particularly effective in restoring the continuity of the ureter following its partial resection. Polyethylene tubing inserted into the lumen not only

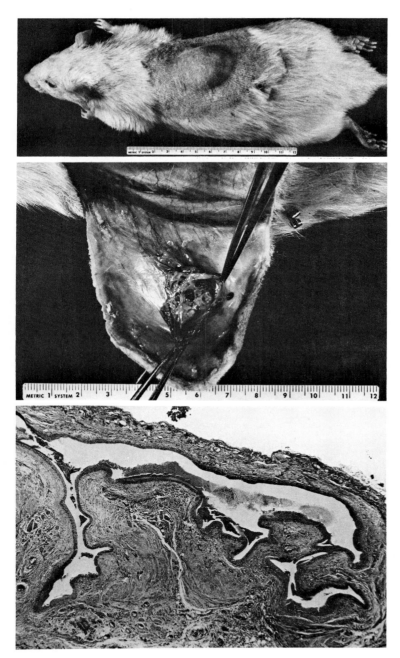

Fig. 143. Above, rat 2 months following the subcutaneous injection of minced urinary bladder plus 15 ml air. The fragments gave rise to the cyst exposed in the middle photograph. Histologically, the fluid-filled cyst is lined with transitional epithelium (below) (35).

provides a conduit for the passage of urine, but acts as a splint over which wound healing can proceed from the margins of the lesion (1). Connective tissue and epithelium restore their continuities with facility. The muscularis may or may not be reconstituted, depending on the size of the animal and how extensive the gap is (7, 24). Nevertheless, the muscularis is at least partially replaced in most cases, although it still remains to be determined whether this is achieved by the proliferation of preexisting smooth muscle cells or the differentiation of new ones *de novo*. Regeneration of the bladder itself can also be assisted by the use of plastic molds *in situ* (20). On the other hand, following the subcutaneous injection of minced bladder in rats, the pieces give rise to a cyst filled with fluid and lined with transitional epithelium (Fig. 143) (35).

In the case of bladder regeneration, autoradiographic studies with ³H-thymidine have shown no uptake of label in the smooth muscle of the bladder (36). This indicates that cellular hypertrophy plays a prominent role in bladder regeneration, but does not rule out the possibility of *de novo* differentiation in the absence of DNA synthesis. The fact remains, however, that when a nearly complete bladder is regenerated from a remnant of the original there is no dearth of smooth muscle cells in its walls. Whether in experimental animals or human patients, the postoperative incontinence that follows is in due course alleviated by a gradual increase in bladder capacity. In the rat, the resulting bladder is anatomically different from the original. The two ureters normally enter the bladder near its neck, in the vicinity of the urethra. During regeneration, the distance between urethra and ureters is increased as the attachment of the latter shifts to the dome of the bladder (30). This affords evidence indicating that the regenerated bladder is produced not by the outgrowth of new tissues from the cut surfaces,

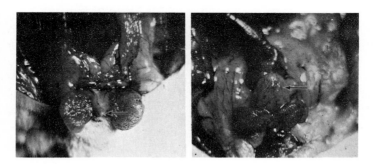

Fig. 144. Atrophy (left) and hypertrophy (right) of the urinary bladder in parabiotic rats 8 weeks after bilateral nephrectomy of one partner. The bladder of the operated rat atrophies to approximately one-third normal mass while that of the opposite intact partner enlarges to an equivalent extent (15).

but by expansion of the remnant. It would appear, therefore, that bladder regeneration is in reality a striking version of tissue hypertrophy coupled with wound healing.

DISTENTION VERSUS CONTRACTION

The bladder and ureters are capable of both atrophy and hypertrophy. Disuse atrophy of the bladder can be induced experimentally by diverting the ureters to the ileum (38) or by allowing them to drain through an abdominal fistula (29). Alternatively, atrophy has been achieved by the bilateral nephrectomy of one partner of a pair of parabiotic rats (Fig. 144) (15). In either case, deprived of the hydrostatic pressure of contained urine the bladders remain in a permanently contracted condition and their mass gradually declines to levels less than one-third of the original. Atrophy is achieved primarily by decreases in the muscularis, while the thickness of the urothelium remains normal (Fig. 145).

These results testify to the importance of hydrostatic pressure in the normal growth and maintenance of the urinary bladder. Regeneration is subject to the same influences, for subtotal cystectomy accompanied by urinary

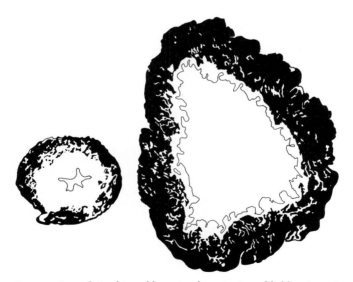

Fig. 145.   Cross sections of atrophic and hypertrophic rat urinary bladders in contracted states (15). Smooth muscle is black, connective tissue white. On the left is an atrophic bladder from an animal deprived of its kidneys and parabiosed to an intact partner for one year. The latter's bladder hypertrophies, as shown on the right. A normal bladder would be of intermediate size.

diversion is not followed by regeneration (29). Nevertheless, the total incapacitation of the bladder, while leading to extensive atrophy, never results
in the total disappearance of the organ (15). Atrophy is also reversible, for the
clinical capacity for hypertrophy following resumption of urine flow may last
for years (40).

In 1921, Carey reported that if increasing quantities of fluid were repeatedly injected into a dog's bladder, its capacity increased to the point
where it could hold heroic quantities of liquid after a number of weeks (2).
This periodic stretching of the bladder wall tripled the weight of the entire
organ and more than doubled the thickness of its walls. Distention of the
guinea pig bladder with saline for only five minutes has been shown to
promote increased DNA synthesis in its epithelial cells one day afterward
(31). A comparable response is found in the ureter under conditions requir-
ing the transport of greater quantities of urine. For example, if one kidney is
removed or its ureter ligated, the output of urine from the contralateral
kidney is increased (22, 39). This is accompanied by elevated rates of DNA
synthesis in the ureter on that side, followed by hypertrophy of its wall.
Similar responses are seen following obstruction of a ureter in which the

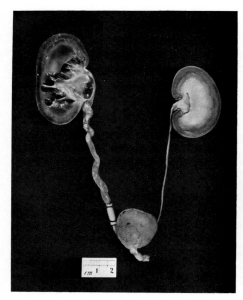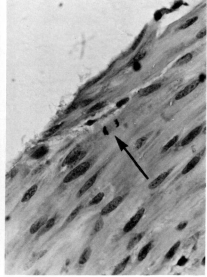

Fig. 146.   Left, dog kidneys 6 days after obstruction of one ureter showing hydronephrosis of
the affected kidney. Right, a mitotic figure in the smooth muscle of the occluded ureter (6).
Reproduced with permission. © 1972 The Williams & Wilkins Co., Baltimore.

upstream increase in hydrostatic pressure brings about hypertrophy of the muscularis (Fig. 146) (6, 11).

Clinical and experimental evidence leads to the inescapable conclusion that the so-called normal size of the bladder is a function of how much urine it is ordinarily called upon to handle. It follows that if the capacity of the bladder were artificially reduced, hypertrophy should occur until the original volume of the organ were restored. These conditions can be achieved by the displacement of urine by injection of a 50:50 mixture of paraffin and Vaseline as a liquid into the lumen of the bladder where it hardens into a solid cast of the vesicle. In order to accommodate to the continued inflow of urine, such bladders hypertrophy during subsequent weeks to weights up to several times their normal mass (34). This evidence confirms other findings that distention promotes growth of the bladder.

It is conceivable, however, that this may not be the only factor involved. If there is an increase in urine flow, not only does the bladder become more distended but it empties itself more frequently. It could be argued that it is the increased frequency of voiding that might be responsible at least in part for the subsequent hypertrophy. To test this possibility, rats have been parabiosed and both kidneys removed from one partner, an operation that leads to atrophy of the ipsilateral bladder (15). Measurements of urine output disclose an 80% increase in diuresis from the intact partner after one week, a level that climbs to considerably more than twice the original amount of

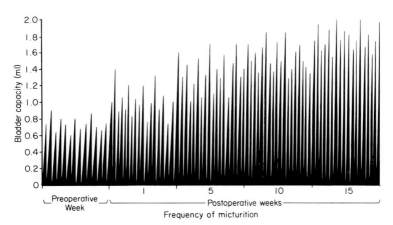

Fig. 147. Effects of increased urine flow on bladder capacity and frequency of micturition in parabiosed rats before and after bilateral nephrectomy of one partner. Each peak represents the filling and emptying of the bladder. The frequency of urination increases approximately 30% in the immediate postoperative intact animal, remaining elevated indefinitely. Meanwhile, the volume of urine voided with each urination gradually rises to twice normal levels after several months (14).

urine during subsequent months. This is caused in part by an immediate increase in the frequency of micturition to levels approaching 40% above normal. Concomitantly, there is a more gradual rise in the capacity of the bladder in intact rats parabiosed to nephrectomized ones, an increase that eventually doubles the amount of urine eliminated at each voiding. These findings show that the urinary bladder adapts to increased urine flow by a combination of more frequent and more copious episodes of urination (Fig. 147) (14). The resulting hypertrophy of such bladders is undoubtedly due in part to hydrostatic distention, but the frequency with which bladders contract remains a potential physiological factor in effecting compensatory growth.

DENERVATION HYPERTROPHY

The physiology of micturition depends upon a spinal reflex. As the bladder becomes filled with urine, proprioceptors in its walls are increasingly stimulated, eventually triggering motor reactions which cause the contraction of smooth muscle in the bladder proper while relaxing the sphincters in its neck. Hence, the capacity of the bladder is determined morphologically by the thickness of its wall and physiologically by the threshold of stretch receptors which trigger the voiding reflex.

If the autonomic innervation to the bladder is interrupted its capacity to fill with urine is not impaired but its emptying reflex is inoperative. Denervation of the bladder thus leads to overflow incontinence, and the chronic distention of such bladders leads inevitably to their hypertrophy (3, 4, 9, 19, 25). Such bladders may grow to sizes many times normal with enormously thickened walls and capacities to store extraordinary quantities of urine (Fig. 148). Pathological consequences inevitably ensue. Urinary stagnation encourages infections in the bladder and concretions frequently develop as a result of the precipitation of salts in the urine. The hypertrophy of such overly distended bladders argues in favor of mechanical stretch as an important stimulus to smooth muscle growth, especially since diversion of the ureters prevents hypertrophy of denervated bladders (41). The explanation may not be this simple, however, in view of the effects of unilateral denervation.

In smaller animals the nerves entering from either side are not restricted to their own halves of the bladder. There is considerable overlap in the pattern of innervation such that if one pudendal nerve is cut the remaining contralateral one is sufficient to trigger the voiding reflex throughout the bladder. In larger animals, such as cats and dogs, there is little or no crossing over of nerves from one side of the bladder to the other. In these animals,

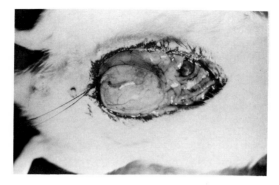

Fig. 148.   Rat bladder after denervation for 2 weeks. Unable to contract, the bladder becomes greatly distended to the point of overflow incontinence and its walls hypertrophy. This specimen, ligated in the distended condition, is several times larger than a maximally inflated normal bladder.

unilateral denervation likewise does not prevent the bladder from emptying, a response achieved by contraction of smooth muscle on the innervated side alone. Nevertheless, the denervated sides of such bladders undergo hypertrophy of their walls, a unilateral reaction that occurs even in the absence of overflow incontinence (*19, 25*). This curious phenomenon cannot be explained in terms of bladder distention since the side that has remained innervated does not hypertrophy. Although the explanation is for the future to provide, this reaction is reminiscent of the hypertrophy of the denervated hemidiaphragm, a growth response believed to be attributable to the passive stretching caused by contractions of the opposite functional half (see Chap. 8). A similar situation may prevail in the bladder where the denervated side is subjected to periodically elevated hydrostatic pressures when the intact half contracts. The inability of the denervated side to contract against the forces of distension may lead to its hypertrophy as a result of the tensions of being excessively stretched. The further analysis of this interesting phenomenon may yield fundamental information not only about the growth of the bladder itself but that of other hollow organs also wrapped in layers of smooth muscle.

REFERENCES

1. Boyarsky, S., and Duque, O. Ureteral regeneration. *J. Urol.* **73**, 53–61 (1955).
2. Carey, E.J. Studies in the dynamics of histogenesis. Tension of differential growth as a stimulus to myogenesis. VII. The experimental transformation of the smooth bladder muscle of the dog, histologically into cross-striated muscle and physiologically into an organ manifesting rhythmicality. *Am. J. Anat.* **29**, 341–378 (1921).

3. Carpenter, F.G. Histological changes in parasympathetically denervated feline bladder. *Am. J. Physiol.* **166**, 692–698 (1951).

4. Carpenter, F.G., and Root, W.S. Effect of parasympathetic denervation on feline bladder function. *Am. J. Physiol.* **166**, 686–691 (1951).

5. Clayson, D.B., and Pringle, J.A.S. The influence of a foreign body on the induction of tumours in the bladder epithelium of the mouse. *Br. J. Cancer* **20**, 564–568 (1966).

6. Cussen, L.J., and Tymms, A. Hyperplasia of ureteral muscle in response to acute obstruction of the ureter. A quantitative study. *Invest. Urol.* **9**, 504–508 (1972).

7. Dalley, B.K., Bartone, F.F., and Gardner, P.J. Smooth muscle regeneration in swine ureters: A light and electron microscopic study. *Invest. Urol.* **14**, 104–110 (1976).

8. Dicker, S.E., and Haslam, J. Effects of exteriorization of the ureters on the water metabolism of the domestic fowl. *J. Physiol. (London)* **224**, 515–520 (1972).

9. Elliott, T.R. The innervation of the bladder and urethra. *J. Physiol. (London)* **35**, 367–445 (1907).

10. Farsund, T. Cell kinetics of mouse urinary bladder epithelium. I. Circadian and age variations in cell proliferation and nuclear DNA content. *Virchows Arch. B.* **18**, 35–49 (1975).

11. Gee, W.F., and Kiviat, M.D. Ureteral response to partial obstruction. Smooth muscle hyperplasia and connective tissue proliferation. *Invest. Urol.* **12**, 309–316 (1975).

12. Ghidoni, J.J., and Campbell, M.M. Fine structure of metaplastic cornified squamous epithelium in the urinary bladder of rats. *J. Pathol.* **97**, 665–670 (1969).

13. Goldstein, A.B.M., Harp, G.E., Jones, W.G., and Morrow, J.W. Epithelial seeding: A possible mechanism in the regeneration of the urothelium—a preliminary report. *Anat. Rec.* **178**, 512 (1974).

14. Goss, R.J., Liang, M.D., Weisholtz, S.J., and Peltzer, T.J. The physiological basis of urinary bladder hypertrophy. *Proc. Soc. Exp. Biol. Med.* **142**, 1332–1335 (1973).

15. Goss, R.J., and Singleton, S.D. Disuse atrophy of the urinary bladder after bilateral nephrectomy. *Proc. Soc. Exp. Biol. Med.* **138**, 861–864 (1971).

16. Hansen, R.I., Lund, F., and Wanstrup, J. Regeneration of bladder epithelium after mucosal denudation (stripping) in guinea pigs. *Scand. J. Urol. Nephrol.* **3**, 208–213 (1969).

17. Hicks, R.M. Nature of the keratohyalin-like granules in hyperplastic and cornified areas of transitional epithelium in the vitamin A-deficient rat. *J. Anat.* **104**, 327–339 (1969).

18. Hicks, R.M. The mammalian urinary bladder: An accommodating organ. *Biol. Rev.* **50**, 215–246 (1975).

19. Jacobson, C.E., Jr. Neurogenic vesical dysfunction. An experimental study. *J. Urol.* **53**, 670–695 (1945).

20. Johnson, A.J., Kinsey, D.L., and Rehm, D.A. Observations on bladder regeneration. *J. Urol.* **88**, 494–502 (1962).

21. Kelami, A. Alloplastic replacement of the urinary bladder wall with lyophilized human dura. *Eur. Surg. Res.* **2**, 195–202 (1970).

22. Kiviat, M.D., Ross, R., and Ansell, J.S. Smooth muscle regeneration in the ureter: Electron microscopic and autoradiographic observations. *Am. J. Pathol.* **72**, 403–416 (1973).

24. Kozak, J.A., Deniz, E., and McDonald, J.H. Reparative processes of lower third of ureter following total segment resection. *J. Urol.* **91**, 509–514 (1964).

25. Langworthy, O.R., and Kolb, L.C. Histologic changes in the muscle following injury of the peripheral innervation. *Anat. Rec.* **71**, 249–263 (1938).

26. Lawson, T.A., Dawson, K.M., and Clayson, D.B. Acute changes in nucleic acid and protein synthesis in the mouse bladder epithelium induced by three bladder carcinogens. *Cancer Res.* **30**, 1586–1592 (1970).

27. Leblond, C.P., Vulpé, M., and Bertalanffy, F.D. Mitotic activity of epithelium of urinary bladder in albino rat. *J. Urol.* **73**, 311–313 (1955).

28. Levi, P.E., Cowen, D.M., and Cooper, E.H. Induction of cell proliferation in the mouse bladder by 4-ethylsulphonyl-naphthalene-1-sulphonamide. *Cell. Tissue Kinet.* **2**, 249–262 (1969).
29. Liang, D.S. Mechanical force as a factor in regeneration of the bladder: Experimental studies. *J. Urol.* **96**, 304–307 (1966).
30. Liang, D.S., and Goss, R.J. Regeneration of the bladder after subtotal cystectomy in rats. *J. Urol.* **89**, 427–430 (1963).
31. Martin, B.F. Cell replacement and differentiation in transitional epithelium: A histological and autoradiographic study of the guinea-pig bladder and ureter. *J. Anat.* **112**, 433–455 (1972).
32. McMinn, R.M.H., and Johnson, F.R. The repair of artificial ulcers in the urinary bladder of the cat. *Br. J. Surg.* **43**, 99–103 (1955).
33. Mulcahy, J.J., Farrow, G.M., Furlow, W.L., and Leary, F.J. The effect of intravesical formalin on the destruction and regeneration of the canine bladder. *Invest. Urol.* **14**, 177–181 (1976).
34. Peterson, C.M., Goss, R.J., and Atryzek, V. Hypertrophy of the rat urinary bladder following reduction of its functional volume. *J. Exp. Zool.* **187**, 121–126 (1973).
35. Roberts, D.D., Leighton, J., Abaza, N.A., and Troll, W. Heterotopic urinary bladders in rats produced by an isograft inoculum of bladder fragments and air. *Cancer Res.* **34**, 2773–2778 (1974).
36. Ross, G., Thompson, I.M., Keown, K.K., Jr., Judy, B.B., and Gammel, G.E. Further observations on the role of smooth muscle regeneration. *J. Urol.* **102**, 49–52 (1969).
37. Schein, C.J., and Sanders, A.R. The epithelial morphology of autogenous grafts when utilized as ureteral and vesical substitutes in the experimental animal: A collective review. *J. Urol.* **75**, 659–664 (1956).
38. Schmaelzle, J.F., Cass, A.S., and Hinman, F., Jr. Effect of disuse and restoration of function on vesical capacity. *J. Urol.* **101**, 700–705 (1969).
39. Swartzberg, J.E., Heilbron, D., and Hinman, F., Jr. Disuse and increased function of the dog ureter. II. Effect on length. *Urol. Int.* **26**, 51–64 (1971).
40. Tanagho, E.A. Congenitally obstructed bladders: Fate after prolonged defunctionalization. *J. Urol.* **111**, 102–108 (1974).
41. Veenema, R.J., Carpenter, F.G., and Root, W.S. Residual urine, an important factor in interpretation of cystometrograms, an experimental study. *J. Urol.* **68**, 237–241 (1952).
42. Winter, W.A. Induction of DNA synthesis by isoproterenol in the rat urinary bladder epithelium. *Histochemistry* **41**, 141–143 (1974).

19

Nonreproductive Endocrine Glands

Thyroid: Gr., *thyros*, shield shaped

THYROID

The distinction between endocrine and exocrine glands is based upon the presence or absence of a duct and whether or not the secretory products are released into the bloodstream. According to both of these criteria, the thyroid is definitely endocrine, but judging from its histology, as well as its phylogeny, one wonders if it might not have originally served an exocrine function. In the larval lamprey, for example, the subpharyngeal gland, or endostyle, consists of epithelium that takes up iodine, yet it is not rounded up into follicles until after metamorphosis. In the sense that iodine and thyroglobulin are secreted into the follicular colloid, the thyroid resembles an exocrine gland. The resorption of thyroid hormone elaborated in the colloid and its secretion into the bloodstream is diagnostic of the endocrine component of the thyroid's physiology.

The growth of the thyroid, like its physiological activity, is under the influence of functional demands. In a classical negative feedback loop, the output of thyroxine is monitored by the hypothalamus which in turn produces a releasing factor to stimulate the secretion by the anterior pituitary of the thyroid stimulating hormone (TSH). It is this hormone that mediates the secretion of thyroid hormones and upon which the size of the thyroid gland depends. Indeed, virtually every experimental and pathological condition that affects the physiological activity and morphological integrity of the thyroid operates via TSH.

The effects of hypophysectomy illustrate the dependence of the thyroid upon the pituitary gland. The ensuing atrophy of the thyroid, however, is reversible upon subsequent treatment with TSH. As with other target organs of pituitary tropic hormones, the thyroid does not disappear completely in the absence of this physiological stimulation. Moreover, the thyroid of a fetus continues its development in the absence of the pituitary following decapitation *in utero*, although colloid fails to accumulate in the follicles (49, 50). Exogenous TSH will accelerate the growth of the thyroid in such fetuses (94). As in adults, it stimulates proliferation of the thyroid cells and increases the height of the follicular epithelium, a reliable criterion for hyperthyroidism.

Nowhere is the role of the pituitary in thyroid maintenance more clearly illustrated than in the case of compensatory thyroid hypertrophy. Surgical excision of 50% or 75% of a thyroid gland leads to a prompt increase in mitotic activity in the remnant together with an increase in the epithelial height (65). There is a concomitant increase in iodine metabolism by the thyroid as indicated by its increase in ^{131}I uptake. Depending upon the extent of thyroid ablation, the residual tissue may increase its mass to something approaching the original size of the gland. This it does by increasing the dimensions of preexisting follicles as well as forming new ones. These may be produced *de novo* by differentiation of interfollicular cells, or by reorganization from follicles already present, a process achieved either by budding or subdivision (28).

No compensatory growth of the thyroid occurs if the part which is removed is grafted elsewhere in the body. Even intrasplenic grafts continue to function normally—in contrast to the stimulation of steroid-secreting tissues in this site owing to the breakdown of their hormones in the liver and the resulting overproduction of tropic hormones from the pituitary. The failure of this reaction in the case of the thyroid testifies to the fact that thyroxine is not normally degraded in the liver.

It can only be concluded that compensatory growth of the thyroid following its reduction in mass is due to the production of increased amounts of TSH from the pituitary in response to the decline in the levels of circulating thyroxine (14). Not until the output by the thyroid returns to normal following the production of more cells (or more follicles) does the secretion of TSH return to preoperative levels. It follows that any physiological alteration that upsets the balance between the supply of and demand for thyoxine will be compensated by hypothalamic and hypophyseal reactions to restore homeostasis. A case in point is the frizzle fowl. Owing to a genetic defect, these birds tend to lose their feathers (57). Such denuded chickens develop high basal metabolic rates and enlarged thyroids, a reaction prevented by keeping the birds warm and simulated by plucking normal fowl.

The thyroid is also sensitive to the seasons. Animals native to the temperate zone typically exhibit an annual alternation between fertile and infertile phases of their life cycles. During the mating season the thyroids are hyperactive compared with other times of year. Under both conditions there is a balance between thyroxine and TSH, except that the level of this equilibrium shifts from one season to another presumably in response to changes in the photoperiod. In polyestrous animals it is well established that the mitotic activity of the thyroid is maximal in estrus, a reaction consistent with the growth of the thyroid after pinealectomy (45). In the absence of the pineal, and of the melatonin which it produces, gonadal activity is increased. Curiously, the enlargement of the thyroid following pinealectomy is not prevented by hypophysectomy, suggesting that the response of the thyroid is not mediated by TSH (43). If this possibility is confirmed, one might look for direct effects of sex hormones on the thyroid gland.

Certain pharmacological agents have been useful in analyzing the physiological factors by which thyroxine secretion is controlled. One of these is phenobarbital, which upon repeated administration has been shown to promote thyroid hyperplasia and eventual enlargement of the gland to several times its normal weight (47). The goitrogenic effect of phenobarbital is believed to operate by stimulating TSH secretion. Most goitrogens exert their effects by virtue of their antithyroid activities. Such compounds as thiourea, thiouracil, or propylthiouracil (PTU) are commonly used to stimulate thyroid growth (31, 42, 50, 89). By interfering with thyroxine production, these compounds are ultimately responsible for the secretion of excess TSH which, in a vain attempt to rectify the incorrigible, succeeds only in racing the thyroid's motor and eventually producing a goiter. A similar situation prevails in cases of iodine deficiency.

There are two kinds of goiters, parenchymatous and colloid. The former are characterized by hyperactive glands in which the follicular epithelium is heightened and the colloid is relatively sparse. The growth of parenchymatous goiters is due more to cytoplasmic augmentation than to colloid accumulation, for thyroxine is secreted as fast as it is produced. Should the conditions responsible for the original hyperthyroidism be corrected, the parenchymatous goiter may be converted into a colloid goiter. Such a gland is less active, as a result of which its follicular epithelium converts to a low cuboidal type as large quantities of colloid accumulate. This is responsible for even further enlargement of the goiter. Successive rounds of hyper- and hypoactivity may eventually be responsible for the prodigious sizes to which goiters may sometimes grow.

Experimentally, the administration of goitrogens stimulates proliferation of both thyroid cells and follicles, but this is prevented by hypophysectomy (2). If PTU or thiouracil is administered to pregnant rats, it causes the enlargement of both the maternal and fetal thyroids (31, 42). The latter effect, however, does not occur in the absence of the fetal pituitary (50), which may be taken to indicate that although thyroxine itself is a small enough molecule to cross the placenta, TSH is not.

The offspring of mothers made hypothyroid by treatment with thiouracil are born as cretins. If thiouracil treatment continues postnatally, cretinism persists and the thyroids may be three times as large as normal (42). Owing to the retarded overall growth of such infant rats, coupled with the marked increase in absolute size of their thyroids, the relative thyroid weights of such animals may reach nine times normal. It has been shown that if half of such an extra-large gland is removed there is no compensatory hypertrophy of the remaining portion (42). This again underscores the importance of TSH in promoting both normal and compensatory growth of the thyroid. One can only conclude that under conditions of maximal stimulation there remains no capacity for further compensatory growth.

Effects opposite from those produced by goitrogens can be demonstrated by the administration of exogenous thyroxine. Prolonged treatment causes a decrease in the weight of the thyroid as it undergoes involution. Following partial thyroidectomy, no compensatory hypertrophy occurs in the presence of replacement therapy (5). Thyroxine abolishes the hyperplasia which is otherwise caused by goitrogens and prevents the production of goiters in animals treated with antithyroid drugs (31).

PARATHYROIDS AND ULTIMOBRANCHIAL BODIES

Animals that sequester large amounts of calcium in their skeletons, such as crustaceans and vertebrates, require mechanisms for extracting it when necessary. When a crustacean molts it first withdraws calcium from its exoskeleton, stores it as gastroliths in the stomach and deposits it in the new cuticle after molting. Although crustaceans lack parathyroids, they presumably have analogous glands responsible for these remarkable shifts in calcium reserves.

Evolution of the vertebrate endoskeleton necessitated the development of glands to regulate the deposition and dissolution of calcium. Calcium resorption is promoted by parathyroid hormone, its deposition by calcitonin, a hormone produced by the ultimobranchial bodies of lower vertebrates and their homologous C cells in higher forms. Parathyroid glands are absent in fishes and neotenous amphibians, and first appear in larval amphibians before metamorphosis when the skeleton is still cartilaginous (106). Their extirpation from tadpoles or urodeles does not upset the balance of calcium in the body enough to cause the convulsions and subsequent lethal effects seen in frogs and other terrestrial forms *(80, 106)*. It is difficult to ignore the correlation between the vital necessity of the parathyroid glands and the evolution of bony skeletons rigid enough to support animals on land.

In frogs, calcium balance is a seasonal phenomenon. In the winter, both the ultimobranchial bodies and the parathyroid glands are atrophic (87, 105). In the spring they regenerate. If winter frogs are immersed in a solution of $CaCl_2$ and injected with vitamin D_2, the rise in blood Ca^{2+} levels is accompanied by cellular proliferation and hypertrophy of the normally atrophic ultimobranchial bodies (Fig. 149) (87). This response to hypercalcemia is accompanied by a severalfold enlargement of the ultimobranchial bodies, increased vascularization, and depletion of their secretory granules. This may be taken to indicate that calcitonin is secreted by the ultimobranchial bodies in response to heightened levels of blood Ca^{2+}. Ultimobranchialectomy in frogs causes hypercalcemia. It also depletes the paravertebral lime sacs of $CaCl_2$ and doubles the excretion of calcium in the urine, effects that are reversed by the transplantation of ultimobranchial bodies (88). The decrease in bone density due to osteoclastic activity in frogs deprived of their ultimobranchial bodies suggests that the role of the parathyroid hormone in shifting calcium from the skeleton to the blood is normally counteracted by the action of calcitonin secreted by the ultimobranchial bodies.

Not surprisingly, the parathyroid glands of birds and mammals are also sensitive to calcium levels. Diets deficient in calcium typically promote enlargement of the parathyroid glands due primarily to cell hyperplasia (95). This effect is reversible when dietary calcium is restored. Artificial elevation of blood phosphate levels causes parathyroid hyperplasia also (108). Physiological conditions that cause the calcium level of the blood to fall or phosphate levels to rise tend to promote parathyroid hyperplasia. This is seen in the laying hen when the need to deposit calcium in the eggshells causes the mineral in the skeleton to be drawn upon under the influences of parthyroid hormone (104). In pregnant mammals the maternal parathyroid enlarges (97). This is associated with decreases in maternal blood Ca^{2+} levels and a relative hypercalcemia in the fetal blood. If pregnant rats are fed a diet rich in calcium and poor in phosphate, the growth of the maternal parathyroid is prevented. Conversely, it is enhanced by a low calcium, high phosphate diet (97). Although hyperparathyroidism in pregnancy is associated with fetal hypoparathyroidism, presumably owing to heightened levels of calcium in the fetal blood, maternal hypoparathyroidism has more serious consequences (1). The resulting hypocalcemia in the blood of both the mother and fetus tends to cause the latter's parathyroid glands to become hyperactive, a condition that promotes demineralization of the prenatal skeleton. Thus, although the parathyroid hormone is too large a molecule to

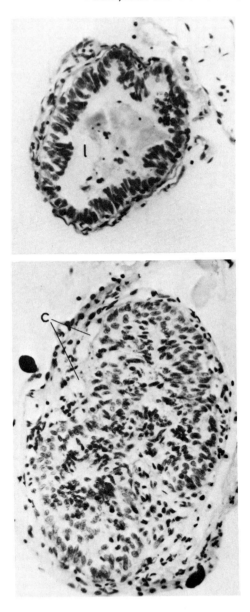

Fig. 149.   Ultimobranchial bodies from a control frog (above) and one treated with vitamin D_2 and maintained in 0.8% $CaCl_2$ for 2 weeks (below). Note occlusion of the central cavity and enlargement of the capillaries (c) in the enlarged gland (87).

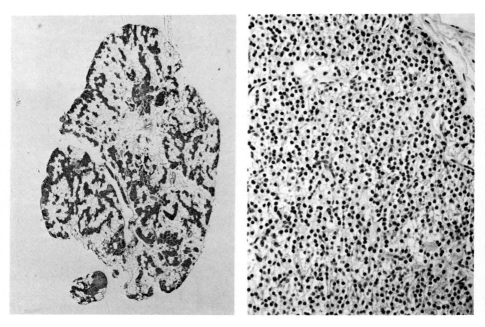

Fig. 150.   Comparison between normal (left) and hyperplastic (right) human parathyroid glands. From Paloyan, E., Lawrence, A.M., and Straus, F.H., "Hyperparathyroidism," pp. 65 and 94 (1973). By permission of Grune & Stratton, Inc., New York.

cross the placenta, maternal–fetal relationships are profoundly affected by the facility with which Ca^{2+} is exchanged between the two bloodstreams and the consequent effects of blood calcium fluctuations on the activity of the parathyroid glands (Fig. 151). This is illustrated by the effects of removing the parathyroids from pregnant rats. Although maternal hypocalcemia ensues, the blood calcium levels of her fetuses remain approximately normal, an effect presumably attributable to placental transfer of Ca^{2+} as well as hyperparathyroidism in the fetuses (85). Despite the vicarious compensation of the fetal parathyroids for the maternal deficiency, the drain on the maternal calcium reserves can be retarded only at the expense of the fetal skeleton, the decalcification of which is hardly enough to prevent maternal hypocalcemia.

It is the function of the parathyroid hormone to maintain normal blood calcium levels. This it can do in two ways. One is to promote the resorption of calcium from mineralized bone. The other is to prevent its excretion via the kidneys. Accordingly, the growth and activity of the parathyroid glands are significantly affected by the condition of the bones and kidneys. For example, such bone diseases as osteomalacia are characteristically associated

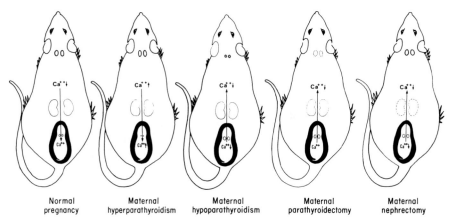

| Normal pregnancy | Maternal hyperparathyroidism | Maternal hypoparathyroidism | Maternal parathyroidectomy | Maternal nephrectomy |

Fig. 151. Maternal–fetal relationships in the regulation of parathyroid gland size and calcium balance in rats. Pregnancy normally causes the mother's parathyroids to enlarge due to the hypocalcemia ($Ca^{2+}\downarrow$) caused by the fetal need for calcium. Maternal hyperparathyroidism promotes hypercalcemia ($Ca^{2+}\uparrow$) which causes disuse atrophy of fetal parathyroids due to elevated fetal blood calcium levels. Maternal hypoparathyroidism or bilateral parathyroidectomy causes hypocalcemia, depleting the fetus of calcium and promoting fetal parathyroid hypertrophy. Bilateral nephrectomy of the mother prevents phosphate excretion thereby lowering the blood calcium levels and enlarging the parathyroids in both mother and fetus. Arrows indicate direction of calcium shift between mother and fetus.

with enlarged parathyroids (3). Chronic nephritis is likewise accompanied by parathyroid hypertrophy (81). Similar effects are produced by removing most or all of the renal tissue, or by ligating the ureters (33, 72). These experimental interventions cause marked cellular proliferation in the parathyroids in the next few days. Nephrectomy of pregnant rats elicits parathyroid growth in both mother and fetuses (32). One might expect that in the absence of the kidneys, hypercalcemia resulting from the failure to eliminate Ca^{2+} would cause hypoparathyroidism. However, there is a concomitant rise in blood phosphate that, combined with elevated blood Ca^{2+}, leads to the deposition of calcium in bone. It is the resultant decline in the blood Ca^{2+} that is responsible for enlargement of the parathyroids. This interpretation is strengthened by the experimental demonstration that when exogenous Ca^{2+} is administered to nephrectomized rats, parathyroid hyperplasia is reduced (100). On the other hand, phosphate administration enhances parathyroid enlargement in nephrectomized rats (108).

As in so many other organs of the body, the remaining parathyroid glands are capable of compensatory hypertrophy following contralateral parathyroidectomy (61). Presumably it is the decline in blood calcium following this operation that triggers the growth response. One might predict that this compensatory hypertrophy would be prevented by administration of exogenous calcium or parathyroid hormone. Indeed, if this hormone is given to intact animals the parathyroids undergo a "functional involutional atrophy due to substitution therapy" (46). This evidence suggests that the size of the parathyroid glands is adjusted to the demands for bone resorption to maintain normal blood calcium levels. It is significant, however, that the parathyroids do not disappear altogether under replacement therapy. Indeed, when as many as 20–80 extra parathyroid glands are grafted to a single rat there results a persistent hypercalcemia relative to the number of transplanted glands (30). Such an embarrassment of endocrine riches is more than the system can cope with. Even when there is no need for their superfluous services, the parathyroid cells are unable to turn off their secretory activities completely. Thus, the indefinite survival of supernumerary tissue is correlated with the persistence of functional activity at minimal levels.

ISLETS OF LANGERHANS

Interposed between the gut and the liver, this endocrine archipelago is in a perfect position to do its job. The insulin and glucagon secreted by the islets are carried with the glucose absorbed from the gut directly to the liver where the balance between the blood sugar level and the storage of glycogen is adjusted to the needs of the body. Insulin converts glucose into glycogen, not only in the liver but in muscle and other tissues. The hyperglycemia that results from the absence of insulin is diagnostic of diabetes. Glucagon deficiency is less serious because other hormones, notably growth hormone and glucocorticoids, also promote glucogenesis and can therefore compensate for the absence of glucagon.

When Langerhans discovered his islets in the mammalian pancreas in 1869, he later found that in cyclostomes, which lack a pancreas, homologous cells were present in the walls of the intestines. In some teleosts, such as the sculpin, goose fish, and catfish, these cells are consolidated into a "principal islet" unassociated with the exocrine pancreas. This anatomical anomaly provides a convenient source of islet hormones uncontaminated, and undigested, by pancreatic enzymes.

Islets are said to differentiate by the seventeenth day in the rat, when granulation of the β cells, which secrete insulin, is initiated (19). They are derived from the duct epithelial cells of the exocrine pancreas. Pancreatic primordia from 16-day rat fetuses can be maintained in culture where they differentiate new islets during the first week (107). In postnatal growth of the pancreas the original number of islets is augmented by the production of new ones from the duct epithelium (7). There is reason to believe that the potential for islet histogenesis persists throughout life (67). Indeed, following partial pancreatectomy of adult rats, acinar cells of the exocrine pancreas can transform into β cells of adjacent islets (67). Evidence for such metaplasia is the existence of cells possessing both zymogen granules and β granulation characteristic of insulin-secreting cells. Fully differentiated β cells can also divide (Fig. 152) (63).

Although total pancreatectomy results in death from diabetes, up to 95% of the pancreas can be removed without fatal consequences owing to the compensatory secretion of insulin by the few remaining islets. During the first week, however, the overworked β cells may show signs of degeneration, and their numbers may be seriously depleted during succeeding weeks. If

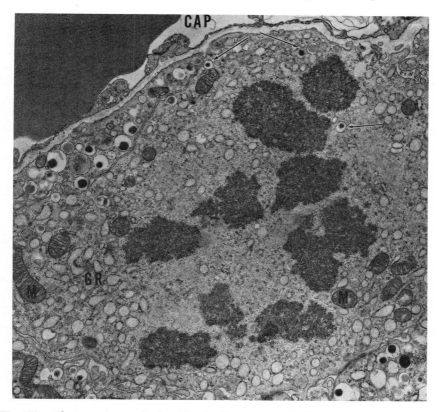

Fig. 152.   Electron micrograph of an islet β cell in mitosis with chromosomes occupying the central portion of the cell. Also shown is the granular endoplasmic reticulum (GR), mitochondria (M), secretory granules (arrows), and a capillary (CAP) (63). From Like, A. A., and Chick, W. L., *Science* **163**, 941–943, 28 February 1969. Copyright 1969 by the American Association for the Advancement of Science.

the animal survives this crisis, new islets may be regenerated in the growing pancreas. Despite some evidence for acino-insular transformation, it is the duct epithelium from which new islets are differentiated (67). Preexisting islets may also enlarge. One can only guess at the stimulus for this regenerative response, but in the absence of evidence for the existence of tropic hormones it is tempting to conclude that islet growth is directly responsive to blood sugar levels.

Growth hormone and ACTH are known to promote mitotic activity in islet cells (12), but the more direct stimulus is in response to the hyperglycemia which these hormones promote. Continuous injections of dextrose into guinea pigs, for example, stimulates secretion by the β cells, depleting their granulation (110). This is followed by β cell hyperplasia coupled with their *de novo* differentiation from duct cells. If pancreatic primordia from fetal rats are cultured in the presence of glucose their β cells become degranulated, secrete insulin, and proliferate (37, 54). Glucose administration to pregnant rats results in hyperglycemia in both the maternal and fetal circulations (93), and maternal diabetes is associated with islet hypertrophy and degranulation of fetal β cells (25). It is uncertain whether the degranulation of β cells under hyperglycemic conditions is due to their depletion when insulin is secreted or inhibition of insulin synthesis. The former explanation would seem to be the more plausible, especially in view of studies on the behavior of isolated islets *in vitro*. If the rat pancreas is incubated in the presence of collagenase the islets can be separated from the exocrine components by differential centrifugation. Maintenance of such islets in culture has yielded evidence that they retain the capacity to secrete insulin (56).

Although heightened levels of blood sugar may stimulate insulin secretion, the possibility remains that the level of insulin itself in the plasma might also constitute a feedback control. This is difficult to test *in vivo* because insulin is inevitably associated with hypoglycemia. Nevertheless, chronic administration of insulin causes involution of the islets, especially of the β cells (59). In the chick embryo it retards the differentiation of β cells and delays the onset of insulin secretion (52). Concomitantly there is an increase in the secretion of glucagon by the α cells. In infant rats, daily injections of insulin decrease proliferation in the islets. When antibodies to insulin are administered, the opposite effects are obtained. The β cells undergo marked degranulation followed by DNA replication and a rise in insulin synthesis (64).

Glucagon affects α cells the way insulin effects β cells. If fetal rat pancreas is grown in culture in the presence of glucagon, the islets that differentiate are composed only of β cells (91). This would suggest that the differentiation of cell types in the islets, if not their functional activity, are as sensitive to hormone levels as they are to the concentration of carbohydrates.

Cases of diabetes are often characterized by the presence of sclerotic islets, the numbers of which are reduced. In acute cases, undamaged islets may be enlarged and their β cells tend to be hyperplastic, evidently in compensation for the deficiency. Diabetes is reversible in rats to which minced isologous fetal pancreas is transplanted intraperitoneally where surviving β cells adhere to the visceral organs (60).

Experimentally, it is easy to produce diabetes by the injection of alloxan. This compound selectively destroys the β cells, resulting in the decreased production of insulin. As a result, α cells predominate in the islet remnants (82). If treatment with alloxan is limited to a single injection, a regenerative response may occur after the effects wear off. New β cells differentiate and proliferate, and the number of islets may eventually approach the original population (8).

If alloxan is administered to pregnant animals, the effects on the offspring are similar to those encountered in babies born to diabetic mothers. The β cells show signs of hydropic degeneration, an indication of physiological exhaustion, and the islets themselves are enlarged (53). Maternal diabetes is accompanied by high mortality, resulting in fetal death and abortion. Infants that survive until birth tend to be considerably larger than normal presumably owing to the exaggerated influx of blood sugars from the maternal circulation. Such infants have increased glucose tolerance, but their islets, having developed in chronic hyperglycemia, are 3–4 times larger than normal (17, 25). The hyperplastic β cells become degranulated in their attempts to secrete enough insulin to reduce the blood sugar level. Hence, such babies suffer from hyperinsulinism and are in jeopardy of dying from hypoglycemia soon after birth when abruptly cut off from the maternal source of blood sugar to which they have adapted. Although it is doubtful that the insulin molecule itself is small enough to be transported across the placental barrier, there is nothing to impede the exchange of monosaccharides. Thus, the infant of a diabetic mother, unable to make up for the maternal deficiency in insulin, can only attempt to reduce the hyperglycemia on both sides of the placenta by increasing its insulin production, an adaptation that can result in fatal postpartum withdrawal symptoms.

JUXTAGLOMERULAR APPARATUS

Homeostasis of the blood pressure is so important that it is affected by a variety of organs. The heart, carotid bodies, and adrenal cortex are important centers of control. The kidney, however, plays the most essential role in maintaining the pressure of the blood within normal ranges. It does so by virtue of the renin secreted by its juxtaglomerular apparatus (JGA).

These cells are smooth muscle fibers modified to perform an endocrine function (Fig. 153). Situated in the tunica media of the arterioles carrying blood to the glomeruli, they synthesize renin and secrete it in response to reductions in blood pressure (76, 102). Hence, these cells are also baroreceptors sensitive to the stretching effects of hemodynamic pressures. It is in response to such stimuli that the growth and proliferation of the JG cells are controlled. The amounts of renin produced are also regulated by local blood pressure.

When Goldblatt discovered that hypertension could be produced by partially clamping one renal artery, he opened the way for the experimental exploration of a condition that had until then been limited to clinical investigations. The development of a histochemical technique to identify the specific granulation of the JG cells made it possible to quantitate a juxtaglomerular index (JGI) for correlation with various experimentally imposed physiological conditions.

It is well known that hypertension is often accompanied by hyperplastic JG cells in a kidney with arterial stenosis (15), and that the condition can be corrected by removing the occlusion in the renal artery if this is surgically feasible, or by unilateral nephrectomy if it is not. Whether the blood pres-

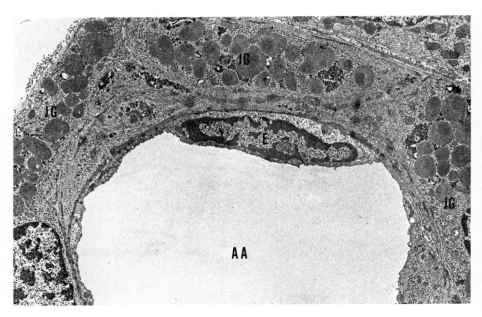

Fig. 153. Electron micrograph of an afferent arteriole (AA) in the rat surrounded by juxtaglomerular cells (JG) containing numerous large secretory granules. E, endothelial cell (83).

sure in one kidney is reduced experimentally or clinically, the hypersecretion of renin by the JG cells in the affected kidney effectively raises the blood pressure in the systemic circulation. Consequently, the opposite intact kidney, subjected to hypertension, reduces its output of renin as reflected in the decline in its JGI (68). Juxtaglomerular granulation can also be reduced in isolated rat kidneys which are perfused at high pressures (103). Clearly, whatever affects the renal circulation exerts an influence on the JGA. The effects of renal arterial constriction are mimicked by wrapping the kidney in cellophane or silk (21). Such encapsulated kidneys lead eventually to hypertension owing to the increased JGI on the affected side. The contralateral degranulation of the JGA is reversed by removing the ischemic kidney. Hydroephrosis, produced by occlusion of one ureter, also results in hypertension. The pressure of accumulating urine in the hydronephrotic kidney effectively interferes with local blood flow thus increasing the JGI on that side. The resulting hypertension then brings about a decrease in JGI in the opposite intact kidney (32). Although removal of the hydronephrotic kidney causes the JGI in the opposite one to revert to normal, ablation of a single healthy kidney has not been shown to result in compensatory responses in the contralateral JGA despite hypertrophy of that kidney. Since unilateral nephrectomy does not cause hypertension, there seems to be little reason to suspect that compensatory renal hypertrophy is attributable to high blood pressure or to the secretion of renin (34).

Renin is an enzyme. When released into the circulation it acts on angiotensinogen, a polypeptide produced by the liver (Fig. 154). Hydrolysis of this molecule results in the production of a decapeptide, angiotensin I, which is subsequently converted to the octapeptide, angiotensin II. This molecule is a vasopressor capable of increasing blood pressure by vascular

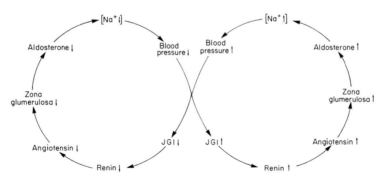

Fig. 154. Diagram outlining the relationships between the juxtaglomerular apparatus, adrenal cortex, electrolyte balance, and blood pressure regulation. Short arrows indicate increases (↑) or decreases (↓) in the factors involved.

constriction. It also stimulates the secretion of mineralocorticoids, such as aldosterone and deoxycorticosterone, adrenal cortical hormones that promote sodium retention and potassium excretion. Like angiotensin II and renin, these hormones exert a hypertensive effect. Their influence on the JGA is to cause degranulation, presumably by raising the blood pressure (109). If salt water is given to drink, there is also a reduction in the JGI (36). This effect is additive to similar responses produced by administration of deoxycorticosterone (84). On the other hand, if this hormone is given to rats on a diet deficient in NaCl, JG granulation is not affected. Bilateral adrenalectomy increases the granulation of the JGA, as does low Na^+ and/or high K^+ (Fig. 155) (66, 83).

The reciprocal relation between Na^+ and K^+ in the body fluids plays a key role in blood pressure regulation. The degree of JG granulation tends to be inversely related to the levels of Na^+ in the plasma. This is clearly demonstrated by the distribution of the JGA in fishes (90). Although absent in cyclostomes and elasmobranchs living in either fresh or salt water (16, 78), the JGA is regularly present in teleosts (10). Here the JG cells are more prominent in marine than freshwater species, even occurring in the aglomerular kidneys of the goose fish where the cells are located in the walls of arterioles just beneath the renal capsule (10, 79). Thus, JGA phylogeny tends

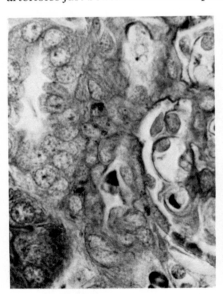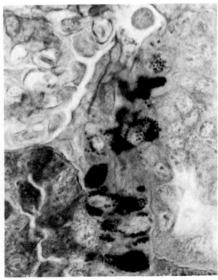

Fig. 155. Juxtaglomerular apparatus in the pig kidney. An animal on a high salt diet has inconspicuous JG cells with few granules (left). One on a sodium deficient diet has increased granulation of the JGA (35). By permission of the American Heart Association, Inc.

to be associated with the need to conserve water by reducing glomerular filtration. Marine teleosts, living as they do in an environment hypertonic to their body fluids, tend to have smaller and fewer glomeruli than do freshwater species which must get rid of the excess water they tend to absorb from their hypotonic environment. Since marine elasmobranchs mimic the osmotic relationships of freshwater teleosts by retaining enough urea in their blood to keep it hypertonic to their salt water environment, they possess relatively large glomeruli adapted for the excretion of excess water. Like marine fishes, terrestrial vertebrates are faced with the problem of water conservation. Their ability to secrete renin from their JGA results in the maintenance of elevated blood pressures attributable in part to increased plasma volumes resulting from water reabsorption in the kidneys. These relationships may account for the anatomical location of the JGA in the kidneys, a location where changes in blood pressure are readily monitored and electrolyte and water balance effectively controlled.

ADRENAL CORTEX

From the beginnings of vertebrate evolution, what we call the adrenal cortex in mammals has been spatially, if not physiologically, associated with the kidneys. Because of its diffuse distribution in fishes and amphibians, it is more properly referred to as interrenal tissue in these animals. The change in prefix of the adrenal or suprarenal of higher vertebrates reflects the consolidation of cortical tissue into a more discrete organ just anterior to the kidneys. This lasting phylogenetic juxtaposition may relate to the functions shared by these two organs in the regulation of electrolyte balance and blood pressure. It is also no coincidence that, as a steroid-secreting organ, the adrenal cortex is situated not in the hepatic portal drainage system but in the systemic circulation where its hormones will not so readily be degraded by the liver.

The association between the adrenal cortex and medulla (a sympathetic ganglion) dates back to the amphibians, but reaches its highest specialization in birds and mammals. That this anatomical intimacy reflects physiological cooperation is difficult to deny, for both tissues play important roles in an animal's adaptation to stress. Wild rats have larger adrenals than tame ones (73). In wild rabbits, the adrenals are larger in the winter than in the summer (75). When rats are forced to run every day their adrenals also enlarge (98). How the cortex and medulla are related in their reactions to stress is not well understood.

There is yet another relationship which may be seen in the interactions between the adrenal cortex and the gonads. Sharing similar embryonic origins, these organs are the only endocrine glands to secrete steroid hormones. Indeed, the adrenal cortex sometimes gets its signals crossed and secretes sex hormones, a phenomenon which accounts for some interesting mix-ups in gender when these hormones are produced in the wrong sex.

At the histological level the mammalian adrenal cortex is subdivided into various zones not seen in other vertebrates. The outermost one, the zona glomerulosa, secretes mineralocorticoids such as aldosterone, steroids which promote sodium retention and potassium excretion. Inside this is the zona fasciculata, made up of radially oriented cords of cells which constitute the source of glucocorticoids. Stimulated by ACTH, these steroids not only regulate carbohydrate balance in the body but are also secreted in the alarm reaction to stress, promoting hyperglycemia and antagonizing protein synthesis. The zona reticularis, sandwiched between the fasciculata and medulla is the least understood of the three zones, but may share some of the functions of the zona fasciculata. Physiologically, therefore, the adrenal cortex may be thought of as an outer zone (zona glomerulosa) responsible for electrolyte balance, and an inner zone (zona fasciculata and reticularis) involved with carbohydrate regulation.

The adrenal cortex has its embryonic origins in the same ridge of mesoderm from which the kidneys and gonads develop. In the mouse embryo, the cortex first appears on the eleventh day, to be invaded by medullary cells during the next day or two, leading to the complete establishment of cortex and medulla by the fourteenth day of gestation (*101*). Subsequent growth gives rise to disproportionately large adrenal glands in the fetus, at least by adult standards (*58*). Both mineralocorticoids and glucocorticoids are known to be secreted prenatally, and the pituitary–adrenal axis is established well before birth (*48*). One suspects that the fetal cortex must play a role quantitatively, if not qualitatively, greater than in the postnatal animal, judging from its inordinately large size.

In view of the discrepancy in pre- and postnatal dimensions, the relative weight of the adrenal declines during the early postnatal period, a reduction which in some cases may even involve atrophy on an absolute scale (*58, 71*). The transient involution during the first week or two after birth may be correlated with the inability of newborn rats and mice to react to stress. This is particularly evident in the case of thermal regulation, for such animals are in fact cold-blooded in the sense that they are incapable of maintaining a typical mammalian body temperature except by virtue of benefiting from their mother's warmth. Removed from the nest, rodent pups may be chilled in the refrigerator, dropping their body temperatures to near-freezing levels. This short-term hibernation is a useful anesthetic substitute for neonatal operations.

The adrenal cortex in the adult undergoes a slow turnover of its cells (70). Other glands are expanding organs, the cells of which tend to proliferate only to keep pace with the enlargement of the body or increases in its functional demands. As such, they possess no growth zones. Although most if not all cortical cells may be capable of division, virtually all of the mitotic activity is confined to a region comprising the outermost portions of the zona fasciculata and the innermost parts of the zona glomerulosa. Autoradiographic studies have shown that cells that synthesize DNA in this zone gradually migrate away, some into the glomerulosa, but most centripetally through the fasciculata, eventually to become incorporated into the zona reticularis (Fig. 156) (6). Though such cells may persist for weeks or months in the adrenal cortex, their eventual disappearance suggests a slow turnover by the production of new cells in the outer fasciculata, their escalation centrally, and their demise in the inner reticularis. The adrenal cortex was therefore originally an expanding organ which, for reasons yet to be explained, has evolved into a renewing tissue.

The growth of the zona glomerulosa is responsive to physiological stimuli different from those affecting the inner zones. Since the function of its hormones is to regulate Na^+ and K^+ balance, its physiological activity and structural integrity are profoundly influenced by electrolyte imbalance. Accordingly, sodium deficiency results in a marked increase in the width of the zona glomerulosa (Fig. 157) (35) in keeping with its enhanced secretion of aldosterone, a steroid that promotes Na^+ retention by the kidneys (75, 86). Interestingly, if the kidneys are removed this reaction fails to occur (24).

Few aspects of adrenal cortical development and function are more intriguing, or less well understood, than the X-zone of the mouse (Fig. 158). Also referred to as the "interlocking zone" or "transitory cortex," the mouse adrenal X-zone is apparently not to be confused with the fetal cortex of humans and other primates (58, 74), nor is it known to have a counterpart in the adrenals of other rodents.

The X-zone makes its appearance during the second week after birth as a layer of cells interposed between the zona reticularis and the medulla (44, 74). This zone may increase considerably in width, and is readily distinguished histologically from the rest of the cortex. In the male mouse, the X-zone reaches its maximum dimensions at around three weeks of age and begins to degenerate about a month later. It can no longer be found at two months of age. In the female, it develops into a more conspicuous layer of tissue than in the male, reaching its maximum proportions at five weeks of age.

Castration of the male before puberty results in the persistence of the X-zone for up to four months (27, 44). Castration of an adult mouse, after the primary X-zone has disappeared, induces the production of a secondary

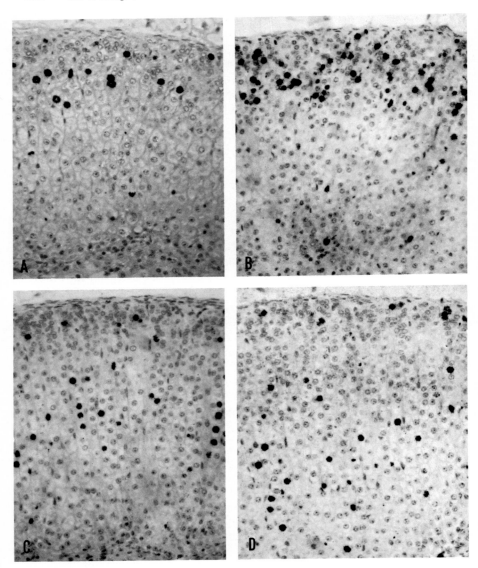

Fig. 156. Escalation of adrenal cortical cells as seen in autoradiographs of mouse adrenals (6). Animals were stressed with CCl₄ injected subcutaneously 3 days prior to administration of ³H-thymidine. A, on day 0 the label is confined to cells in the outer zona fasciculata. After 1–3 weeks (B–D) most of the labeled nuclei are seen to migrate centripetally through the cortex.

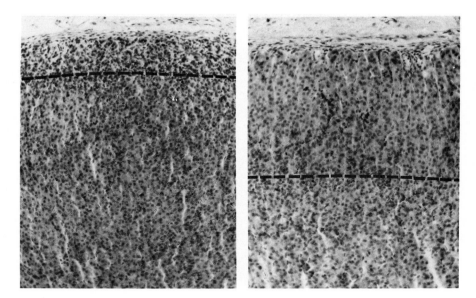

Fig. 157, Adrenal cortex of young pigs illustrating the effects of salt on the width of the zona glomerulosa (35). Left, an animal with a narrow zona glomerulosa after 2 weeks on a high salt diet. Right, hypertrophy of the zona glomerulosa following 2 weeks of sodium deficiency. Broken lines indicate the approximate boundaries between the zona glomerulosa and zona fasciculata. By permission of the American Heart Association, Inc.

X-zone from the cells of the zona reticularis (38). The administration of testosterone suppresses X-zone development in both sexes, and brings about its premature degeneration (Fig. 158) (27, 40).

The fate of the female X-zone depends on whether or not she becomes pregnant (Fig. 159). In virgin females the X-zone may persist for up to four months before undergoing spontaneous degeneration (44). Ovariectomy, like castration in the male, precludes degeneration of the X-zone. When a female becomes pregnant, her X-zone starts to degenerate by the seventh day of gestation, disappearing completely at the end of two weeks (13, 40, 44). Pseudopregnancy has similar effects (40). Indeed, the transplantation of blastocysts to the anterior chamber of the eye or beneath the kidney capsule in otherwise intact nonpregnant mice leads to X-zone degeneration, an effect not seen in ovariectomized females (13, 41, 55). This evidence, coupled with the findings that progesterone administration induces regression of the X-zone, strongly suggests that the ovary or placenta promotes its degeneration by virtue of the secretion of progesterone (13, 40). None of this explains the actual role played by the X-zone, but clearly implies a function during adolescence which relates somehow to sexual maturation.

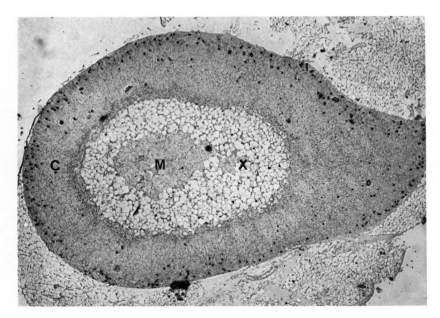

Fig. 158.   X-Zone of a mouse adrenal gland (6). C, cortex; M, medulla; X, X-zone.

Gonadal influences on the adrenal cortex affect tissues other than the X-zone. In the male, fetal castration has no adverse effects on prenatal adrenal growth, but neonatal castration has been found to result in the development of adrenal cancer later in life when the mouse is about a year old (111). Castration after weaning is said to bring about increases in adrenal weight, although adult castration may interfere with compensatory hypertrophy following contralateral adrenalectomy (4, 26).

Ovariectomy at birth in certain strains of mice leads to adrenal hyperplasia after 3–4 months and eventually to adrenal cortical carcinoma at the age of a year or two (111). Ovariectomy after weaning causes a decrease in adrenal cortical weight in rats, promotes adrenal hyperplasia in the mouse and guinea pig, and leads to cell hypertrophy in the zona fasciculata of the gerbil (77, 96). As in the case of male castration, ovariectomy prevents compensatory adrenal hypertrophy after partial adrenalectomy. In the pregnant animal, the adrenals typically become larger than normal, despite the loss of the X-zone under these circumstances. Thus, while progesterone appears to be responsible for degeneration of the X-zone, it may have opposite effects on the rest of the cortex. Administered to the fetus, for example, it can prevent the transient postnatal decrease in adrenal size, something not seen when progesterone is administered to newborn mice (22). Although progesterone

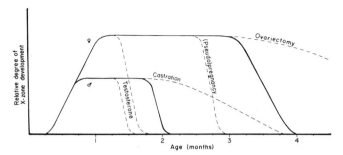

Fig. 159.   Factors affecting survival of the adrenal X-zone in mice. The X-zone normally re-
gresses earlier in males than females, but its disappearance is delayed in both sexes by gonadec-
tomy. Pregnancy accelerates its loss in females, while exogenous testosterone stimulates degeneration of the X-zone in both sexes.

has little effect on adrenal size in pregnant animals, it brings about atrophy of
the zona fasciculata in the male.

No other endocrine gland exceeds the pituitary for its important effects on
the adrenal cortex. Hypophysectomy causes atrophy of the cortex, including
marked decreases in mitotic activity after 2–3 days (*11*). Compensatory
hypertrophy of the adrenal does not occur after hypophysectomy, nor does
regeneration take place from the capsules of enucleated adrenals. The effects
of hypophysectomy can be reversed by ACTH, and to a lesser extent by
growth hormone (*11*). However, it is not the decline in bodily growth attending hypophysectomy that is responsible for the atrophy of the adrenal cortex.
Even in hypophysectomized animals made to grow at normal rates by force-
feeding, atrophy of the cortex is not prevented (*62*). The early development
of the pituitary–adrenal axis in prenatal life is substantiated by the effects of
fetal decapitation. This is not a fatal operation when performed prenatally,
nor does it necessarily result in serious retardation of body growth in the
fetuses of rats and rabbits (see Chap. 24). Nevertheless, relative cortical
atrophy and growth retardation are seen in such animals, as well as in
hypophysectomized chick embryos (*29, 51*). That decapitation is tantamount
to hypophysectomy in fetal mammals is suggested by the fact that replacement therapy, by administration of ACTH, prevents the deleterious effects
of decapitation on fetal adrenal growth (*20*).

The importance of the pituitary in stimulating secretion of glucocorticoids
by the adrenal cortex and in maintaining the structural integrity of the gland
is nowhere more dramatically illustrated than in compensatory hypertrophy
and regeneration of the adrenal cortex, phenomena both depending on
ACTH. When one adrenal gland is removed, the remaining one promptly
enlarges, reaching a size about three-quarters of the combined original
weights of the two organs (*4, 23*). This probably reflects a nearly 100%

enlargement of the cortex when one takes into account that the medulla, while contributing to the weight of the gland, does not undergo compensatory hypertrophy. Even in the unilaterally adrenalectomized rat fetus, the opposite gland will hypertrophy before birth, again emphasizing the prenatal operation of the pituitary–adrenal axis. If one adrenal is removed from a newborn rat very little hypertrophy occurs in the opposite gland, at least during the first postoperative week or two. It will be recalled that this is a period when little or no growth is occurring in the normal adrenal as its relative size adjusts from prenatal to postnatal dimensions (23). It is possible, therefore, that the secretion of ACTH by the pituitary may be held in abeyance during this stage of infancy.

Bilateral adrenalectomy is fatal. Survival of adrenalectomized animals can be insured by replacement therapy with deoxycorticosterone or aldosterone, or by simply allowing access to salt water. Hence, it is the sodium depletion in the absence of the zona glomerulosa that makes the adrenal cortex such a vitally essential organ. In parabiotic rats, bilateral adrenalectomy of one animal is not fatal because hormones secreted by the intact partner cross into the operated one. In this situation, both adrenals of the intact partner undergo hypertrophy. Viviparity is a condition somewhat similar to parabiosis in that at least smaller molecules can cross the placenta between the mother and her fetuses. When a pregnant rat is bilaterally adrenalectomized, instead of dying in a week or so from sodium depletion she may survive twice as long until parturition. In the meantime, the adrenals of her fetuses become enlarged, secreting steroid hormones which cross back into the maternal circulation compensating for her deficiency (39, 69). Presumably the vicarious hypertrophy of the fetal adrenals occurs in response to ACTH from the fetal pituitary, since this glycoprotein is too large to cross into the fetus from the maternal circulation. Only the inner zones of the fetal adrenal enlarge, not the zona glomerulosa which is unresponsive to pituitary tropic hormones. The latter zone may be stimulated to secrete its steroids by the loss of fetal sodium to the mother's circulation. Alternatively, growth and secretion of the fetal zona glomerulosa could be attributed to angiotensin II, an octapeptide produced in response to sodium depletion and small enough to be transferred across the placenta.

In adult rats, it is possible to slit open an adrenal gland, squeeze out the contents, and transplant the enucleated capsule back into the animal. If this is done on one side without removing the opposite gland, there is little regeneration. However, if the other gland is extirpated or similarly enucleated, then regeneration of a new cortex occurs during the following month or so (18, 92). Such regenerated glands may approach the size of the original organs, and the typical zonation of the cortex is restored. The histogenesis of these regenerated cortices is the residual cortical cells adhering to the inner surface of the capsule, a source of tissue different from that responsible for the normal turnover in intact glands. The conclusion is inescapable that, despite the differing functions and differentiated states of the cells in the adrenal cortex, they nevertheless retain a pluripotency which can be expressed under unusual conditions.

The importance of circulatory arrangements in maintaining the adrenal cortex is well illustrated by the regenerative responses of enucleated glands. When left in the systemic circulation, the extent of regeneration is at least in part commensurate with the levels of ACTH secreted, which are in turn responsive to the amounts of hormone secreted by the target organ. On the other hand, if both adrenals are removed and one of them transplanted to the spleen, its secretory products drain into the hepatic portal circulation where with inconvenient efficiency the liver breaks down the steroids it receives from intrasplenic grafts. The incidence of survival of such rats is proportional to the amount of glandular material transplanted (9). Among the survivors, grafts may overshoot the normal mass of cortical tissue in their regeneration (4), a reaction triggered by the chronic and excessive stimulation of ACTH secreted by a pituitary that is itself chronically stimulated by hypothalamic releasing factors. In the absence of artificially reduced serum levels of glucocorticoids it is not known whether such hyperactive grafts of the adrenal cortex may eventually become cancerous as is the case in intrasplenic transplants of ovarian tissue (see Chap. 21). The prospect would seem worth pursuing.

REFERENCES

1. Aceto, T., Jr., Batt, R.E., Bruck, E., Schultz, R.B., and Perez, Y.R. Intrauterine hyperparathyroidism: A complication of untreated maternal hypoparathyroidism. *J. Clin. Endocrinol. Metab.* **26**, 487–492 (1966).
2. Astwood, E.B., Sullivan, J., Bissell, A., and Tyslowitz, R. Action of certain sulphonamides and thiourea upon the function of the thyroid gland of the rat. *Endocrinology* **32**, 210–225 (1943).
3. Bartos, H.R., and Henneman, P.H. Parathyroid hyperplasia in osteomalacia. *J. Clin. Endocrinol. Metab.* **25**, 1522–1523 (1965).
4. Bernstein, D.E., and Biskind, G.R. Autotransplantation of the adrenal of the rat to the portal circulation: Induced adrenal hypertrophy and its prevention by oophorectomy. *Endocrinology* **60**, 575–577 (1957).
5. Bershtein, L.M. Inhibition of compensatory hypertrophy of the thyroid gland by thyroxine in rats of different ages. *Biull. Eksp. Biol. Med.* **80**, 886–888 (1975). (Eng. transl.)
6. Brenner, R.M. Radioautographic studies with tritiated thymidine of cell migration in the mouse adrenal after a carbon tetrachloride stress. *Am. J. Anat.* **112**, 81–96 (1963).
7. Bunnag, S.C. Postnatal neogenesis of islets of Langerhans in the mouse. *Diabetes* **15**, 480–491 (1966).

8. Bunnag, S.C., Warner, N.E., and Bunnag, S. Effect of alloxan on the mouse pancreas during and after recovery from diabetes. *Diabetes* **16**, 83–89 (1967).

9. Butcher, E.O. Adrenal autotransplants with hepatic portal drainage in the rat. *Endocrinology* **43**, 30–35 (1948).

10. Capreol, S.V., and Sutherland, L.E. Comparative morphology of juxtaglomerular cells. I. Juxtaglomerular cells in fish. *Can. J. Zool.* **46**, 249–256 (1968).

11. Cater, D.B., and Stack-Dunne, M.P. The effects of growth hormone and corticotrophin upon the adrenal weight and adrenocortical mitotic activity in the hypophysectomized rat. *J. Endocrinol.* **12**, 174–184 (1955).

12. Cavallero, C., and Mosca, L. Mitotic activity in the pancreatic islets of the rat under pituitary growth hormone and adrenocorticotropic hormone treatment. *J. Pathol. Bacteriol.* **66**, 147–150 (1953).

13. Chester Jones, I. (1952) The disappearance of the X zone of the mouse adrenal cortex during first pregnancy. *Proc. Roy. Soc. London, Ser. B* **139**, 398–410.

14. Clark, O.H., Lambert, W.R., Cavalieri, R.R., Rapoport, B., Hammond, M.E., and Ingbar, S.H. Compensatory thyroid hypertrophy after hemithyroidectomy in rats. *Endocrinology* **99**, 988–995 (1976).

15. Crocker, D.W. Bilateral juxtaglomerular cell counts in renal hypertension. *Arch. Pathol.* **93**, 103–108 (1972).

16. Crockett, D.R., Gerst, J.W., and Blankenship, S. Absence of juxtaglomerular cells in the kidneys of elasmobranch fishes. *Comp. Biochem. Physiol.* **44**, 673–675 (1973).

17. D'Agostino, A.N., and Bahn, R.C. A histopathologic study of the pancreas of infants of diabetic mothers. *Diabetes* **12**, 327–331 (1963).

18. de Groot, J., and Fortier, C. Quantitative and histological aspects of adrenal cortical regeneration in the male albino rat. *Anat. Rec.* **133**, 565–570 (1959).

19. Dixit, P.K., Lowe, I.P., Heggestad, C.B., and Lazarow, A. Insulin content of microdissected fetal islets obtained from diabetic and normal rats. *Diabetes* **13**, 71–77 (1964).

20. Domm, L.V., and Leroy, P. A method for hypophysectomy of rat fetus by decapitation. *Anat. Rec.* **109**, 395–396 (1951).

21. Dunihue, F.W. Effect of cellophane perinephritis on the granular cells of the juxtaglomerular apparatus. *Arch. Pathol.* **32**, 211–216 (1941).

22. Eguchi, Y., and Ariyuki, F. Development of the fetal rat adrenal in prolonged pregnancy. *Endocrinol. Jpn.* **10**, 125–135 (1963).

23. Eguchi, Y., Eguchi, K. and Wells, L.J. Compensatory hypertrophy of right adrenal after left adrenalectomy: Observations in fetal, newborn and week-old rats. *Proc. Soc. Exp. Biol. Med.* **116**, 89–92 (1964).

24. Elema, J.D., Hardonk, M.J., Koudstaal, J., and Arends, A. Acute enzyme histochemical changes in the zona glomerulosa of the rat adrenal cortex. II. The effect of bilateral nephrectomy either alone or followed by peritoneal dialysis with 5% glucose. *Acta Endocrinol.* **59**, 519–528 (1968).

25. Farquhar, J.W. Maternal hyperglycaemia and foetal hyperinsulinism in diabetic pregnancy. *Postgrad. Med. J.* **38**, 612–628 (1962).

26. Fiske, V.M., and Lambert, H.H. Effect of light on the weight of the adrenal in the rat. *Endocrinology* **71**, 667–668 (1962).

27. Garweg, G., Kinsky, I., and Brinkmann, H. Markierung der juxtamedullären X-Zone in der Nebenniere der Maus mit L-Cystein-S^{35}. *Z. Anat. Entwicklungsgesch.* **134**, 186–199 (1971).

28. Gibadulin, R.A. Compensatory hypertrophy of the thyroid gland. *Biull. Eksp. Biol. Med.* **54**, 790–793 (1962). (Eng. transl.)

29. Girouard, R.J., and Hall, B.K. Pituitary-adrenal interaction and growth of the embryonic avian adrenal gland. *J. Exp. Zool.* **183**, 323–332 (1973).

30. Gittes, R.F., and Radde, I.C. Experimental model for hyperparathyroidism: Effect of excessive numbers of transplanted isologous parathyroid glands. *J. Urol.* **95**, 595–603 (1966).

31. Hamburgh, M., Sobel, E.H., Koblin, R., and Rinestone, A. Passage of thyroid hormone across the placenta in intact and hypophysectomized rats. *Anat. Rec.* **144**, 219–225 (1962).

32. Hansson, C.G. Nephrectomy of pregnant rats. Effects of parathyroids in mother and offspring. *Acta Endocrinol.* **54**, 166–172 (1967).

33. Hansson, C.G., Mathewson, S., and Norby, K. Parathyroid cell growth and proliferation in nephrectomised rats. *Pathol. Eur.* **6**, 313–321 (1971).

34. Hartroft, P.M. Studies on renal juxtaglomerular cells. III. The effects of experimental renal disease and hypertension in the rat. *J. Exp. Med.* **105**, 501–508 (1957).

35. Hartroft, P.M. Juxtaglomerular cells. *Circ. Res.* **12**, 525–534 (1963).

36. Hartroft, P.M., and Hartroft, W.S. Studies on renai juxtaglomerular cells. I. Variations produced by sodium chloride and desoxycorticosterone acetate. *J. Exp. Med.* **97**, 415–428 (1953).

37. Hegre, O.D., Wells, L.J., and Lazarow, A. Response of beta cells to different levels of glucose. Fetal pancreases grown in organ culture and subsequently transplanted to maternal hosts. *Diabetes* **19**, 906–915 (1970).

38. Hirokawa, N., and Ishikawa, H. Electron microscopic observations on the castration-induced X-zone in the adrenal cortex of male mice. *Cell Tiss. Res.* **162**, 119–130 (1975).

39. Holland, R.C. The effect of hypoxia on the fetal rat adrenal. *Anat. Rec.* **130**, 177–195 (1958).

40. Holmes, P.V., and Dickson, A.D. X-zone degeneration in the adrenal glands of adult and immature female mice. *J. Anat.* **108**, 159–168 (1971).

41. Holmes, P.V., and Dickson, A.D. The effect on the mouse adrenal X-zone of blastocysts transplanted under the kidney capsule. *J. Reprod. Fert.* **25**, 111–113 (1971).

42. Horn, E.H., and LoMonaco, M.B. Unilateral adrenalectomy in the cretin rat. *Proc. Soc. Exp. Biol. Med.* **98**, 817–820 (1958).

43. Houssay, A.B., and Pazo, J.H. Role of pituitary in the thyroid hypertrophy of pinealectomized rats. *Experientia* **24**, 813–814 (1968).

44. Howard-Miller, E. A transitory zone in the adrenal cortex which shows age and sex relationships. *Am. J. Anat.* **40**, 251–294 (1939).

45. Hunt, T.E. Mitotic activity in the thyroid glands of female rats. *Anat. Rec.* **90**, 133–138 (1944).

46. Jaffe, H.L., and Bodansky, A. Experimental fibrous osteodystrophy (ostitis fibrosa) in hyperparathyroid dogs. *J. Exp. Med.* **52**, 669–694 (1930).

47. Japundzic, M., and Japundzic, I. Observations of the anterior pituitary cytology in the phenobarbital treated rat. *Virchows Arch. B* **7**, 229–235 (1971).

48. Josimovich, J.B., Ladman, A.J., and Deane, H.W. A histophysiological study of the developing adrenal cortex of the rat during fetal and early postnatal stages. *Endocrinology* **54**, 627–639 (1954).

49. Jost, A. Sur le développement de la thyroide chez le foetus de Lapin décapité. *Arch. Anat. Microscop. Morphol. Exp.* **42**, 168–183 (1953).

50. Jost, A. Le problème des interrelations thyréo-hypophysaires chez le foetus et l'action du propylthiouracile sur la thyroide foetale du rat. *Rev. Suisse Zool.* **64**, 821–832 (1957).

51. Jost, A., and Cohen, A. Signification de l'"atrophie" des surrénales foetales du rat provoquée par l'hypophysectomie (décapitation). *Dev. Biol.* **14**, 154–168 (1966).

52. Kalliecharan, R., and Gibson, M.A. Histogenesis of the islets of Langerhans in insulin-treated chick embryos. *Can. J. Zool.* **50**, 265–277 (1972).

53. Kim, J.N. Effects of hyperglycemia on beta granulation in pancreatic islets of fetuses from diabetic rats. *Diabetes* **14**, 137–141 (1965).

54. King, D.L., and Chick, W.L. Pancreatic beta cell replication. Effects of hexose sugars. *Endocrinology* **99**, 1003–1009 (1976).

55. Kirby, D.R.S. The difference in response by the mouse adrenal X-zone to trophoblast derived from transplanted tubal eggs and uterine blastocysts. *J. Endocrinol.* **36**, 85–92 (1966).

56. Lacy, P.E., and Kostianovsky, M. Method for the isolation of intact islets of Langerhans from the rat pancreas. *Diabetes* **16**, 35–39 (1967).

57. Landauer, W. Loss of body heat and disease. *Am. J. Med. Sci.* **194**, 667–674 (1937).

58. Lanman, J.T. The adrenal fetal zone: Its occurrence in primates and a possible relationship to chorionic gonadotropin. *Endocrinology* **61**, 684–691 (1957).

59. Latta, J.S., and Harvey, H.T. Changes in the islets of Langerhans of the albino rat induced by insulin administration. *Anat. Rec.* **82**, 281–296 (1942).

60. Leonard, R.J., Lazarow, A., and Hegre, O.D. Pancreatic islet transplantation in the rat. *Diabetes* **22**, 413–428 (1973).

61. Lever, J.D. Cytological appearances in the normal and activated parathyroid of the rat. A combined study by electron and light microscopy with certain quantitative assessments. *J. Endocrinol.* **17**, 210–217 (1958).

62. Levine, L. Some effects of increased food consumption on the composition of carcass and liver of hypophysectomized rats. *Am. J. Physiol.* **141**, 143–150 (1944).

63. Like, A.A., and Chick, W.L. Mitotic division in pancreatic beta cells. *Science* **163**, 941–943 (1969).

64. Logothetopoulos, J. Electron microscopy of the pancreatic islets stimulated by insulin antibody. *Can. J. Physiol. Pharmacol.* **46**, 407–410 (1968).

65. Logothetopoulos, J.H., and Doniach, I., Compensatory hypertrophy of the rat thyroid after partial thyroidectomy. *Brit. J. Exp. Pathol.* **36**, 617–627 (1955).

66. Marx, A.J., and Deane, H.W. Histophysiologic changes in the kidney and adrenal cortex in rats on a low-sodium diet. *Endocrinology* **73**, 317–328 (1963).

67. Marx, M., Schmidt, W., and Goberna, R. Electron microscopic investigations on the islet regeneration in the rat after subtotal pancreatectomy. *Z. Zellforsch. Mikrosk. Anat.* **110**, 569–587 (1970).

68. Masson, G.M.C., Yagi, S., Kashii, C., and Fisher, E.R. Further observations on juxta-glomerular cells and renal pressor activity in experimental hypertension. *Lab. Invest.* **13**, 321–330 (1964).

69. Milković, K., Paunović, J., Kniewald, Z., and Milković, S. Maintenance of the plasma corticosterone concentration of adrenalectomized rat by the fetal adrenal glands. *Endocrinology* **93**, 115–118 (1973).

70. Mitchell, R.M. Histological changes and mitotic activity in the rat adrenal during postnatal development. *Anat. Rec.* **101**, 161–185 (1948).

71. Moog, F., Bennett, C.J., and Dean, C.M., Jr. Growth and cytochemistry of the adrenal gland of the mouse from birth to maturity. *Anat. Rec.* **120**, 873–891 (1954).

72. Morrison, A.B. Experimentally induced chronic renal insufficiency in the rat. *Lab. Invest.* **11**, 321–332 (1962).

73. Mosier, H.D. Comparative histological study of the adrenal cortex of the wild and domesticated Norway rat. *Endocrinology* **60**, 460–469 (1957).

74. Müntener, M., and Theiler, K. Development of the adrenal glands of the mouse. II. Postnatal development. *Z. Anat. Entwicklungsgesch.* **144**, 205–214 (1974).

75. Myers, K. Morphological changes in the adrenal glands of wild rabbits. *Nature (London)* **213**, 147–150 (1967).

76. Nairn, R.C., Fraser, K.B., and Chadwick, C.S. The histological localization of renin with fluorescent antibody. *Br. J. Exp. Pathol.* **40**, 155–163 (1959).

77. Nickerson, P.A. Stimulation of the adrenal gland of the Mongolian gerbil after ovariectomy. *Cell Tissue Res.* **165**, 135–139 (1975).

78. Nishimura, H., Ogwi, M., Ogawa, M., Sokabe, H., and Imai, M. Absence of renin in kidneys of elasmobranchs and cyclostomes. *Am. J. Physiol.* **218**, 911–915 (1970).

79. Oguri, M., Ogawa, M., and Sokabe, H. Juxtaglomerular cells in aglomerular teleosts. *Bull. Jpn. Soc. Sci. Fish.* **38**, 195–197 (1972).

80. Oguro, C. Are the parathyroid glands necessary to the life of the newt, *Cynops pyrrhogaster? Endocrinol. Jpn.* **16**, 555–556 (1969).

81. Pappenheimer, A.M., and Wilens, S.L. Enlargement of the parathyroid glands in renal disease. *Am. J. Pathol.* **11**, 73–91 (1935).

82. Patent, G.J., and Alfert, M. Histological changes in the pancreatic islets of alloxan-treated mice, with comments on β-cell regeneration. *Acta Anat.* **66**, 504–519 (1967).

83. Peter, S., Lazar, J., Gross, F., and Forssmann, W.G. Studies on the juxtaglomerular apparatus. II. Quantitative morphology after adrenalectomy. *Cell. Tissue Res.* **151**, 457–469 (1974).

84. Peter, S., Lazar, J., Gross, F., and Forssmann, W.G. Studies on the juxtaglomerular apparatus. III. Quantitative morphology after treatment with deoxycorticosterone (DOC). *Cell Tissue Res.* **151**, 471–480 (1974).

85. Pic, P. Maintien d'une calcémie foetale élevée en l'absence des parathyroïdes maternelles et foetales chez le Rat. *C.R. Soc. Biol.* **162**, 1043–1047 (1968).

86. Pohanka, D.G., and Pike, R.L. Effects of dietary sodium restriction during pregnancy on the histochemistry of the rat zona glomerulosa. *Proc. Soc. Exp. Biol. Med.* **133**, 246–251 (1970).

87. Robertson, D.R. The ultimobranchial body in *Rana pipiens*. IV. Hypercalcemia and glandular hypertrophy. *Z. Zellforsch. Mikrosk. Anat.* **85**, 441–452 (1968).

88. Robertson, D.R. The ultimobranchial body in *Rana pipiens*. IX. Effects of extirpation and transplantation on urinary calcium excretion. *Endocrinology* **84**, 1174–1178 (1969).

89. Santler, J.E. Growth in the cell populations of the thyroid gland of rats treated with thiouracil. *J. Endocrinology* **15**, 151–161 (1957).

90. Sauer, J.C., and Hartroft, P.M. The effects of oxygen and salinity on juxtaglomerular cells in kidneys of goldfish, *Carassius auratus. Anat. Rec.* **175**, 434 (1973).

91. Schweisthal, M.R., and Thompson, J.F. Effect of glucagon on the islets of embryonic rat pancreases grown in organ culture. *Anat. Rec.* **154**, 487 (1966).

92. Seki, M., Sekiyama, S., Miyahara, H., and Ichii, S. Studies on regenerating adrenal cortex. 2. Autoradiographic and electron microscopic observations. *Endocrinol. Jpn.* **16**, 361–377 (1969).

93. Seller, M.J. The effect of glucose and insulin on the pregnant rat and foetus. *J. Physiol. (London)* **172**, 353–357 (1964).

94. Sethre, A.E., and Wells, L.J. Accelerated growth of thyroid in normal and "hypophysectomized" fetal rats given thyrotrophin. *Endocrinology* **49**, 369–373 (1951).

95. Sevastikoglu, J.A., and Larsson, S.E. Osteoporosis and parathyroid glands. I. The effect of prolonged calcium deficiency on the parathyroids of the adult rat. *Clin. Orthop. Relat. Res.* **85**, 163–170 (1972).

96. Sharawy, M., Larke, V., and Liebelt, A. Fine structural study of postcastrational adrenocortical carcinomas in female CE-mice. *Anat. Rec.* **184**, 526 (1976).

97. Sinclair, J.G. Size of the parathyroid glands of albino rats as affected by pregnancy and controlled diets. *Anat. Rec.* **80**, 479–496 (1941).

98. Song, M.K., Ianuzzo, C.D., Saubert, C.W., IV and Gollnick, P.D. The mode of adrenal gland enlargement in the rat in response to exercise training. *Pfluegers Arch.* **339**, 59–68 (1973).

99. Straus, F.H., II, and Paloyan, E. The pathology of hyperparathyroidism. *Surg. Clin. N. Am.* **49**, 27–42 (1969).

100. Talmage, R.V., and Toft, R.J. The problem of the control of parathyroid secretion. *In* "The Parathyroids" (R.O. Greep and R.V. Talmage, eds.), pp. 224–240. Thomas, Springfield, Illinois, 1961.

101. Theiler, K., and Muntener, M. Development of the adrenal glands of the mouse. I. Prenatal development. *Z. Anat. Entwicklungsgesch.* **144**, 195–203 (1974).

102. Tobian, L. Interrelationship of electrolytes, juxtaglomerular cells and hypertension. *Physiol. Rev.* **40**, 280–312 (1960).

103. Tobian, L., Tomboulian, A., and Janecek, J. The effect of high perfusion pressures on the granulation of juxtaglomerular cells in an isolated kidney. *J. Clin. Invest.* **38**, 605–610 (1959).

104. Urist, M.R. Avian parathyroid physiology: Including a special comment on calcitonin. *Am. Zool.* **7**, 883–895 (1967).

105. Waggener, R.A. A histological study of the parathyroids in the Anura. *J. Morphol. Physiol.* **48**, 1–44 (1929).

106. Waggener, R.A. An experimental study of the parathyroids in the Anura. *J. Exp. Zool.* **57**, 13–55 (1930).

107. Wells, L.J., Erlandsen, S.L., and Eguchi, Y. Development of pancreases and adrenals of rat embryos in organ culture. *Anat. Rec.* **160**, 201–205 (1968).

108. Whitaker, R.H. The experimental induction of parathyroid hyperplasia in rats, and the use of toluidine blue O for *in vivo* identification of parathyroid tissue. *Calc. Tissue Res.* **8**, 133–141 (1971).

109. Wiedman, M.L., Dunihue, F.W., and Robertson, W. Van B. Effect of sodium chloride intake and mineralocorticoid level on the granularity of juxtaglomerular cells. *J. Endocrinol.* **17**, 261–264 (1958).

110. Woerner, C.A. Studies of the islands of Langerhans after continuous intravenous injection of dextrose. *Anat. Rec.* **71**, 33–58 (1938).

111. Woolley, G.W., and Little, C.C. The incidence of adrenal cortical carcinoma in gonadectomized female mice of the extreme dilution strain. III. Observations on adrenal glands and accessory sex organs in mice 13–24 months of age. *Cancer Res.* **5**, 321–327 (1945).

20

Experimental Regulation of the Testis

Testis: L., *testis*, a witness (i.e., proof of a man's gender)

The testis is two glands in one. It is partly exocrine and partly endocrine. The exocrine component, made up of the seminiferous tubules, may be thought of as a holocrine gland secreting spermatozoa. The endocrine moiety comprises the testosterone-secreting interstitial or Leydig cells interspersed among the seminiferous tubules. It is probably no coincidence that each of these two components of the testis has not evolved as a separate organ. There is reason to believe that the intimacy of their anatomical association reflects a subtle interplay responsible for regulating the form, if not the function of these two tissues.

Seminiferous tubules do not exist as such in fishes. Instead, sperm are released into lobules lined with germinal epithelium (53). The familiar seminiferous tubules of mammals may be thought of as elongated versions of the testicular lobules of lower vertebrates. Each tubule is basically a double-ended loop, the two ends of which open into the rete testis as tubuli recti (Fig. 160). The number of tubules per testis is species-specific and is fixed during early stages of gonadal development. It has been estimated that there are about 8–13 seminiferous tubules in the mouse testis, 20–30 in the rat, and about 16 in the rabbit (8, 16, 20). Such figures are only approximations, however, because many loops branch and anastomose, especially in larger species (25, 31).

The testis is more than just a bag of spaghetti. Its tubules are systematically packed within the confines of the tunica albuginea often folding back and forth upon themselves in an orderly manner (Fig. 161) (8). In smaller mammals the loops tend to be unbranched entities. Only about 0.2 mm in diameter, their lengths may reach 30 cm in the rabbit and 55 cm in

369

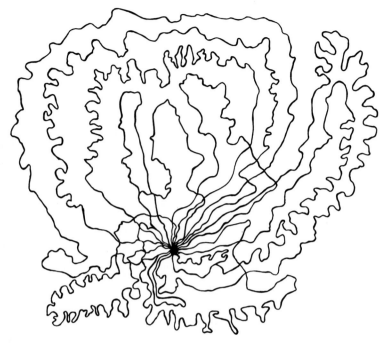

Fig. 160.   Seminiferous tubules of a mouse testis after being teased apart. Although the tubules are arranged as loops with two branches emptying into the rete testis, side branches and joined loops are not uncommon (*16*).

the rat (*8, 17*). In large animals, such as man, there may be considerable branching and anastomoses such that the tubules, if teased apart, form a spider web of interconnecting strands. Their total length has been estimated at 244 m in man (*25*).

KINETICS OF SPERMATOGENESIS

Along the length of the seminiferous tubule, spermatogenesis goes on in an orderly manner (*7*). It takes a long time, about 48 days in the rat, for a mature sperm to differentiate from its spermatogonium at the periphery of the tubule (*9*). If spermatogonia continuously embarked upon the path of differentiation it would be possible to visualize every step in the process by inspecting the concentric layers of cells in a single cross section of the tubule. Such is not the case, for there are so many steps in the process that the

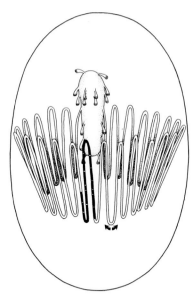

Fig. 161.   Arrangement of a seminiferous tubule in the testis of a rat (8). Tubular loops are folded on themselves in a regular fashion, the two ends opening into the rete testis. The segments drawn in black represent the 14 cell associations of a spermatogenic wave propagated from the "site of reversal" (arrows) toward the rete testis.

tubule would have to be considerably larger than it is if all stages of sper-matogenesis were to be accommodated at one level.

To avoid such cytological congestion, the spermatogonia at any one site initiate differentiation intermittently, about every 12 days in a rat (9). The onset of differentiation is staggered along the length of the tubule as succes-sive rounds of cellular differentiation commence. In other words, instead of continuously sending out cells into the differentiated compartment, as is typical of most other renewing tissues, the basal layers of spermatogonia dispatch cohorts of daughter cells at regular intervals, each generation giving the preceding one a lengthy headstart. This is the reason why, like the student body of a four-year college, a cross section through the seminiferous tubule is represented by only four evenly spaced stages in the development of the final product. However, if one moves along the length of the tubule, successive phases of these four stages are to be seen. Therefore, owing to the remarkable regularity of this process, the same stages of cellular matura-tion are invariably found in association. It is the stepwise occurrence of these cellular associations in space and time that enables us to disentangle the rather complicated kinetics of spermatogenesis by analyzing the events at successive levels along a seminiferous tubule.

The length of a seminiferous tubule in the rat is divided into 14 different kinds of segments, each representing a stage in the cycle of spermatogenesis. Because of the asynchrony with which spermatonia at different levels commence differentiation, each segment is slightly ahead of or behind adjacent ones. All 14 of these cell associations, each confined to its own segment, are laid out in succession along the tubule for a distance of about 2.6 cm. This represents one spermatogenic wave (Fig. 161). Although the participating cells stay put, their consecutive relays of spermatogenesis sweep down the tubule toward the rete testis at a rate of about 2 mm/day. Inasmuch as these waves are propagated along both arms of the tubular loop toward the rete testis, there is a "site of reversal" about midway between the two ends where the waves that are generated move away from each other.

In the rat, there are about 12 such waves along the length of each loop of seminiferous tubule, and each wave is made up of 14 segments of cell associations. In the mouse, each wave is divided into 12 segments (40), and in man there are six (6). It is not possible to define spermatogenic waves in the human seminferous tubule, however, because the six cell associations of the cycle are not consecutively arranged. Notwithstanding the mosaic nature of such segments in man, spermatogenesis is still the orderly process it is in rodents except that it defies the precise histological analysis that is possible in the latter animals.

The spermatogenic wave is not to be imagined as the actual passage of a stimulus along the length of the tubule, however much it may appear that way from the orderly onset of spermatogenesis in successive locations. Nor does it appear to be a domino effect in which each activated set of spermatogonia triggers the next on down the line. Instead, the timing of this process seems to be carefully coordinated with the development of the final product. As the sperm in each relay undergo their final stages of maturation they send back a signal to the basal layer of spermatogenic epithelium thereby initiating the next round of differentiation (29). The nature of this signal may be sought in the maturing spermatozoa themselves. One of their last acts of differentiation is to shed fragments of their cytoplasm. These fragments, or residual bodies, are rich in lipids, mitochondria, and RNA. Gravitating toward the periphery of the tubule, they are phagocytosed by Sertoli cells (13). It is following this event that a new generation of spermatogonia in the neighborhood begins to differentiate. Although the evidence is circumstantial, it is possible that residual bodies, produced at an opportune time for initiating a new generation of differentiating cells, may stimulate the Sertoli cells to induce the next cycle of spermatogenesis among the spermatogonia with which they are so closely associated. So it is that every 12 days the graduating class of spermatozoa may initiate the next round of differentiation and thus be responsible for spacing their replacements in lockstep fashion.

HORMONAL CONTROL OF SEMINIFEROUS TUBULES

The coordination of spermatogenic cycles is, as we have seen, under the control of internal feedbacks within the seminiferous epithelium. The promotion of spermatogenesis and the maintenance of the tubule itself is under the influence of the follicle stimulating hormone (FSH). Hypophysectomy brings about testicular atrophy not only of the tubules for lack of FSH, but of the Leydig cells in the absence of the interstitial cell stimulating hormone (ICSH). The administration of estrogen to a male inhibits spermatogenesis presumably by turning off the secretion of FSH (30). Cyproterone, an antiandrogen compound, likewise inhibits spermatogenesis and brings about tubular atrophy (37). This compound, believed to compete with testosterone for receptor sites, may also interfere with FSH secretion.

Another way to interfere with the pituitary–gonadal axis is to castrate an animal and transplant portions of the testis into the spleen. In this site all of the testicular hormones drain into the hepatic portal circulation and are therefore subject to degradation by the liver. In the apparent absence of testosterone, the pituitary secretes increasing amounts of ICSH, the hormone that normally stimulates testosterone secretion by the interstitial cells. The latter cells in the intrasplenic testis graft, subjected to chronic stimulation by ICSH, may in due course give rise to tumors (49). The seminiferous tubules in such grafts, however, are not known to undergo excessive development, presumably owing to the decreased levels of FSH secretion in the absence of testosterone.

It is FSH that is responsible in large measure for the seasonal changes in fertility in those species of animals, mostly native to temperate zones, which tend to breed only at certain times of year. In the vast majority of cases, such animals lock onto changes in day lengths as the most reliable environmental cue to the changes in seasons. In some species of birds and mammals the testicular size varies tremendously from one season to another. The pineal gland plays a pivotal role in mediating the influence of photoperiod on reproductive activity (44).

In an animal such as the hamster the testes tend to atrophy on short days and hypertrophy on long ones. Pinealectomy prevents the atrophy that normally occurs when the days are short and can even bring about increases in testicular weight (18). Such findings can be explained in terms of melatonin, a compound elaborated by the pineal and responsible for the inhibition of gonadal growth and activity. Hamsters treated with melatonin fail to undergo testicular recrudescence under the influence of long days (19). Melatonin is normally produced by the pineal in the dark, while light inhibits its synthesis, which explains why the testes atrophy on short days and enlarge on long ones. Hence, exogenous melatonin can override the otherwise stimulating influence of long photoperiods. The effect of the pineal on the gonads, however, is not a direct one. Evidence indicates that melatonin affects the growth and activity of the testes by inhibiting the secretion of gonadotropins from the pituitary.

It is the pituitary that also appears to mediate compensatory growth of the testis after unilateral castration. In most species, a remaining testis undergoes only a modest degree of hypertrophy believed to be attributable primarily to increases in the interstitial tissues (14, 18, 32, 39, 51). In seasonal breeders, compensatory testicular hypertrophy usually fails to occur during nonreproductive times of year, presumably owing to the inhibitory effects of melatonin secreted by the pineal. Pinealectomy alone can counteract the effects of short days which in small mammals normally decrease testicular weight and prevent compensatory growth (18). Hence, compensatory testicular hypertrophy following unilateral castration may be the result of increasing the ratio of gonadotropin to target organ such that what remains of the latter is enlarged until its normal equilibrium with the former is restored.

COMPENSATORY SPERMATOGENESIS?

The ovary is endowed with an impressive capacity for compensatory ovulation following unilateral ovariectomy (see Chap. 21). The possibility of compensatory spermatogenesis in the hemicastrated male, however, remains to be confirmed. The problem is that techniques for monitoring the rate of sperm production are yet to be perfected. Nevertheless, it is established that sperm counts in ejaculates of rats and rabbits fall to half their preoperative levels after unilateral orchidectomy (34, 41, 52). In the ram, this is accompanied by a doubling in the flow rate of seminal fluid which, with no change in sperm concentration, amounts to twice the normal rate of sperm production (51). In the rat, sperm counts have been observed to rise to normal values about a month after hemicastration, indicating a twofold increase in sperm production by the remaining testis (34). How this can be achieved so soon when it takes about seven weeks for a sperm to differentiate from a spermatogonium is not clear. Other results have been obtained with a different method of collecting sperm, a method based on the fact that sperm are often found in samples of urine where they are remarkably resistant to degeneration. By grafting the vasa deferentia to the bladder it is possible to recover sperm in each day's output of urine collected in metabolism cages. This technique is plagued with problems not easily overcome (e.g., bladder stones, occluded vasa), and has failed to yield data confirming compensatory spermatogenesis within at least 90 days following unilateral castration in rats (52). Until reliable information is forthcoming on the normal day-to-day rate of sperm production and their delivery via the vas deferens, it will be difficult to establish whether or not the remaining testis of a hemicastrated animal doubles its sperm production. If so, does this involve doubling the number of participating cells by increasing the circumference and length of each seminiferous tubule, or is it possible to accelerate the rate of spermatogenesis itself?

CRYPTORCHIDISM

It is possible to destroy selectively the spermatogenic component of the testis while leaving the interstitial cells intact by transferring the testis from the scrotum to the abdominal cavity, a condition known as cryptorchidism. The cryptorchid testis is subjected to extensive degeneration of its spermatogenic tissues (Fig. 162) (5, 35). Such testes may shrink to less than one-third their normal size and the diameters of their seminiferous tubules become reduced to one-half their former dimensions. The cycle of spermatogenesis becomes disorganized, spermatocytes degenerate, and eventually the spermatogonia are destroyed (although a small percentage of them may persist for many months). The Sertoli cells seem unaffected by cryptorchidism. It is the survival of these cells together with some of the spermatogonia that is responsible for the reversibility of cryptorchidism. If such a testis is returned to the scrotum (an operation known as orchidopexy) the tubules may recover and spermatogenesis may resume (2, 5). The extent of such recovery, however, diminishes in proportion to the duration of cryptorchidism.

The deleterious effects of the abdominal environment are attributable to the exposure of the testis to normal body temperatures rather than the cooler scrotal temperatures upon which spermatogenesis depends. Correlated with this is the evidence that amino acid incorporation into testicular proteins occurs optimally at 32°C in the rat, presumably because testicular enzymes are more heat labile than are those in other organs (11). What advantage there may be in this state of affairs defies speculation, but it provides a convenient experimental approach to destroy selectively the spermatogenic moiety of the testis. Just as the cryptorchid testis exposed to 37.5°C undergoes collapse of its seminiferous tubules, experimental exposure of scrotal

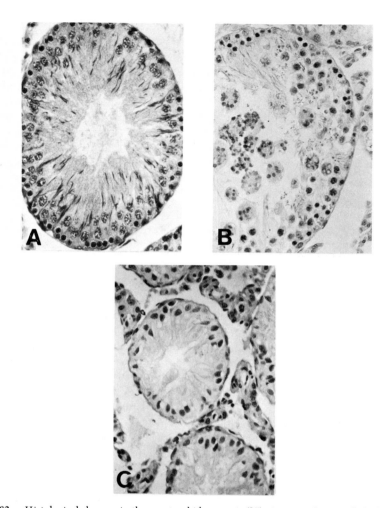

Fig. 162.   Histological changes in the cryptorchid rat testis (35). A, seminiferous tubule from a scrotal testis showing normal spermatogenesis. B, 5-day cryptorchid testis with cellular debris in the tubular lumen and disorganized germinal epithelium. C, atrophy of the seminiferous tubule with only the Sertoli cells remaining after 20 days' cryptorchidism.

testes to water baths of 41–43°C for varying periods of time likewise destroys the spermatogenic epithelium (10). Exposure to these temperatures for half an hour brings about extensive damage to the seminiferous rubules, reducing the sperm count to one-third of normal levels a month later (5). Such organs are capable of recovery, however, with spermatogenesis restored by seven weeks and the sperm count normal after three months.

It might be concluded from the foregoing evidence that the testis is sensitive to absolute temperature *per se*. There is reason to believe, however, that the relative temperature may also be involved (48). When rats are held at ambient temperatures of 35°C for three months their deep body temperatures average about 1° above those measured in rats exposed abruptly to a hot environment without prior acclimatization. In both groups, however, the scrotal temperatures are at least 1° below body temperature. Hence, the testes of the heat-acclimatized rats are at a temperature approximately the same as the body temperatures of the nonacclimatized animals. Nevertheless, these testes do not undergo dengenerative changes even though their temperatures are commensurate with those to which cryptorchid testes are normally exposed. Although the fertility of such animals is somewhat reduced, these findings may be taken to indicate that, so long as a temperature differential is maintained between the testis and the rest of the body, spermatogenesis may be exempt from the damaging effects of too much warmth.

CHEMICAL DESTRUCTION OF THE TESTIS

Certain chemical agents have also been found to damage the testis. One such compound is ethionine, an analogue of the essential amino acid methionine. When administered to rats, ethionine brings about the degeneration of the tubular epithelium within a few weeks, while sparing the Sertoli cells and interstilial tissue (4). The testis is protected from the deleterious influences of ethionine if methionine is simultaneously administered. The ethionine-injured testis, however, is capable of recovery when treatment is discontinued.

More severe damage is caused by treatment with cadmium chloride. A single subcutaneous injection suffices to cause the complete degeneration of the testes, an effect that can be counteracted by concomitant treatment with zinc acetate. Indeed, chronic zinc deficiencies alone can bring about degenerative changes in the testes. The cadmium effect is not altered by testosterone, but can be prevented by treatment with estradiol (15). The testis may also be protected from cadmium damage by reserpine or adrenalectomy, suggesting a dependence on catecholamines (42).

Testicular damage by cadmium would appear to be correlated with the normal scrotal position of the testes. It has been demonstrated in the mouse, rat, bat, hamster, rabbit, goat, and monkey but does not occur in the frog, pigeon, fowl, armadillo, hedgehog, or opossum. None of the animals in the latter group has a scrotum, except for the opossum. The lack of effect in nonscrotal animals appears not to be attributable to the failure of their testes to concentrate cadmium, for radioactive cadmium becomes localized just as

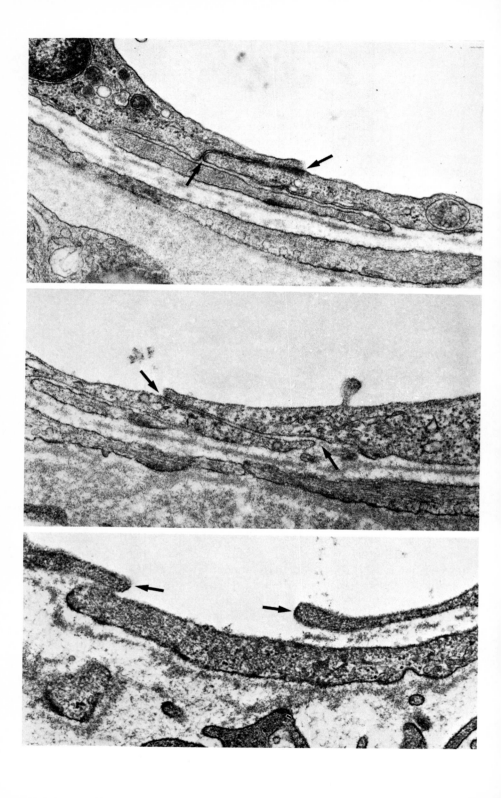

much in the testes of the fowl as in scrotal mammals, yet does not cause degeneration of the avian testis (24).

The immediate target of cadmium is the vasculature of the testis where the permeability of the blood vessels is increased (46). This is attributable to disruption of junctions between endothelial cells lining capillaries in the testes (Fig. 163) (12). Within a few hours after treatment, the testicular blood supply is interrupted and necrosis of the ischemic testis ensues (38). Thus, the effects of cadmium simulate those that take place in the testis after surgical ligation of its blood vessels. The specific sensitivity of the testicular vasculature to cadmium may relate to the role of the blood supply in maintaining the below-normal temperatures of the scrotum (23). During the second hour after injection of $CdCl_2$, the temperature of the testes rises, falling to normal thereafter. Zinc acetate prevents this transient elevation in testicular temperature, and even in the absence of cadmium, zinc reduces the temperature of the testes. Thus, zinc not only counteracts the effects of cadmium on testicular temperature but also protects against the interruption in blood flow normally caused by cadmium, a correlation not altogether without significance.

The failure of cadmium to damage the testes of nonscrotal animals, coupled with its effect on testicular temperature, suggests that cadmium might have no deleterious effects on cryptorchid testes. Such is not the case, however, for when $CdCl_2$ is administered to a rat immediately after rendering it unilaterally cryptorchid, the extent of damage to the abdominal testis is more severe than in the contralateral scrotal organ (42). One can only conclude that the location of the testicular blood vessels, or the temperature at which they are held, affects their susceptibility to cadmium poisoning (23).

Once the testis has been damaged by cadmium (or blood vessel ligation) the animal undergoes a typical castration syndrome whereby the accessory sex organs atrophy and castration cells appear in the anterior pituitary. In due course, however, limited regenerative changes may be seen owing to the survival of some of the interstitial cells beneath the tunica albuginea. These residual Leydig cells proliferate extensively in the absence of tubular regeneration. What is created, therefore, is an unprecedented organ made up of androgen-secreting cells among the defunct remnants of seminiferous tubules. As these cells resume their endocrine functions the symptoms of castration disappear in several weeks. After some months (27), hyperplastic Leydig cells may give rise to interstitial cell tumors in many cases (26, 43).

Fig. 163.   Influence of $CdCl_2$ on the vascular endothelium of the rat testis (12). Top, normal endothelial cell junction (arrows) in a control capillary. Middle, 3 hours after $CdCl_2$ injection the electron dense material normally associated with the junctional membranes has been depleted. Bottom, separation of the endothelial cell junction 6 hours after treatment with $CdCl_2$.

TESTICULAR EFFECTS OF VASECTOMY

Recent years have witnessed a revival of interest in vasectomy as a means of male birth control, and a heightened concern for the possible effects this operation might have on the testis (22). Virtually every other exocrine gland undergoes degenerative changes when their ducts have been obstructed. What happens in the testis depends very much on the species of animal involved.

Two kinds of occlusion have been studied, one of the vasa efferentia, the ducts leading directly from the testis to the epididymis, the other of the vas deferens which conducts sperm from the epididymis to the outside. In the former case, the testis is invariably damaged (45, 47). Studies in rats and mice have shown that the sizes of such testes increase during the first few days after operation due to the retention of sperm and the distension of tubules by accumulated fluid. This leads to a decline in spermatogenesis, varying degrees of degeneration of the germinal epithelium, and a decrease in the size of the testis to as little as one-half its original dimensions.

Occlusion of the vas deferens by ligation and/or transection has been claimed to cause no damage to the testis in cats, rabbits, gerbils, and rats (33, 36), although contradictions in the literature abound (3). In the dog, on the other hand, there is overwhelming evidence that vasectomy leads to extensive degeneration of the seminiferous tubules during the succeeding few months (28, 50). Eventually, however, surviving spermatogonia, in association with Sertoli cells, may resume spermatogenesis. Why it is that the

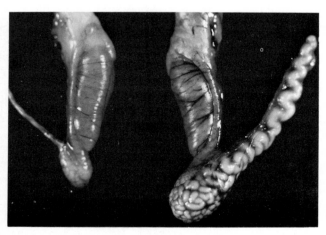

Fig. 164.   Effect of vasectomy on the rabbit (3). Left, unoperated control testis. Right, 18 weeks after unilateral vasectomy the vas deferens and cauda epididymidis are swollen. The testis itself is rarely affected by this operation.

canine testis undergoes transient degeneration of its spermatogenic epithelium, while those of other species do not, remains to be determined.

Although vasectomy may be surgically reversible, there is reason to suspect that fertility might be impaired as a result of the production of antibodies against one's own sperm. It is well known that injection of spermatozoa elicits the production of spermagglutinins, and mounting evidence suggests that antibodies are commonly produced against sperm after vasectomy, presumably due to the inadvertent release of sperm from their immunologically privileged sites within the testicular tubules into the surrounding connective tissue. However, the fact that unilateral vasectomy does not necessarily interfere with fertility suggests that autoimmunological reactions are probably not so important as might have been anticipated (1, 3).

A number of experimental interventions, including cryptorchidism, heat treatment, ethionine, $CdCl_2$, and sometimes vasectomy, all have profound effects on the testis. The seminiferous tubules are especially sensitive to such treatments while the interstitial tissue is either unaffected, or can regenerate from a few surviving cells, as in the case of cadmium poisoning and testicular ischemia. This regeneration of the interstitial tissue is especially interesting because it reveals a hitherto unsuspected relationship between the seminiferous tubules and the interstitial cells among them. In the cryptorchid testis, the interstitial cells often are seen to become hyperplastic as the tubules degenerate (21). A comparable reaction might be expected in testes exposed to elevated temperatures for brief periods, although this has yet to be reported in such cases. Testes damaged by administration of ethionine typically exhibit hyperplastic interstitial tissues (4). In the vasectomized dog testis, Leydig cells may also become more prominent (28). Finally, idiopathic testicular atrophy is frequently accompanied by hyperplastic Leydig cells. Clearly, there appears to be a reciprocal relationship between the condition of the seminiferous tubules and the mass of interstitial tissue such that atrophy or degeneration of the former leads to the proliferation of the latter. Indeed, when the tubules are devitalized altogether, as after cadmium poisoning and ischemia, the Leydig cells eventually become neoplastic (26, 43). Such evidence would predict that the seminiferous tubules might normally exert a restraining influence on the growth of interstitial cells. If true, this would provide a logical explanation for the intimate association between these two tissues, without which the cancerous potential of the Leydig cells perhaps could not be held in abeyance.

REFERENCES

1. Alexander, N.J. Vasectomy and vasovasostomy in Rhesus monkeys: The effect of circulating antisperm antibodies on fertility. *Fertil. Steril.* **28**, 562–569 (1977).

2. Atkinson, P.M. The effects of early experimental cryptorchidism and subsequent orchidopexy on the maturation of the guinea-pig testicle. *Br. J. Surg.* **60**, 253–258 (1973).

3. Bedford, J.M. Adaptations of the male reproductive tract and the fate of spermatozoa following vasectomy in the rabbit, rhesus monkey, hamster and rat. *Biol. Reprod.* **14**, 118–142 (1976).

4. Benson, W.R., and Clare, F.S. Regenerative changes and spermatic granulomas in the rat testis after treatment with DL-ethionine. *Am. J. Pathol.* **49**, 981–995 (1966).

5. Blackshaw, A.W., and Massey, P.F. The recovery of spermatogenesis in the rat testis after heat-induced degeneration. *J. Reprod. Fertil.* **28**, 142–143 (1972).

6. Clermont, Y. The cycle of the seminiferous epithelium in man. *Am. J. Anat.* **112**, 35–51 (1963).

7. Clermont, Y. Kinetics of spermatogenesis in mammals: Seminiferous epithelium cycle and spermatogonial renewal. *Physiol. Rev.* **52**, 198–236 (1972).

8. Clermont, Y., and Huckins, C. Microscopic anatomy of the sex cords and seminiferous tubules in growing and adult male albino rats. *Am. J. Anat.* **108**, 79–98 (1961).

9. Clermont, Y., Leblond, C.P., and Messier, B. Durée du cycle de l'épithélium séminal du rat. *Arch. Anat. Microscop.* **48**, 37–55 (1959).

10. Collins, P., and Lacy, D. Studies on the structure and function of the mammalian testis. II. Cytological and histochemical observations on the testis of the rat after a single exposure to heat applied for different lengths of time. *Proc. Roy. Soc. London, Ser. B* **172**, 17–38 (1969).

11. Davis, J.R., Firlit, C.F., and Hollinger, M.A. Effect of temperature on incorporation of L-lysine-U-C^{14} into testicular proteins. *Am. J. Physiol.* **204**, 696–698 (1963).

12. Fende, P.L., and Niewenhuis, R.J. An electron microscopic study of the effects of cadmium chloride on cryptorchid testes of the rat. *Biol. Reprod.* **16**, 298–305 (1977).

13. Firlit, C.F., and Davis, J.R. Morphogenesis of the residual body of the mouse testis. *Q. J. Microscop. Sci.* **106**, 93–98 (1965).

14. Gomes, W.R., and Jain, S.K. Effect of unilateral and bilateral castration and cryptorchidism on serum gonadotrophins in the rat. *J. Endocrinol.* **68**, 191–196 (1976).

15. Gunn, S.A., Gould, T.C., and Anderson, W.A.D. Protective effect of estrogen against vascular damage to the testis caused by cadmium. *Proc. Soc. Exp. Biol. Med.* **119**, 901–905 (1965).

16. Hirota, S. The morphology of the seminiferous tubules. I. The seminiferous tubules of the mouse. *Kyushu Mem. Med. Sci.* **3**, 121–128 (1952).

17. Hirota, S. The morphology of the seminiferous tubules. II. The seminiferous tubules of the monkey. *Kyushu Mem. Med. Sci.* **3**, 129–136 (1952).

18. Hoffman, R.A., and Reiter, R.J. Influence of compensatory mechanisms and the pineal gland on dark-induced gonadal atrophy in male hamsters. *Nature (London)* **207**, 658–659 (1965).

19. Hoffmann, K. The influence of photoperiod and melatonin on testis size, body weight, and pelage colour in the Djungarian hamster (*Phodopus sungorus*). *J. Comp. Physiol.* **85**, 267–282 (1973).

20. Huber, G.C., and Curtis, G.M. The morphology of the seminiferous tubules of mammalia. *Anat. Rec.* **7**, 207–219 (1913).

21. Iturriza, F.C., and Irusta, O. Hyperplasia of the interstitial cells of the testis in experimental cryptorchidism. *Acta Physiol. Lat. Am.* **19**, 236–242 (1969).

22. Jhaver, P.S., and Ohri, B.B. The history of experimental and clinical work on vasectomy. *J. Int. Coll. Surg.* **33**, 482–486 (1960).

23. Johnson, A.D., Gomes, W.R., and VanDemark, N.L. Early actions of cadmium in the rat and domestic fowl testis. I. Testis and body temperature changes caused by cadmium and zinc. *J. Reprod. Fertil.* **21**, 383–393 (1970).

24. Johnson, A.D., and Sigman, M.B. Early actions of cadmium in the rat and domestic fowl testis. IV. Autoradiographic location of ^{115m}cadmium. *J. Reprod. Fertil.* **24**, 115–117 (1971).

25. Johnson, F.P. Dissections of human seminiferous tubules. *Anat. Rec.* **59**, 187–199 (1934).

26. Knorre, D. Induction of interstitial cell tumors by cadmium chloride in albino rats. *Arch. Geschwulstforsch* **38**, 257–263 (1971).

27. Kormano, M. Microvascular supply of the regenerated rat testis following cadmium injury. *Virchows Arch. A* **349**, 229–235 (1970).

28. Kothari, L.K., and Mishra, P. Histochemical changes in the testis and epididymis after vasectomy. *Int. J. Fertil.* **18**, 119–125 (1973).

29. Lacy, D. The seminiferous tubule in mammals. *Endeavour* **26**, 101–108 (1967).

30. Lacy, D., and Lofts, B. Studies on the structure and function of the mammalian testis. I. Cytological and histochemical observations after continuous treatment with oestrogenic hormones and the effects of F.S.H. and L.H. *Proc. Roy. Soc. London, Ser. B* **162**, 188–197 (1965).

31. Liang, D.S. Anatomical structure of the testicular tubule. *Invest. Urol.* **4**, 285–287 (1966).

32. Liang, D.S., and Liang, M.D. Testicular hypertrophy in rats. *J. Reprod. Fertil.* **21**, 537–540 (1970).

33. Lohiya, N.K., and Dixit, V.P. Biochemical studies of the testes and sex accessory organs of the desert gerbil after vasectomy. *Fertil. Steril.* **25**, 617–620 (1974).

34. Mauss, J., and Hackstedt, G. The effect of unilateral orchidectomy and unilateral cryptorchidism on sperm output in the rat. *J. Reprod. Fertil.* **30**, 289–292 (1972).

35. Nagy, F. Tritiated leucine incorporation by Sertoli cells in scortal and cryptorchid rat testes. *Fertil. Steril.* **24**, 805–810 (1973).

36. Neaves, W.B. The rat testis after vasectomy. *J. Reprod. Fertil.* **40**, 39–44 (1974).

37. Neumann, F. Use of cyproterone acetate in animal and clinical trials. Hormones and antagonists. *Gynecol. Invest.* **2**, 150–179 (1971–1972).

38. Niemi, M., and Kormano, M. An angiographic study of cadmium-induced vascular lesions in the testis and epididymis of the rat. *Acta Pathol. Microbiol. Scand.* **63**, 513–521 (1965).

39. Noller, D.W., Howards, S.S., and Panko, W. Metabolic effects of unilateral orchiectomy on the guinea pig testis. *Fertil. Steril.* **28**, 186–190 (1977).

40. Oakberg, E.F. A description of spermiogenesis in the mouse and its use in analysis of the cycle of the seminiferous epithelium and germ cell renewal. *Am. J. Anat.* **99**, 391–413 (1956).

41. Paufler, S.K., and Foote, R.H. Semen quality and testicular function in rabbits following repeated testicular biopsy and unilateral castration. *Fertil. Steril.* **20**, 618–625 (1969).

42. Ray, P., and Chatterjee, A. Reserpine, adrenalectomy and the reversal of the early action of cadmium on scrotal and cryptorchid testes in the rat. *J. Reprod. Fertil.* **33**, 523–526 (1975).

43. Reddy, J., Svoboda, D., Azarnoff, D., and Dawar, R. Cadmium-induced Leydig cell tumors of rat testis: Morphologic and cytochemical study. *J. Nat. Cancer Inst.* **51**, 891–903 (1973).

44. Reiter, R.J. Exogenous and endogenous control of the annual reproductive cycle in the male golden hamster: Participation of the pineal gland. *J. Exp. Zool.* **191**, 111–120 (1975).

45. Ross, M.H. The organization of the seminiferous epithelium in the mouse testis following ligation of the efferent ductules. A light microscope study. *Anat. Rec.* **180**, 565–578 (1974).

46. Setchell, B.P., and Waites, G.M.H. Changes in the permeability of the testicular capillaries and of the blood-testis barrier after injection of cadmium chloride in the rat. *J. Endocrinol.* **47**, 81–86 (1970).

47. Smith, G. The effects of ligation of the vasa efferentia and vasectomy on testicular function in the adult rat. *J. Endocrinol.* **23**, 385–399 (1962).

48. Sod-Moriah, U.A., Goldberg, G.M., and Bedrak, E. Intrascrotal temperature, testicular histology and fertility of heat-acclimatized rats. *J. Reprod. Fertil.* **37**, 263–268 (1974).
49. Twombly, G.H., Meisel, D., and Stout, H.P. Leydig cell tumours induced experimentally in the rat. *Cancer* **2**, 884–892 (1949).
50. Vare, A.M., and Bansal, P.C. Changes in the canine testes after bilateral vasectomy—an experimental study. *Fertil. Steril.* **24**, 793–797 (1973).
51. Voglmayr, J.K., and Mattner, P.E. Compensatory hypertrophy in the remaining testis following unilateral orchidectomy in the adult ram. *J. Reprod. Fertil.* **17**, 179–181 (1968).
52. Vreeburg, J.T.M., van Andel, M.V., Kort, W.J., and Westbroek, D.L. The effect of hemicastration on daily sperm output in the rat as measured by a new method. *J. Reprod. Fertil.* **41**, 355–359 (1974).
53. Weisel, G.F. A histological study of the testes of the sockeye salmon (*Oncorhynchus nerka*). *J. Morphol.* **73**, 207–229 (1943).

21

Cycles of Ovarian Growth

Estrus: Gr., *oistros*, gadfly (bothersome as the sexual urge)

In the eighteenth century there were many who believed that sex was determined by whether the ovum was produced by the left or the right ovary. This was not an illogical idea in the era before chromosomes were discovered, although it was not known which ovary gave rise to which sex. To answer this question, the renowned English anatomist, John Hunter, removed one ovary from a sow to learn whether the resulting piglets would be all male or all female. It took a man of Hunter's imagination to devise such an ingeniously simple experiment, and a surgeon of his skill to carry out the operation on a squealing pig half a century before the discovery of anesthetics. The results of his experiment, published in 1792, were disappointing (18). When the sow became pregnant and delivered her litter their sexes were mixed, thus disproving the "one ovary:one sex" theory. However, the size of the litter produced by the unilaterally ovariectomized mother was equal to the number of piglets normally produced by sows with both ovaries intact. Hunter was not unaware of the significance of this observation, for in disproving one theory he discovered the important phenomenon of compensatory ovulation, the explanation for which was not to be discovered for almost two more centuries.

Another popular misconception is that the two ovaries shed their eggs alternately. There is no evidence for this in humans, and it obviously cannot prevail in polytocous animals which routinely become pregnant in both uterine horns. There is one exception. Monotocous marsupials are said to ovulate from alternate ovaries, becoming pregnant first in one side of the uterus, then in the other (12).

In the vast majority of animals both ovaries share equally in the job of egg production and hormone secretion, and readily compensate for their missing counterparts should one ovary be lost or incapacitated. There are some interesting exceptions to this rule, the most notable of which is the case of the avian ovary which is present only on the left side in many birds (the right gonad remaining rudimentary). Likewise, in the Indian vampire bat only the left ovary is functional, pregnancy always occurring in the left horn of the uterus. The so-called American chamelion (*Anolis carolinensis*) has two ovaries but the left one is the larger (*19*).

In other kinds of animals the ovarian asymmetry is dextral. The garter snake, for example, has a right ovary that is the larger of the two. In the human fetus, the right ovary averages 17% heavier than the left one (*27*). In many species of bats only the right ovary is functional. One wonders if the unilateral asymmetry of ovaries in bats, like that in birds, is an adaptation to their aerial existence. The most extreme case of reproductive lopsidedness is found in the mountain viscacha, a hystricomorph rodent native to the Peruvian Andes. Although immature females have gonads of equal size, the adult animal ovulates only from the larger right ovary (*28, 38*). What selective advantage this imbalance affords remains a mystery.

The growth and structure of the ovary are dominated by the cyclicity of its function, cycles that are punctuated by periods of estrus recurring as frequently as every 4–5 days in small rodents (hamster, rat, mouse) or once a month or more as in many primates. Some mammals are polyestrous, coming into heat at regular intervals until pregnancy supervenes. Some are monestrous, coming into heat twice a year, for example. Nontropical species tend to have seasonally synchronized reproductive periods while those inhabiting equatorial regions may be polyestrous throughout the year. The human species falls in the latter category.

There is a three-cornered feedback by which the rhythms of ovarian activity are regulated. Under appropriate hormonal conditions, the hypothalamus secretes a releasing factor which stimulates the secretion of follicle stimulating hormone (FSH) from the anterior pituitary. This hormone is responsible for promoting the growth and maturation of follicles in the ovary, the theca interna of which secretes estrogen. Rising levels of estrogen in the system, monitored by cells in the hypothalamus, turn off the hypothalamic secretion of releasing factors until the onset of the next cycle. Concomitantly, luteinizing hormone (LH) is then secreted by the pituitary, inducing ovulation and interrupting further estrogen secretion. The granulosa cells of the now collapsed follicles become organized into corpora lutea and secrete progesterone. If pregnancy has not occurred, this activity subsides and the corpora lutea regress as a new cycle is initiated. Hence, the life history of the ovary is characterized by alternations in follicular and luteal development. The regulation of ovarian growth involves the determination of how many follicles are to develop and how long the corpora lutea are to survive.

COMPENSATORY OVARIAN HYPERTROPHY

The ovary is very sensitive to reductions in its mass. Indeed, considerable regeneration is possible from small fragments of the original organ, but only if the opposite ovary is missing (29). Following unilateral ovariectomy, compensatory hypertrophy of the remaining gonad occurs, sometimes doubling its size. The fact that little or no compensatory ovarian hypertrophy takes place in prepubertal animals (21), coupled with its absence in anovulatory females in the nonbreeding season (24, 28), implicates the hypothalamo-hypophyseal axis as responsible for the physiological and morphological accommodation of ovarian remnants. This possibility is strengthened by the fact that compensatory growth of the remaining ovary does not occur in rabbits isolated from males (5). In this species, the nervous stimulation of copulation triggers the release of LH thus stimulating ovulation at the most propitious moment so that mating is tantamount to conception. In bred and pregnant does, compensation occurs with its usual efficiency.

Adaptive growth of the ovary is achieved primarily by an increase in the number of follicles that mature and corpora lutea into which they develop. Expansion of the connective tissue stroma in the ovary makes only a minor contribution to hypertrophy of the organ. It is significant that the Graafian follicles increase in number instead of size, for only in this way can the number of eggs ovulated be restored to normal, a phenomenon that has been documented in frogs, mice, rats, hamsters, rabbits, and pigs. In order for compensatory ovulation to occur, however, ovariectomy must be performed at least one day before estrus in rats and hamsters, leaving only a remarkably short time in which to double the number of follicles completing their maturation (14, 30). It is now established that the number of mature follicles to develop in an ovary is dependent upon the levels of gonadotropin secreted by the pituitary, not by the mass of the target organ. Experiments in the mouse, for example, have shown that the number of eggs produced by both ovaries may be increased over sixfold following the administration of sufficiently high doses of FSH (32). This superovulation results in excessively gravid females, although the extent of this superfecundity is limited to not more than triple the normal litter size, presumably owing to limitations in uterine capacity.

Compensatory ovulation also occurs if the Graafian follicles are selectively destroyed in one ovary, in the absence of removing the entire organ. Following such operations in ewes, new follicles mature to replace the missing ones, but they do so more in the operated ovary than in the opposite intact one (11). Thus, there is no dearth of immature follicles from which to recruit those destined for maturation. In compensatory ovulation the number of ova and immature follicles that differentiate at each cycle remains unchanged:

the proportion completing development is increased (33). How the lucky ones are selected from an apparently homogeneous pool of candidates is not known.

In attempting to fathom the interplay of hormones regulating ovarian growth, the responses of transplanted ovaries are particularly instructive. In general, a grafted ovary, or fragment thereof, will fail to grow until the rest of the host's ovaries are removed. The latter's estrogenic output restricts the secretion of gonadotropins to normal levels and the *in situ* ovaries apparently have priority over ectopic grafts. Even in the absence of host ovaries, grafts in the systemic circulation do not grow beyond normal dimensions. Transplants in sites drained by the hepatic portal circulation, however, behave differently. The liver is the principal organ involved in estrogen degradation, so hormones secreted by ovarian grafts in the spleen seldom get past the liver intact. In hosts lacking other ovarian tissue, the hypothalamus detects little or no circulating estrogens and the pituitary therefore heightens its secretion of gonadotropins. Chronic stimulation of intrasplenic ovarian grafts by tropic hormones results in the excessive enlargement of the transplanted tissue, including hyperplasia and luteinization. After some months neoplastic transformation may occur, giving rise to luteomas and eventually granulosa cell tumors (13, 16). None of these changes occurs if one of the host's own ovaries remains or if estrogen is injected. Moreover, if an Eck fistula is established (by anastomosis of the hepatic portal vein to the inferior vena cava), the liver is bypassed, the graft's estrogen reaches the hypothalamus, and in the absence of gonadotropins the graft fails to grow (22). On the other hand, if one ovary is transplanted to the spleen, the *in situ* one can still undergo compensatory hypertrophy, something it cannot do if the excised ovary is grafted to a site drained by the systemic circulation (39).

It seems undeniable that compensatory follicular multiplication, like the production of the normal crop of Graafian follicles with each estrous cycle, is under the control of FSH. It has been a matter of contention, however, whether the increase in number of follicles produced by a solitary ovary was due to the production of increased FSH by the pituitary or the prolongation of exposure to normal circulating levels of the hormone. Despite lack of unanimity on this important point, improved techniques have now confirmed that following unilateral ovariectomy there is indeed an increase in FSH secretion by the pituitary, a response that occurs within hours after operation (3, 9, 40). It is this transient flood of FSH that doubles the number of follicles reaching maturity in the remaining ovary.

Unilateral ovariectomy creates an asymmetry in a normally symmetrical pair of organs, an imbalance that elicits the compensatory reactions described above. In view of this tendency to correct ovarian inequities, how is the normal asymmetry occurring in the gonads of certain animals to be explained? It will be recalled that the left ovary of *Anolis carolinensis* is normally larger than the right one. When the smaller of the two ovaries is removed, no compensatory growth occurs in the remaining larger one (*19*). In the mountain viscacha (Fig. 165), if the functional right ovary is removed, subsequent pregnancies occur in the left uterine horn (*28*). There is no compensatory hypertrophy of the left ovary in nonbreeding animals, however.

The situation is more complicated in birds which have a left ovary paired with only a rudimentary gonad on the right side. If the left ovary is removed the rudimentary right gonad then develops, but in the large majority of cases it becomes a testis (*20, 36*). In only a few instances does the right gonad develop into an ovary, not infrequently differentiating both ovarian and testicular characteristics. The development of a testis in a genetic female results in the reversal of secondary sexual characters, including plumage and

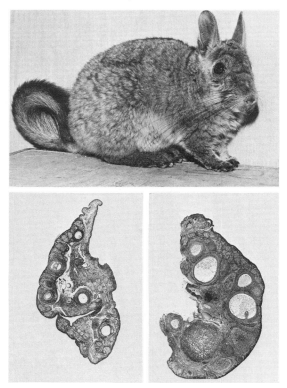

Fig. 165.   Above, the mountain viscacha. Below, the left and right ovaries from a pregnant viscacha, illustrating the lack of corpora lutea in the nonovulating left one and the presence of corpora lutea and large follicles on the right side where ovulation occurs (*38*).

behavior. Although the testes of these birds produce mature sperm, abnormalities in their efferent ducts are responsible for their infertility. The administration of estrogen or androgen to such birds decreases the percentage of right gonads that develop into testes (20, 36). This may be attributable to the resulting decrease in gonadotropin secretion, for hypophysectomy of left ovariectomized birds reduces the hypertrophy of the right gonad (20). In the avian embryo, both gonads are originally equal, and it is noteworthy that extirpation of the left one at this stage of development prevents the remaining right one from developing into a testis (15). Therefore, the potential for sex reversal that characterizes adult birds would seem to depend upon the establishment of gonadal asymmetry during early stages of maturation.

In other animals, sex reversal may be a normal component of the life cycle. Certain species of fishes, for example, begin life as females, but as they grow older their ovaries degenerate and are replaced by testes. In male frogs there is a prospective ovary (Bidder's organ) located anterior to the testes. Should the latter be removed, Bidder's organs differentiate into ovaries capable of egg production.

THE CORPUS LUTEUM

It is not sufficient for an ovary just to produce eggs. The nature of motherhood lies in the care of one's offspring. In oviparous animals this is expressed in nest building and associated behavioral patterns. In viviparous forms, nidification occurs in the uterus, the preparation of which depends upon progesterone secreted by the corpora lutea. These glands, derived from the remnants of the Graafian follicles after ovulation, are indispensable for the success of gestation. Their days are numbered, however, by the transient nature of pregnancy, if indeed it occurs at all.

In compensatory ovarian hypertrophy, there is an increase in the number of corpora lutea commensurate with the rise in the population of follicles (26). Each corpus luteum, however, grows to normal size, except that the output of progesterone may increase when one ovary has been removed in pregnancy or pseudopregnancy (9, 35). If the corpora lutea are selectively excised from an ovary, the decline in progesterone secretion brings about increased gonadotropin production leading to an eventual rise in the number of mature follicles subsequently produced (6).

The physiological basis for the rise and demise of the corpora lutea is not fully understood. Their initial production, triggered by ovulation under the influence of LH, may also be enhanced by the luteotropic action of prolactin. It is altogether fitting that this hormone, which promotes the "water drive" in adolescent newts, nest building in birds, and milk production in mammals and in pigeons, should also play an important role in the control of pregnancy. Indeed, the preservation of corpora lutea during lactation, presumably owing to the luteotropic action of prolactin, effectively postpones the resumption of estrous cycles until after weaning in some species. Unlike other hormones produced by the anterior pituitary, the secretion of prolactin is regulated by inhibitory factors from the hypothalamus. Thus, if the anterior pituitary is grafted elsewhere in the body its separation from the hypothalamus effectively prevents the secretion of most other tropic hormones while permitting the uncontrolled production of prolactin. This may explain why, following hypothalamic lesions in newborn rats, the ovaries grow to excessive dimensions due to the accumulation and persistence of innumerable corpora lutea (Fig. 166) (25).

Whether the eventual destruction of the corpora lutea is a case of murder or suicide remains to be determined. There is reason to believe that the persistence of luteal tissue depends upon its ability to synthesize and secrete progesterone. Since the target organ of progesterone is the uterus, particularly the endometrium, it is not unreasonable that this should play a role in determining the fate of the corpora lutea. It is well established that pregnancy prolongs the survival of corpora lutea (which in the absence of conception would begin to regress). In this way the source of progesterone needed to maintain pregnancy is preserved, at least until the placentas can take over this function. It would appear that the uterus is more important than the embryos in preventing luteal regression, for pseudopregnancy has an effect similar to pregnancy. Indeed, if the pseudopregnant uterus is traumatized to

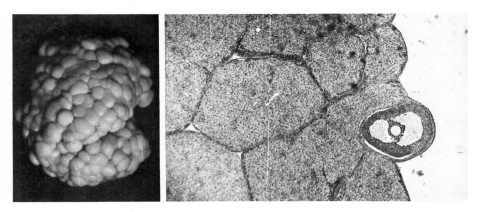

Fig. 166. Ovary from a rat subjected to hypothalamic lesion on its day of birth and unilaterally ovariectomized later. At 10 months of age, the remaining ovary is greatly enlarged (left) due to the accumulation of large numbers of corpora lutea which, for reasons not clearly understood, fail to degenerate as normal ones do. On the right is a histological section of the same ovary composed almost exclusively of corpora lutea but with occasional follicles (25).

induce the development of a deciduoma (see Chap. 22), the corpora lutea survive for the equivalent of nearly a full gestation period in the rat, a week or so longer than in the absence of the decidual reaction. If glass beads are introduced into the nongravid horn of the guinea pig uterus 5–8 days after mating, the regression of the corpora lutea in the ipsilateral ovary is prevented (4). If implanted before or after this period, the corpora lutea regress on the operated side. In either case, they persist in the contralateral ovary associated with the gravid horn of the uterus. It is between the fifth and eighth days that the endometrium is capable of decidualization in response to endometrial trauma. Thus, it is the reaction of the endometrium to implantation of embryos (or other kinds of traumatization) that is responsible for sparing the corpora lutea from regression. This effect is not to be explained in terms of the production of luteotropic substances by the decidual reaction, but by withholding the production of substances involved in luteal destruction. Such an interpretation is necessitated by the observation that hysterectomy results in the prolonged persistence of corpora lutea in a variety of animals (1, 7, 8, 17), presumably because this removes the source of the agent responsible for promoting the regression of corpora lutea. This agent is luteolysin.

Luteolysin is normally produced by the nonpregnant, nontraumatized uterus. Its activity is present in extracts of the endometrium and is carried to the ovaries via the blood or lymphatic vessels. In some species, such as the pig, the effects of luteolysin are so strong as to terminate pregnancy when there are only a few embryos in one horn and the opposite one is sterile (23). In rats, parabiosis of a hysterectomized pseudopregnant animal to an intact one leads to the termination of pseudopregnancy when luteolysin from the latter animal crosses into her partner (2).

The deleterious effects of luteolysin on the corpora lutea are accompanied by interference with the production of progesterone. Under the influence of luteolysin, there is an increase in 20α-hydroxysteroid dehydrogenase (20α-OH-SDH), an enzyme that catalyzes the conversion of progesterone into its 20α-dihydro derivative, an inactive form. The alkaloid drug ergocornine mimics this effect (41), being active in pseudopregnant rats or pregnant ones during the first week of gestation before placental prolactin secretion. These reactions can be prevented by the exogenous administration of prolactin presumably owing to its luteotropic action. However, since ergocornine inhibits prolactin secretion by the pituitary, its action on the corpus luteum may be mediated by a decline in prolactin, a hormone that normally interferes with the activity of 20α-OH-SDH. It is this body of evidence that suggests that the sustenance of the corpora lutea depends on their capacity to produce progesterone and that inactivation of this hormone leads to luteolysis.

PINEAL CONTROL OF BREEDING SEASONS

Since most animals have recurrent breeding seasons, their gonads are especially sensitive to photoperiods. It is the change in day length that serves as the environmental cue that the majority of organisms monitor to signal, if not to anticipate, changes in the seasons. Accordingly, the growth and activity of the ovaries are profoundly affected by changes in the light cycle. The pineal gland plays a pivotal role in coordinating such cycles with the environment. Studies on laboratory rodents have confirmed that the pineal secretes melatonin in the dark, an activity inhibited by exposure to illumination. Melatonin in turn blocks the secretion of gonadotropins by the pituitary thereby exerting an inhibitory effect on gonadal growth and activity (37). These relationships are consistent with the fact that birds and small mammals tend to breed when the days are long and melatonin secretion by the pineal is minimal.

Pinealectomy promotes growth of the gonads owing to diminished melatonin secretion (10, 31). It also enhances compensatory ovarian hypertrophy after unilateral ovariectomy. If an animal is blinded, however, the magnitude of compensatory ovarian hypertrophy declines, an effect attributable to increased secretion of melatonin by the pineal (10, 34).

Larger animals with longer gestation periods tend to breed in the autumn when the days are decreasing in length. This insures that birth will occur in the spring. Therefore, the relationship between pineal and gonads must be different in these forms from that in rodents in which short gestations permit breeding in the spring and summer. If the pineal is responsible for the seasonal coordination of reproduction in ungulates, it would seem that either melatonin must be secreted in response to light instead of darkness, or that its action is stimulatory instead of inhibitory on the production of gonadotropins. Some such modification of the regulatory mechanism would be required in order to inusre that the gonads would become functional during short photoperiods. The alternative would be to have adopted a mechanism altogether independent of that known to exist in small mammals and birds, an intriguing possibility worthy of investigation.

REFERENCES

1. Anderson, L.L. Effects of hysterectomy and other factors on luteal function. In "Handbook of Physiology," Endocrinology II, Part 2 (R.O. Greep and E.B. Astwood, eds.), pp. 69–86. Williams & Wilkins, Baltimore, Maryland, 1973.
2. Anderson, R.R. Uterine luteolysis in the rat: Evidence for blood borne and local actions. J. Reprod. Fertil. 16, 423–431 (1968).

3. Benson, B., Sorrentino, S., and Evans, J.S. Increase in serum FSH following unilateral ovariectomy in the rat. *Endocrinology* **84**, 369–374 (1969).

4. Blatchley, F.R., and Donovan, B.T. The relationship between the uterus and ovaries during early pregnancy in the guinea-pig. *J. Endocrinol.* **53**, 295–301 (1972).

5. Bond, C.J. An inquiry into some points in uterine and ovarian physiology and pathology in rabbits. *Br. Med. J.* **2**, 121–127 (1906).

6. Brinkley, H.J., and Young, E.P. Effects of unilateral ovariectomy or the unilateral destruction of ovarian components on the follicles and corpora lutea of the nonpregnant pig. *Endocrinology* **84**, 1250–1256 (1969).

7. Caldwell, B.V., Moor, R.M., and Lawson, R.A.S. Effects of sheep endometrial grafts and extracts on the length of pseudopregnancy in the hysterectomized hamster. *J. Reprod. Fertil.* **17**, 567–569 (1968).

8. Clemens, J.A., Minaguchi, H., and Meites, J. Relation of local circulation between ovaries and uterus to lifespan of corpora lutea in rats. *Proc. Soc. Exp. Biol. Med.* **127**, 1248–1251 (1968).

9. De Greef, W.J., Dullaart, J., and Zeilmaker, G.H. Serum luteinizing hormone, follicle-stimulating hormone, prolactin and progesterone concentrations and follicular development in the pseudopregnant rat after unilateral ovariectomy. *J. Endocrinology* **66**, 249–256 (1975).

10. Dickson, K., Benson, B., and Tate, G., Jr. The effect of blinding and pinealectomy in unilaterally ovariectomized rats. *Acta Endocrinol.* **66**, 177–182 (1971).

11. Dufour, J., Ginther, O.J., and Casida, L.E. Compensatory hypertrophy after unilateral ovariectomy and destruction of follicles in the anestrous ewe. *Proc. Soc. Exp. Biol. Med.* **138**, 1068–1072 (1971).

12. Flynn, T.T. The uterine cycle of pregnancy and pseudopregnancy as it is in the diprotodont marsupial *Bettongia cuniculus*. *Proc. Linn. Soc. N.S.W.* **55**, 506–531 (1930).

13. Gardner, W.U. Development and growth of tumors in ovaries transplanted into the spleen. *Cancer Res.* **15**, 109–117 (1955).

14. Greenwald, G.S. Temporal relationship between unilateral ovariectomy and the ovulatory response of the remaining ovary. *Endocrinology* **71**, 164–166 (1962).

15. Groenendijk-Huijbers, M.M. The right ovary of the chick embryo after early sinistral castration. *Anat. Rec.* **153**, 93–106 (1965).

16. Horvarth, E., Kovacs, K., Barg, B.D., and Tuchweber, B. Ultrastructure of the rat ovary transplanted into the spleen. *Gynecol. Invest.* **4**, 73–83 (1973).

17. Howe, G.R. The uterus and luteal activity in the rat, hamster, and rabbit. *Fertil. Steril.* **19**, 936–944 (1968).

18. Hunter, J. "Observations on Certain Parts of the Animal Oeconomy." London, 1972.

19. Jones, R.E., Gerrard, A.M., and Roth, J.J. Endocrine control of clutch size in reptiles. II. Compensatory follicular hypertrophy following partial ovariectomy in *Anolis carolinensis*. *Gen. Comp. Endocrinol.* **20**, 550–555 (1973).

20. Kornfeld, W., and Nalbandov, A.V. Endocrine influences on the development of the rudimentary gonad of fowl. *Endocrinology* **55**, 751–761 (1954).

21. Labhsetwar, A.P. Age-dependent changes in the pituitary-gonadal relationship: A study of ovarian compensatory hypertrophy. *J. Endocrinol.* **39**, 387–393 (1967).

22. Lee, S. The effect of Eck fistula upon intrasplenic ovarian neoplasm formation. *Surg. Forum* **23**, 110–112 (1972).

23. Longenecker, D.E., and Day, B.N. Maintenance of corpora lutea and pregnancy in unilaterally pregnant gilts by intrauterine infusion of embryonic tissue. *J. Reprod. Fertil.* **31**, 171–177 (1972).

24. Mallampati, R.S., and Casida, L.E. Absence of ovarian compensatory hypertrophy after unilateral ovariectomy during the anestrous season in the ewe. *Proc. Soc. Exp. Biol. Med.* **134**, 237–240 (1970).

25. Martinovitch, P.N., Ivansievic, O.K., and Martinovic, J.V. Induction of hyperluteinization and precocious opening of the vagina in rats with a transverse cut in the hypothalalmus made shortly after birth. *Nature (London)* **217**, 866–867 (1968).

26. McLaren, A. Mechanism of ovarian compensation following unilateral ovariectomy in mice. *J. Reprod. Fertil.* **6**, 321–322 (1963).

27. Mittwoch, U., and Kirk, D. Superior growth of the right gonad in human foetuses. *Nature (London)* **257**, 791–792 (1975).

28. Pearson, O.P. Reproduction of a South American rodent, the mountain viscacha. *Am. J. Anat.* **84**, 143–173 (1949).

29. Peppler, R.D. Effect of removing one ovary and a half on ovulation number in cycling rats. *Experientia* **31**, 243 (1975).

30. Peppler, R.D., and Greenwald, G.S. Effects of unilateral ovariectomy on ovulation and cycle length in 4- and 5-day cycling rats. *Am. J. Anat.* **127**, 1–8 (1970).

31. Reiter, R.J. Pineal-mediated regression of the reproductive organs of female hamsters exposed to natural photoperiods during the winter months. *Am. J. Obstet. Gynecol.* **118**, 878–880 (1974).

32. Sato, A. Some observations on induced polyovulation and superpregnancy in mature mice. *Embryologia* **7**, 285–294 (1963).

33. Short, R.E., Peters, J.B., First, N.L., and Casida, L.E. Effect of unilateral ovariectomy at three stages of the estrous cycle on the activity of the remaining ovary and pituitary gland. *J. Anim. Sci.* **27**, 691–696 (1968).

34. Sorrentino, S., Jr., and Benson, B. Effects of blinding and pinealectomy on the reproductive organs of adult male and female rats. *Gen. Comp. Endocrinol.* **15**, 242–246 (1970).

35. Staigmiller, R.B., First, N.L., and Casida, L.E. Compensatory growth and function of luteal tissue following unilateral ovariectomy during early pregnancy in pigs. *J. Anim. Sci.* **39**, 752–758 (1974).

36. Taber, E., Knight, J., Flowers, J., Gambrell, D., and Clayton, M. Hormonal inhibition of medullary development of the right gonad in sinistrally ovariectomized brown Leghorn fowl. *Anat. Rec.* **122**, 451–452 (1955).

37. Vaughan, M.K., Vaughan, G.M., Blask, D.E., and Reiter, R.J. Influence of melatonin, constant light, or blinding on reproductive system of gerbils (*Meriones unguiculatus*). *Experientia* **32**, 1341–1342 (1976).

38. Weir, B.J. Some notes on reproduction in the Patagonian mountain viscacha, *Ligidum boxi* (Mammalia: Rodentia). *J. Zool.* **164**, 463–467 (1971).

39. Welschen, R. Compensatory ovarian growth and compensatory ovulation after unilateral ovariectomy in rats with an ovarian autograft in the region of the portal vein. *Acta Endocrinol.* **65**, 509–516 (1970).

40. Welschen, R., and Dullaart, J. Serum concentrations of follicle-stimulating hormone and luteinizing hormone after unilateral ovariectomy in the adult rat. *J. Endocrinol.* **63**, 421–422 (1974).

41. Zeilmaker, G.H. Effect of ergocornine methanesulphonate on the luteotrophic activity of the ectopic mouse trophoblast. *Acta Endocrinol.* **59**, 442–446 (1968).

22

The Pregnant Uterus

Metrium: Gr., *metron*, measure (as of the menstrual cycle)

Pregnancy is the essence of motherhood. Its success depends upon the adaptive reactions of the uterus, reactions that are both chemical and mechanical. Dominated by endocrine control from the ovaries, the dimensions of the uterus wax and wane with each reproductive cycle. The hormonal interplay between uterus and ovary coordinates the receptivity of the endometrium to coincide with the arrival of the blastocyst, thus optimizing successful implantation. Next, the embryonic trophoblast and the uterine endometrium join forces to produce a placenta. Meanwhile, the conceptus and the fluids within which it is immersed distend the uterus to the hypertrophic proportions for which it is famous (Fig. 167). Finally, postpartum involution restores the uterus to its original size.

The uterus is especially sensitive to estrogen, a hormone that promotes proliferation of its endometrium and stimulates hypertrophy of the smooth muscle cells in the myometrium (15). Pinealectomy, or exposure to constant illumination, likewise promotes uterine hypertrophy by stimulating estrogen secretion from the ovaries (17). It is the prior conditioning of the uterus under the influence of estrogen that enhances the subsequent effects of progesterone in preparing the uterus for implantation by the fertilized egg. Ovariectomy tends to cause uterine atrophy, an effect stimulated by exposure to shortened days, total darkness, or blinding (16). The decidual reaction to trauma is also inhibited or reversed in spayed animals (21). On the other hand, ovariectomy does not prevent growth of the distended uterus nor interfere with the healing of wounds (8).

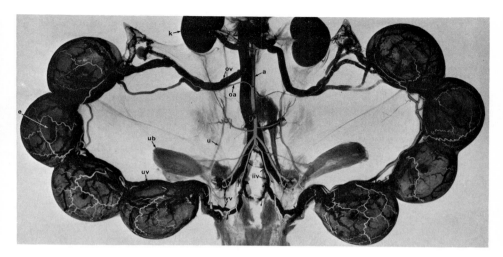

Fig. 167.   Pregnant cat uterus with vasculature injected with latex (arteries white, veins dark) (7). a, aorta; e, embryo; iiv, internal iliac vein; k, kidney; oa, ovarian artery; ov, ovarian vein; u, ureter; ub, urinary bladder; uv, uterine vein; vv, vaginal vein.

DISTENTION AND REGRESSION

The uterus is famous for its growth in response to being stretched. The distention of pregnancy is responsible for hyperplasia of the smooth muscle fibers, especially during the first trimester, and particularly in the vicinity of implantation sites (2, 22). During the latter two-thirds of pregnancy, smooth muscle hypertrophy predominates, the diameters of the muscle fibers increasing as much as 50% in the gravid uterus. Although these effects normally follow upon the hormonally induced progestational changes in the uterus, they can be mimicked by strictly mechanical interventions in the absence of pregnancy. One way to achieve this is to tie off a segment of uterus with double ligations to allow fluid to accumulate in the lumen (18). The resulting hydrometria distends the uterus and stimulates growth of the myometrium. Similarly, implantation of paraffin pellets (18) or inflation of a balloon inserted in the lumen (8) likewise stimulates myometrial mitosis and hypertrophy in the distended horn, effects that are observed even after ovariectomy. It is noteworthy that the atrophy that normally takes place following delivery can also be prevented by artificial distention of the postpartum uterus (4).

The resilience of the uterus is illustrated by its involution after birth. The mass of the uterus is abruptly reduced during the first week or two following parturition (13). Excess collagen disappears and the size of smooth muscle fibers is decreased (26). Although uterine involution is due in large measure to the release of intraluminal pressure, it is not unaffected by less mechanical stimuli. For example, if one horn of a postpartum uterus is injured by repeated crushing, the resulting wound healing response retards the loss in weight which normally occurs after birth (27). Hormones are also involved. Active lactation promotes involution owing to the depression of ovarian follicle activity (5), while postpartum atrophy of the uterus is slower to occur in mothers deprived of their litters.

THE DECIDUAL REACTION

The uterus is naturally endowed with an ability to repair injuries. This is expressed in the recurrent healing of the endometrium following the loss of its decidual layers in species that menstruate. Even in those that do not, the implantation of the blastocyst is achieved by local traumatization of the endometrial epithelium, an event that elicits a growth response in the decidual tissues which give rise to the maternal components of the placenta. In the absence of a fertilized ovum, injury to the endothelium at this stage of the cycle stimulates massive overgrowth in the endometrial connective tissue. In view of these regenerative responses, it is not surprising that the uterus can also repair more extensive injuries with considerable alacrity. For example, if the uterus is slit lengthwise, the opening is closed in zipper-like fashion from either end by the ingrowth of connective tissue stroma and epithelium from the two cut edges (20). Transversely sectioned uteri, the cut edges of which become separated, can heal only to themselves, not one another.

The reaction to trauma is especially interesting in the case of the pseudopregnant uterus. Pseudopregnancy can be induced by mating to a vasectomized male, or simply by stimulation of the cervix mechanically or electrically at the time of estrus. In the rat, for example, estrus occurs every four or five days as indicated by the presence of cornified epidermal cells in vaginal smears (instead of leukocytes or nucleated epidermal cells). If the cervix is stimulated at this time, neural transmission to the hypothalamus is responsible for the surge of LH secreted from the pituitary which triggers ovulation. The development of corpora lutea in the ovaries and the resultant secretion of progesterone prepares the uterus for implantation. Since mating in rats is normally tantamount to conception, the lack of fertilized ova is an unprecedented situation for which the uterus is not prepared. The corpora lutea persist and the uterus remains progestational for about 13 days in the pseudopregnant rat, 11 in the mouse, or 9 in the hamster, by which time the termination of pseudopregnancy is characterized by uterine regression, destruction of the corpora lutea, and resumption of normal reproductive cycles. Pseudopregnancy can usually be prolonged, however, if deciduomata are induced to develop in the uteri (Fig. 168). These are transient overgrowths of the endometrium which swell the uterus to dimensions approaching those attained in pregnancy.

To induce the production of deciduomata, it is necessary only to simulate implantation in the progestational uterus. This can be done in a variety of ways, all of which traumatize the endometrium. The most common methods involve piercing the uterus with needle and thread, or scraping the endometrium with a sharp instrument. However, the injection into the lumen of the pseudopregnant uterus of a variety of substances, including certain oils, salt solutions, or biologically active materials (e.g., trypsin, histamine), has also been shown to be effective (6, 11, 12, 19). The nature of the trauma would therefore appear to be relatively nonspecific.

What is more important is the timing. The sensitivity of the endometrium to trauma is synchronized with when the blastocysts arrive in the uterus at a stage when they are ready to implant. This is 4–6 days after ovulation in the rat, or between the fifth and eighth days in the guinea pig and rabbit. The incidence of deciduoma formation drops off sharply on either side of the peak of responsiveness. When successful, however, the decidual reaction leads to the production of a massive tumescense of the endometrium. The resulting deciduoma rapidly develops to its maximum size in approximately one week, after which it undergoes a gradual degeneration (Fig. 168). Its presence, like that of the placenta, is responsible for the prolongation of pseudopregnancy by interfering with the production of luteolysin from the uterus (see Chap. 21). Consequently, the corpora lutea survive and continue to secrete progesterone which in turn preserves the pseudopregnant condition of the uterus. Eventually, however, the charade must end, which it does when the degenerating deciduoma is eventually sloughed into the uterine lumen. Although the duration of pseudopregnancy is proportional to the strength of the decidual reaction, in the rat it generally lasts for a period approximately equal to the normal gestation period, after which the uterus reverts to its original dimensions (24, 25).

Concomitant with the development of deciduomata or placentae is the differentiation of the metrial gland. Situated on the mesometrial side of the uterus, this little-known organ is derived from mesodermal cells of the endometrium and myometrium (10). It reaches its maximum development in the rat about two weeks after ovulation and may survive into the period of lactation. Nevertheless, it is a transient organ the function of which remains

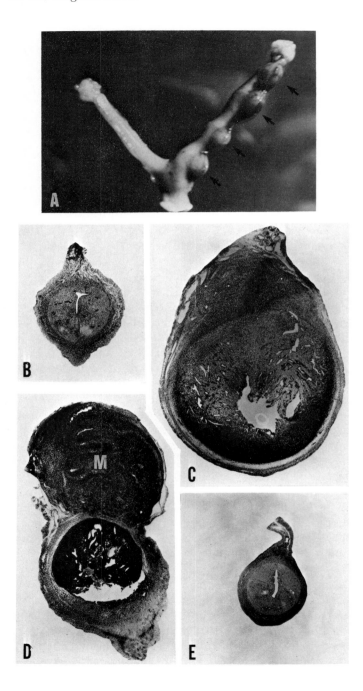

unknown. Its life span is said to be lengthened by progesterone and shortened by estrogen.

The reaction of the uterus to traumatization, not to mention the presence of developing embryos, is important to our understanding of how intrauterine devices (IUD) interfere with conception. A variety of foreign bodies in the uterine lumen, from silk or nylon sutures to plastic or metal wires, has been shown not to prevent fertilization but to interfere with implantation of the embryo (9, 14, 28). In animals with two uterine horns, one of which contains an IUD, pregnancy is not necessarily prevented in the contralateral side (1, 9). Although the physical presence of an IUD is not without its effects, its composition is an important consideration. Some metals, for example, prevent implantation more than others, copper being one of the most effective, while inert metals such as silver and gold are poor contraceptives (3, 28). Although IUD's have been shown in some cases to induce the production of deciduomata (23), presumably owing to endometrial trauma, their presence has also been claimed to interfere with the development of deciduomata (1). Much remains to be learned about the interactions between the uterus and the contents of its lumen, not only to enhance our knowledge of how to prevent pregnancy but to improve its chances of success when it occurs.

REFERENCES

1. Bartke, A. Effect of an IUD on implantation and the decidual reaction in different strains of mice. *J. Reprod. Fertil.* 15, 185–190 (1968).
2. Bulmer, D., and Peel, S. An autoradiographic study of cellular proliferation in the uterus and placenta of the pregnant rat. *J. Anat.* 117, 433–441 (1974).
3. Chang, C.C., Tatum, H.J. and Kincl, F.A. The effect of intrauterine copper and other metals on implantation in rats and hamsters. *Fertil. Steril.* 21, 274–278 (1970).
4. Csapo, A., Erdos, T., de Mattos,, C.R., Gramss, E., and Moscowitz, C. Stretch-induced uterine growth, protein synthesis and function. *Nature (London)* 207, 1378–1379 (1965).
5. Dawson, A.B. The effects of lactation on the postpartum involution of the uterus of the cat. *Am. J. Anat.* 79, 241–265 (1946).
6. De Feo, V.J. Determination of the sensitive period for the induction of deciduomata in the rat by different inducing procedures. *Endocrinology* 73, 488–497 (1963).

Fig. 168. Above, deciduomata in the rat uterus (A). The animal was first made pseudopregnant by cervical stimulation at estrus, and 4 days later her right uterine horn was pierced in 4 locations by a needle and thread. After one week a deciduoma (arrows) had developed at each of the traumatized loci. Below, cross sections through rat uteri 5, 10, 18, and 20 days (B–E) after onset of pseudopregnancy and traumatization of the endometrium on the fourth day. The decidual reaction reaches its peak about a week after trauma (A, C), then degenerates and is sloughed into the lumen (D) as the uterus reverts to normal (E). The metrial gland (M) is conspicuous 2 weeks after trauma (D). (Courtesy of Dr. Richard A. Ellis, Brown University.)

7. Del Campo, C.H., and Ginther, O.J. Arteries and veins of uterus and ovaries in dogs and cats. *Am. J. Vet. Res.* **35**, 409–415 (1974).

8. De Mattos, C.E.R., Kempson, R.L., Erdos, T., and Csapo, A. Stretch-induced myometrial hypertrophy. *Fertil. Steril.* **18**, 545–556 (1967).

9. Doyle, L.L., and Margolis, A.J. The effect of an IUFB on reproduction in mice. *J. Reprod. Fertil.* **11**, 27–32 (1966).

10. Ellis, R.A. Histochemistry of the cellular components of the metrial gland of the rat during prolonged pseudopregnancy. *Anat. Rec.* **129**, 39–52 (1957).

11. Elton, R.L. The decidual cell responses in rabbits. *Acta Endocrinol.* **51**, 543–550 (1966).

12. Humphrey, K.W., and Martin, L. Attempted induction of deciduomata in mice with mast-cell, capillary permeability, and tissue inflammatory factors. *J. Endocrinol.* **42**, 129–141 (1968).

13. Maibenco, H.G. Connective tissue changes in postpartum uterine involution in the albino rat. *Anat. Rec.* **136**, 59–67 (1960).

14. Marston, J.H., and Chang, M.C. Contraceptive action of intra-uterine devices in the rabbit. *J. Reprod. Fertil.* **18**, 409–418 (1969).

15. Martin, L., Finn, C.A., and Trinder, G. Hypertrophy and hyperplasia in the mouse uterus after oestrogen treatment: An autoradiographic study. *J. Endocrinol.* **56**, 133–144 (1973).

16. Reiter, R.J., Hoffman, R.A., and Hester, R.S. The effects of thiourea, photoperiod and the pineal gland on the thyroid, adrenal and reproductive organs of female hamster. *J. Exp. Zool.* **162**, 263–268 (1966).

17. Reiter, R.J., and Klein, D.C. Observations on the pineal gland, the Harderian glands, the retina, and the reproductive organs of adult female rats exposed to continuous light. *J. Endocrinol.* **51**, 117–125 (1971).

18. Reynolds, S.R.M., and Kaminester, S. The rate of uterine growth resulting from chronic distention. *Anat. Rec.* **69**, 281–286 (1937).

19. Salvatore, C.A. Significance of the myometrial cell hypertrophy during pregnancy. (Critical review.) *Ergeb. Allg. Pathol. Anat.* **42**, 148–169 (1962).

20. Selye, H., and McKeown, T. On the regenerative power of the uterus. *J. Anat.* **69**, 79–81 (1934).

21. Selye, H., and McKeown, T. Studies of the physiology of the maternal placenta in the rat. *Proc. Roy. Soc. London, Ser. B* **119**, 1–31 (1935).

22. Shelesnyak, M.C. Decidualization: The decidua and the deciduoma. *Perspect. Biol. Med.* **5**, 503–518 (1962).

23. Tobert, J.A. Induction of deciduomata by intrauterine copper in the rabbit. *J. Reprod. Fertil.* **45**, 197–200 (1975).

24. Velardo, J.T., Dawson, A.B., Olsen, A.G., and Hisaw, F.L. Sequence of histological changes in the uterus and vagina of the rat during prolongation of pseudopregnancy associated with the presence of deciduomata. *Am. J. Anat.* **93**, 273–305 (1953).

25. Velardo, J.T., Olsen, A.G., Hisaw, F.L., and Dawson, A.B. The influence of decidual tissue upon pseudopregnancy. *Endocrinology* **53**, 216–220(1953).

26. Woessner, J.F., and Brewer, T.H. Formation and breakdown of collagen and elastin in the human uterus during pregnancy and post-partum involution. *Biochem. J.* **89**, 75–82 (1963).

27. Woessner, J.F., Jr., and Celio, J.R. Effect of injury on collagen resorption in the involuting rat uterus. *Proc. Soc. Exp. Biol. Med.* **147**, 475–478 (1974).

28. Zipper, J., Mendel, M., and Prager, K. Suppression of fertility by intrauterine copper and zinc in rabbits. *Am. J. Obstet. Gynecol.* **105**, 529–534 (1969).

23

The Placental Connection

Placenta: Gr., *plakous*, a flat cake

By definition, viviparity is the establishment of an intimate association between mother and offspring for the exchange of nutrients and waste products. The placenta not only facilitates communication, but also serves as a barrier to the transfer of cells, pathogenic organisms, and large molecules. It is an organ of respiration, nutrition, and excretion. It is also an endocrine gland, secreting gonadotropins and progesterone. Above all, the placenta protects the antigenic individuality of the embryo and guards against its immunological rejection by the mother.

ON THE NATURE OF VIVIPARITY

Mammals did not invent the placenta, for its origins, at least in the uterus, date back to certain adaptations in some of the reptiles in which the morphological complementariness between mother and offspring in the oviduct are suggestive of physiological interchange (9). In lower forms there are even stranger adaptations. The Surinam toad is famous for its habit of depositing its eggs in integumental pockets on the back of the mother. Here the embryos develop disproportionately enlarged tail fins, the vascularity of which suggests they may be involved in the absorption of nutrients secreted by the maternal tissues. In certain species of fishes that give birth to live young, the fertilized ovum develops in a cavity within the ovary where it is nourished by maternal secretions. There is a shark in which the inner lining of the oviduct develops long finger-like projections which grow through the spiracles down into the stomach of the developing young where nutrient secretions are produced. Viviparity also occurs in various orders of insects, perhaps the most famous case being that of the aphids, the offspring in some species being so precocious as to be pregnant with the next generation before they themselves are born. Perhaps the most outstanding example among the invertebrates is to be found in *Peripatus*, that hard-to-classify inhabitant of the tropics which shares as many attributes with annelids as it does with arthropods (Fig. 169). In the absence of specifically designed copulatory organs, fertilization is said to occur by the direct deposition of semen onto the body surface of the female where the sperm make their way inside her body to fertilize the eggs. The embryos then establish placental connections with the oviduct along which they progress during the subsequent course of gestation (5). Thus, there are many arrangements between mothers and their embryos to insure the survival of the next generation, and whether they are classified as oviparous or viviparous depends entirely on whether the nutrients of maternal origin are delivered before or after fertilization.

In mammals the placenta is usually consolidated into a single organ. Ungulates, however, tend to have multiple placentas, or cotyledons, each one plugged into the uterine wall. In either case, the intimacy with which the maternal and fetal circulations are associated varies from species to species (4). In some cases the embryonic trophoblast is in apposition with the endometrial epithelium, diffusion between the two bloodstreams being required to traverse these epithelia, the two endothelial linings of the maternal

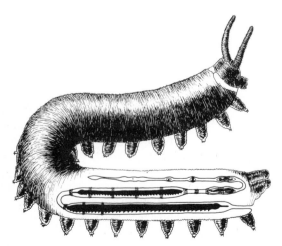

Fig. 169.   Schematic view of the gravid uterus of a peripatus (5). Ova produced from the ovary at the upper end of the uterus establish placentas for the nourishment of early embryos. As development progresses the embryos (black) move distally, eventually losing their placental relationships with the uterus.

and fetal blood vessels as well as the intervening connective tissues. At the opposite extreme the two circulatory systems are separated only by a single layer of fetal endothelium, an arrangement that reduces the placental barrier to an absolute minimum. Nature has never overstepped this limitation by allowing the maternal and fetal circulations to intermingle. To have done so would presumably have created more problems, both hemodynamic and immunological, than would have been worth the advantages of establishing confluent blood supplies.

Although all mammals with the exception of the echidna and playtypus are viviparous, birds are exclusively oviparous. Ovoviviparity is not uncommon among reptiles but has never been adapted by the birds, presumably as a concession to their aerial existence. Whether flightless birds would ever evolve in this direction, or even become viviparous, is an interesting matter for conjecture. It is even more interesting to contemplate the effects that oviparity might have had on higher mammals such as man. Our habitat would have been confined to warm climates unless we became seasonal breeders, and we might have retained a coat of fur with which to line the nest and assist in incubation. Aside from its obvious effects on our social structure, oviparous reproduction probably would have caused us to gain an egg tooth and to lose our navels.

RELATIONS TO THE EMBRYO AND FETUS

Throughout most of gestation the weight of the placenta surpasses that of the embryo. Eventually, however, the exponential growth of the latter overtakes the more linear enlargement of the former (15, 24). Although the growth of the placenta may become more hypertrophic than hyperplastic during the final phases of gestation (50), it undergoes a fivefold increase in weight during the last week of pregnancy in the rat (17).

Almost any condition affecting the size of the fetus likewise influences the dimensions of the placenta. For example, the number of offspring in the litter is inversely proportional to both fetal and placental size (35, 52). Overcrowding may even result in the fusion of adjacent placentas (36). There is evidence that systemic influences may also be operating, for overcrowding in one horn of the uterus is sometimes accompanied by reductions in the sizes of fetuses and placentas in the opposite horn. Nevertheless, the effects of crowding do not become evident until later in gestation when a critical mass is attained. In humans, singletons and twins grow at equivalent rates throughout the first two trimesters of pregnancy, and not until after the thirtieth week of gestation does the size of each twin fetus and placenta drop below that of singles (34). In the case of identical twins, both of which share a

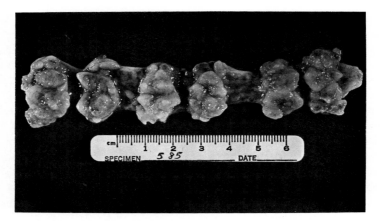

Fig. 170.   Persistent maternal placentas in a rabbit uterus after 18 days of gestation following removal of the embryos by aspiration on day 8. The uterus has been slit open antimesometrially (23).

common placenta, its size is correlated with the combined weights of the two twins, being somewhat larger than that of single infants but smaller than the combined placental weights of fraternal twins.

The possibility that placental growth might be regulated by that of the fetus can be tested by experiments on the fetus itself. A variety of investigations on rats, rabbits, and monkeys involving the removal or destruction of the embryo or fetus has yielded similar results (Fig. 170). In general, fetectomy does not interfere with the survival of the placenta, nor does it prevent its continued growth (23, 39, 40, 43). Despite the deprivation of blood flow to the fetal component of the placenta, the trophoblast persists, although in the case of the monkey the villi may degenerate (32). Nevertheless, it is remarkable that the placenta, in the absence of its fetus, continues to enlarge during the remainder of the gestation period, finally to be expelled at term (40), but sometimes surviving even longer (43). Although the rate of placental growth under these circumstances may fall short of that occurring when the fetus remains intact, these results emphasize that the development of the placenta is considerably more autonomous than one might otherwise be led to believe.

MATERNAL FACTORS CONTROLLING PLACENTAL GROWTH

The development of the embryonic trophoblast into a placenta normally occurs in association with the endometrium. However, it is possible for a

5. Anderson, D.T., and Manton, S.M. Studies on the Onychophora. VIII. The relationship between the embryos and the oviduct in the viviparous placental onychophorans *Epiperipatus trinidadensis* Bouvier and *Macroperipatus torquatus* (Kennel) from Trinidad. *Phil. Trans. Roy. Soc.* **264**, 161–189 (1972).

6. Avery, G.B., and Hunt, C.V. Giant cell formation in ectopic mouse trophoblast. *Exp. Cell Res.* **74**, 3–8 (1972).

7. Beer, A.E., and Billingham, R.E. Implantation, transplantation, and epithelial-mesenchymal relationships in the rat uterus. *J. Exp. Med.* **132**, 721–736 (1970).

8. Beer, A.E., and Billingham, R.E. The embryo as a transplant. *Sci. Am.* **230**(4), 36–46 (1974).

9. Bellairs, A. "The Life of Reptiles." Weidenfeld and Nicholson, London, 1969.

10. Billington, W.D. Influence of immunological dissimilarity of mother and foetus on size of placenta in mice. *Nature (London)* **202**, 317–318 (1964).

11. Boshier, D.P., and Moriarty, K.M. Some effects on the conceptus of prior immunological sensitization of ewes to the sire. *J. Reprod. Fertil.* **21**, 495–500 (1970).

12. Bright, A.S., and Maser, A.H. Advanced abdominal pregnancy. *Obstet. Gynecol.* **17**, 316–324 (1961).

13. Bruce, N.W. The effect of ligating a uterine artery on fetal and placental development in the rat. *Biol. Reprod.* **14**, 246–247 (1976).

14. Bruce, N.W., and Norman, N. Influence of sexual dimorphism on foetal and placental weights in the rat. *Nature (London)* **257**, 62–63 (1975).

15. Bruce, N.W., and Abdul-Karim, R.W. Relationship between fetal weight, placental weight and maternal placental circulation in the rabbit at different stages of gestation. *J. Reprod. Fertil.* **32**, 15–24 (1973).

16. Butterstein, G.M., and Leathem, J.H. Ribosomes in normal and giant placentae. *J. Endocrinol.* **48**, 473–474 (1970).

17. Butterstein, G.M., and Leathem, J.H. Placental growth modification during pregnancy in the rat. *Endocrinology* **95**, 645–649 (1974).

18. Csapo, A.I., and Wiest, W.G. An examination of the quantitative relationship between progesterone and the maintenance of pregnancy. *Endocrinology* **85**, 735–746 (1969).

19. Csapo, A.I., and Wiest, W.G. Plasma steroid levels and ovariectomy-induced placental hypertrophy in rats. *Endocrinology* **93**, 1173–1177 (1973).

20. Chang, M.C., Casas, J.H., and Hunt, D.M. Development of ferret eggs after 2 to 3 days in the rabbit Fallopian tube. *J. Reprod. Fertil.* **25**, 129–131 (1971).

21. Dawes, G.S. The physiological determinants of fetal growth. *J. Reprod. Fertil.* **47**, 183–187 (1976).

22. Fawcett, D.W., Wislocki, G.B., and Waldo, C.M. The development of mouse ova in the anterior chamber of the eye and in the abdominal cavity. *Am. J. Anat.* **81**, 413–443 (1947).

23. Hoffman, L.H., and Davies, D. Production of the maternal placenta in rabbits following aspiration of conceptuses. *J. Reprod. Fertil.* **26**, 255–257 (1971).

24. Ibsen, H.L. Prenatal growth in guinea-pigs with special reference to environmental factors affecting weight at birth. *J. Exp. Zool.* **51**, 51–91 (1928).

25. James, D.A. Effects of antigenic dissimilarity between mother and foetus on placental size in mice. *Nature (London)* **205**, 613–614 (1965).

26. James, D.A., Acierto, S., and Murphy, B.D. Growth of mouse trophoblast transplanted to syngeneic and allogeneic testes. *J. Exp. Zool.* **180**, 209–216 (1972).

27. Joshi, S.G., and Kraemer, D.C. Development of mouse embryos in uterine washings of rats and baboons bearing an intrauterine foreign body. *Contraception* **2**, 353–359 (1971).

28. Joubert, D.M., and Hammond, J. A crossbreeding experiment with cattle, with special reference to the maternal effect in South Devon-Dexter crosses. *J. Agr. Sci.* **51**, 325–341 (1958).

29. Kirby, D.R.S. Development of the mouse blastocyst transplanted to the spleen. *J. Reprod. Fertil.* **5**, 1–12 (1963).

30. Koshy, T.S., Sara, V.R., King, T.L., and Lazarus, L. The influence of protein restriction imposed at various stages of pregnancy on fetal and placental development. *Growth* **39**, 497–506 (1975).

31. Lawson, R.A.S., Adams, C.E., and Rowson, L.E.A. The development of sheep eggs in the rabbit oviduct and their viability after re-transfer to ewes. *J. Reprod. Fertil.* **29**, 105–116 (1972).

32. Lewis, J., Jr., and Hertz, R. Effects of early embryectomy and hormonal therapy on the fate of the placenta in pregnant Rhesus monkeys. *Proc. Soc. Exp. Biol. Med.* **123**, 805–809 (1966).

33. McKeown, T., and Record, R.G. Influence of pre-natal environment on correlation between birth weight and parental height. *Am. J. Hum. Genet.* **6**, 457–463 (1952).

34. McKeown, T., and Record, R.G. The influence of placental size on foetal growth in man, with special reference to multiple pregnancy. *J. Endocrinol.* **9**, 418–426 (1953).

35. McLaren, A. Genetic and environmental effects on foetal and placental growth in mice. *J. Reprod. Fertil.* **9**, 79–98 (1965).

36. McLaren, A., and Michie, D. The spacing of implantations in the mouse uterus. *Mem. Soc. Endocrinol.* **6**, 65–75 (1959).

37. Nuti, K.M., Ward, W.F., and Meyer, R.K. Embryonic survival and development in unilaterally ovariectomized rats. *J. Reprod. Fertil.* **26**, 15–24 (1971).

38. Oh, W., D'Amodio, M.D., Yap, L.L., and Hohenauer, L. Carbohydrate metabolism in experimental intrauterine growth retardation in rats. *Am. J. Obstet. Gynecol.* **108**, 415–421 (1970).

39. Petropoulos, E.A. Maternal and fetal factors affecting the growth and function of the rat placenta. *Acta Endocrinol.* **73**, (Suppl. 176), 9–69 (1973).

40. Pritchard, J.J., and St.G. Huggett, A. Experimental foetal death in the rat: Histological changes in the membranes. *J. Anat.* **81**, 212–224 (1947).

41. Pulkkinen, M.O., and Csapo, E. Placental hypertrophy in the rat induced by ovariectomy. *J. Reprod. Fertil.* **18**, 125–127 (1969).

42. Rogers, J.F., and Dawson, W.D. Foetal and placental size in a *Peromyscus* species cross. *J. Reprod. Fertil.* **21**, 255–262 (1970).

43. Shintani, S., Glass, L.E., and Page, E.W. Studies on induced malignant tumors of placental and uterine origin in the rat. I. Survival of placental tissue following fetectomy. *Am. J. Obstet. Gynecol.* **95**, 542–549 (1966).

44. Simmons, R.L. Histoincompatibility and the survival of the fetus: Current controversies. *Transplant. Proc.* **1**, 47–52 (1969).

45. Simmons, R.L., and Weintraub, J. Transplantation experiments on placental ageing. *Nature (London)* **208**, 82–83 (1965).

46. Tarkowski, A.K. Inter-specific transfers of eggs between rat and mouse. *J. Embryol. Exp. Morphol.* **10**, 476–495 (1962).

47. van der Vies, J., and Feenstra, H. The effects of ovarian hormones on the placenta of rats. *Acta Endocrinol.* **119**, 235 (1967).

48. Walton, A., and Hammond, J. The maternal effects on growth and conformation in Shire horse-Shetland pony crosses. *Proc. Roy. Soc. London, Ser. B* **125**, 311–335 (1938).

49. Warwick, B.L., and Berry, R.O. Inter-generic and intra-specific embryo transfers in sheep and goats. *J. Hered.* **40**, 297–303 (1949).

50. Winick, M., Coscia, A., and Noble, A. Cellular growth in human placenta. I. Normal placental growth. *Pediatrics* **39**, 248–251 (1967).
51. Winick, M., and Noble, A. Cellular growth in human placenta. II. Diabetes mellitus. *J. Pediat.* **71**, 216–219 (1967).
52. Zambrana, M.A., and Greenwald, G.S. Effects of fetal, ovarian, and placental weight of various number of fetuses in the rat. *Biol. Reprod.* **4**, 216–233 (1971).

24

From Embryo to Adult

Umbilicus: L., *umbo*, the protuberance from a shield

The distinction between organ and organism becomes obscure in the viviparous condition. It is customary to think of the mammalian embryo as an organism with its own intrinsic rate of growth no more influenced by its mother than a parasite is affected by its host. On the other hand, if the embryo is thought of as just another of its mother's organs, then one would expect its development to be affected by a variety of physiological influences impinging on it in the intrauterine ecology. Since neither of these perspectives is entirely incorrect, the problem is to determine the extent to which the developing embryo shares with its mother the control of its own growth (25).

MATERNAL INFLUENCES ON PRENATAL DEVELOPMENT

Maternal influences on embryonic and fetal growth fall into three categories: hormonal, nutritional, and physical. The single most important endocrine gland for fetal development is the maternal ovary, the loss of which early in gestation may interfere with implantation and bring about fetal death (5). Ovariectomy later in pregnancy, if accompanied by estrogen and progesterone therapy, tends to cause an increase in fetal weight (2). This effect, however, is prevented by hypophysectomy of the mother (7). Indeed, in the absence of the maternal pituitary, pregnancy is terminated in the first half of gestation unless ovarian hormones are administered. Fetal growth is accelerated under conditions of maternal diabetes, presumably owing to the nutritional effects of hyperglycemia.

It is well known that if a pregnant animal is sustained on a diet low in protein or caloric content her fetuses do not grow at normal rates and their postnatal weights are reduced (*18, 36*). Growth retardation in the malnourished fetus is accompanied by decreases in cell number, an effect that may be attributed to abnormally low rates of mitosis during early stages of development. This cannot be entirely made up once cellular differentiation has occurred, particularly in those organs in which proliferation is incompatible with specialization (*45*). Nevertheless, catch-up growth is still possible prenatally, as it is postnatally, upon refeeding normal diets to malnourished mothers (*18*).

Although mammalian embryos are capable of continuing their development to a considerable, but not unlimited, degree when cultured in nutrient medium in the absence of placental connections (Fig. 172) (*31*), it is the blood supply to the uterus which delivers the nourishment necessary for placental and fetal development *in vivo*. If this blood flow is reduced, the fetuses which depend upon it exhibit intrauterine growth retardation (*32, 34, 44*). The uterus receives arterial blood from two sources, the abdominal aorta and the iliac arteries, the former also giving rise to the ovarian arteries. In its course along the length of the uterus, the uterine artery gives off branches to each placenta, the vascular architecture of which resembles an arteriovenous shunt. If the uterine artery is ligated at its cervical end, those fetuses closest to the site of operation may fail to survive the effects of ischemia (*44*). Farther away, where arterial blood from the ovarian end of the uterus is available, the growth of viable fetuses exhibits varying degrees of retardation depending upon the extent to which their blood supplies have been curtailed (Fig. 173). The decrease in fetal growth under these circumstances may be as much as 50%, with concomitant reductions in DNA, RNA, and protein contents (*32*). Although the brains of such dysmature fetuses may be reduced in size, their DNA contents tend to be less affected than in other organs when the uterine blood flow is not restricted until the later stages of gestation when the hyperplastic phase of neural development is mostly completed (*32, 34*). In view of the dependence of fetal growth on blood supply, the correlation between fetal and placental size is not unexpected. Indeed, lesions in the placenta or ligation of umbilical vessels retards development of the fetus (*12, 30*).

It is difficult to separate hormonal and nutritional influences from the effects of position in the uterus on the growth of the fetus. In the mouse and guinea pig, the smallest fetuses are located in the middle of the uterine horn, the largest ones on either end. This distribution is believed to reflect the fact that in the terminal locations the fetuses are exposed to more copious blood flows at each end of the uterine arterial loop, while those in the middle of the

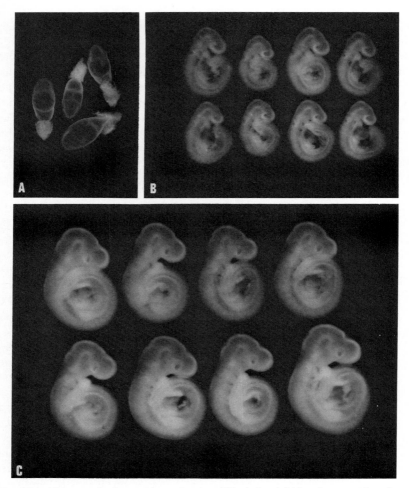

Fig. 172. The growth of rat embryos *in vitro* (*30*). A, head-fold embryos, within their embryonic membranes, at 9½ days of gestation when they were explanted. B, embryos grown for 32 hours *in vivo* (above) and *in vitro* (below). C, after 48 hours *in vitro* (below) the embryos have kept pace with those held *in vivo* (above) for the same period. All embryos at same magnification.

horn are farthest away from the source of supply (*28*). Indeed, it has been noted in the case of the mouse that the fetus at the ovarian end of the uterine horn is disproportionately smaller than the one next to it, its reduction being correlated with the fact that the anterior branch of the uterine artery is divided between the ovary and the placenta of the nearest fetus, to the disadvantage of the latter (Fig. 174). Such a size distribution among litter-

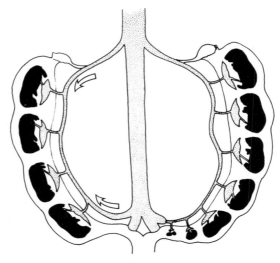

Fig. 173. Effects of ligating the uterine artery on fetal development in the rat (43). The uterine artery receives blood from both ends (arrows), and if it is tied off near the cervical end of the pregnant uterus (right) the nearest fetus may die. Others farther away from the occlusion experience growth retardation according to the degree of placental ischemia.

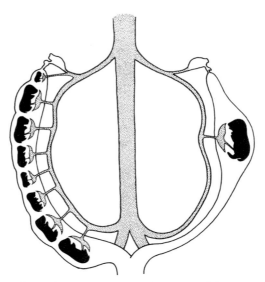

Fig. 174. Fetal growth in relation to position in the uterus of the mouse (28). There is normally a size gradient along the uterine horn such that the larger fetuses are at either end. The one adjacent to the ovary, with which it must share its blood supply, tends to be smaller. The single fetus on the right illustrates maximal growth in the absence of intrauterine crowding.

mates in the mouse argues in favor of hemodynamic control of fetal and placental growth (29). In the rat, however, a different size distribution obtains despite the similarity in vascular anatomy. Here, the largest fetuses are typically to be found in the center of each uterine horn, the smallest ones being at either end (1). This discrepancy is not likely to be explained until more data are gathered from additional species, including physiological studies of the hemodynamic properties along the uterine artery to determine how this might correlate with the development of fetuses and placentas.

The size of the litter is another factor in the intrauterine environment which affects fetal growth. Birth weights in large litters are generally smaller than those from small litters (11). If the size of a litter is reduced by cauterizing some of its members, the remaining ones grow larger (41). These effects of litter size on fetal development may be due in part to the physical crowding within the uterine horn (28). Competition between neighboring fetuses for available blood supply is probably another important factor in regulating growth. In addition to such local influences, it is also recognized that systemic effects are also important, albeit to a lesser degree. In guinea pigs, for example, it has been shown that when a single fetus occupies one horn and several others are present in the contralateral one, the size of the solitary fetus is slightly reduced as the number of littermates in the opposite horn increases (11). This may be the result of prenatal competition for limited resources of maternal origin, but whether these are nutritional or hormonal remains to be determined.

Whatever may be the relationship between fetal growth and litter size, it is not expressed until the latter stages of gestation. In other words, embryos develop at the same rate during earlier phases of pregnancy, and not until the effects of crowding are felt is the optimal rate of growth reduced. The onset of this reduction is earlier in larger litters, and varies with the species. For example, the inhibitory effects of multiple births become evident by the seventeenth day of gestation in the rat and the fiftieth day in the guinea pig. Studies on humans have shown that the growth of twins begins to lag behind that of singletons after 30 weeks of gestation while the growth of triplets and quadruplets keeps pace with that of single fetuses for only 27 and 26 weeks, respectively (Fig. 175) (26). Although the average birth weights of such infants are reduced approximately 25% with each increment from singleton to quadruplets, the combined weights of the offspring in multiple births exceeds that of single infants, though seldom rising above the normal upper range for individual newborns.

The aforementioned evidence testifies to the fact that, despite its remarkable distensibility, the uterus itself is a major limiting factor in the regulation of prenatal growth. Inasmuch as the size of the uterus is proportional to that of the mother, it might be predicted that maternal size should play an

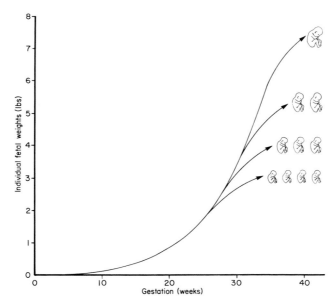

Fig. 175.   Prenatal growth curves of human fetuses as a function of litter size. The larger the litter the earlier in pregnancy growth slows down, and the shorter the gestation period (arrows). Individual birth weights decrease while combined litter weights increase with size of litter (25).

important role in regulating the growth of her offspring. To test this possibility, one must contrive to mismatch the sizes of mothers and their unborn infants to determine if the maternal environment can alter the expression of the fetal genotype with respect to bodily growth rates. In the case of humans, statistical analyses have confirmed that the birth weights of infants are correlated with the heights of their mothers, not their fathers (6). Experimentally, it has been possible to breed large and small animals together in reciprocal crosses to determine if offspring of the same genotype develop to dimensions commensurate with the body sizes of large versus small mothers. The classic experiment along these lines was that of Walton and Hammond in 1938 (42), who, by artificial insemination, crossed Shetland ponies with Shire horses, there being approximately a fourfold discrepancy in parental body weights. Normally, the foals of horses weigh about three times that of newborn ponies. The offspring of crossbred animals were intermediate between these extremes, but those from Shire mares were larger at birth than were those from Shetland mothers, a difference that persisted throughout postnatal maturation (Fig. 176). Since the discrepancies in birth weights occurred despite similar gestation periods, it must be concluded that maternal size can have a significant modulating effect on the rate of prenatal

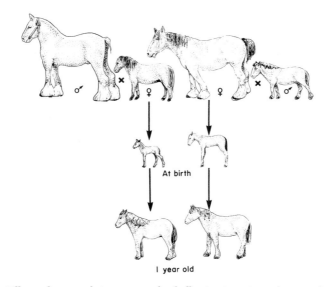

Fig. 176. Effects of maternal size on growth of offspring in reciprocal crosses between Shire horses (large) and Shetland ponies (small). The foal of a Shetland mare (left) is considerably smaller than is that produced by a Shire mare (right), a size difference that persists for up to one year after birth (below).

growth. These striking effects in the horse have been confirmed in reciprocal crosses between small and large breeds of cattle and sheep (8, 19, 22). In each case, the birth weights are greater when the mother is of the larger breed.

It could be argued that such effects might be an expression of cytoplasmic inheritance rather than maternal environment. This possibility can be tested only by transferring purebred eggs from one strain into another, an approach that has been achieved between ewes of the smaller Welsh strain and the larger Lincoln breed. As in the cases of crossbreeding experiments, the eggs of the genetically small strain developed into larger lambs in the uteri of large mothers, while the reverse was true in the reciprocal situation (9).

Crossbreeding experiments between large and small strains of mice and rabbits have not yielded the same results as in the cases of larger animals. Although the offspring are of intermediate size between the parental strains, they do not necessarily reflect differences in maternal size (15, 33). It is worth nothing, however, that these species are multiparous, while larger animals tend normally to have single births. Therefore, it may be the limited capacity of the uterus in larger species that affects the growth of individual fetuses, an influence obscured by the superfluity of uterine capacity adapted to multiple births in smaller species.

INHERENT CONTROL OF FETAL GROWTH

However important maternal influences may be on fetal growth, the fact remains that it is the genotype of the individual that exerts the overriding influence on prenatal, as well as postnatal, growth. This may be mediated at least in part by hormones of fetal origin. The simplest, though not necessarily the most elegant, kind of deletion experiment is intrauterine decapitation of the fetus, a procedure tantamount at least to hypophysectomy. Anencephaly is a teratological model for this, and has been shown to retard fetal growth in humans (*17*). In rats and rabbits, whether decapitated by cervical ligation or decerebrated by aspiration of the brain *in utero*, postoperative somatic growth is slower than normal, but nevertheless continues at a reduced rate in the total absence of the pituitary (Fig. 177) (*16, 20, 21, 38*). If growth hormone is administered to the pregnant mother, or given directly to the fetus, the subsequent growth of the unborn rat is not accelerated (*16, 36*). Anti-α-melanocyte stimulating hormone, however, decelerates prenatal growth when injected directly into 19-day rat fetuses (*39*), suggesting that MSH may normally promote the growth of the fetus.

Apropos of the possible role of the pituitary in prenatal development, some interesting results have come to light in the alpine meadows of the Rocky Mountains (*4*). Here there grows a species of skunk cabbage (*Veratrum californicum*) upon which sheep sometimes feed, much to the disadvantage

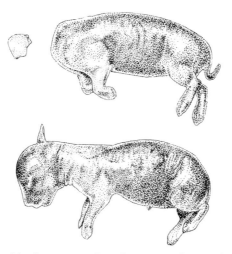

Fig. 177.   Littermate rabbit fetuses at 28 days of gestation. The one above was decapitated *in utero* at 19 days (the head of which is shown at the left). The headless fetus continues to grow at a nearly normal rate compared with its intact control (below) (*21*).

Fig. 178.   The head of a cyclopic lamb born of a ewe allowed to eat *Veratrum californicum* early in gestation (*4*). Such monsters also lack pituitaries, as a result of which their birth may be greatly delayed.

of their unborn lambs which tend to develop into cyclopian monsters (Fig. 178). Fusion of the cerebral hemispheres precludes development of the pituitary glands in such fetuses, yet their growth continues unabated. Indeed, parturition fails to occur at the expected time leading to the production of fetuses weighing over 7 kg by 230 days' gestation, compared with 3 kg lambs at normal term of 149 days. Therefore, the pituitary would seem to be necessary for the normal initiation of birth. The adrenal cortex appears to have an inhibitory effect on fetal growth. *In utero* administration of ACTH to the rabbit fetus decreases the rate of prenatal growth (*20*). On the other hand, bilateral adrenalectomy of fetal lambs permits growth to continue (*10*).

Considerable significance is attached to the length of gestation, yet little is understood about the physiological factors that trigger parturition. Indeed, it is not known for certain whether birth is initiated by the mother or the fetus, despite evidence that the fetal pituitary is involved in the stimulation of parturition. Decapitation or hypophysectomy of fetal lambs and pigs has been shown to lengthen the gestation period (*4, 23, 37*). Bilateral adrenalectomy of fetal lambs also delays labor (*10*). Parabiosis of pregnant rats when both partners are in the same stage of gestation has no effect on the time of birth. When they are in different stages of pregnancy the delivery of the more advanced litter is delayed, suggesting that if fetuses control their own birth they may do so by inhibiting the onset of parturition until the appropriate stage of development has been attained (*24*).

One of the most important factors in determining the length of gestation is the size of the litter. Naturally larger litters are usually born earlier than smaller ones, a phenomenon more related to intrauterine crowding than to absolute litter size (*26*). For example, following unilateral ovariectomy in the rat, pregnancy occurs in one horn where twice the normal number of offspring implant owing to compensatory ovulation in the ipsilateral ovary (*3*).

Such crowding results in a shortened period of gestation compared with litters of the same size distributed between both horns of the uterus. On the other hand, if one Fallopian tube is occluded, no compensatory ovulation occurs and a litter of half the normal size develops in the opposite horn of the uterus. Under these conditions, the length of gestation is equivalent to that of a normal full-sized litter.

Still another factor affecting the length of gestation is the genetic disparity between the parents. Crossbreeding tends to produce heavier offspring during prenatal development, resulting in a shortened period of gestation (27). This may be logically explained in terms of intrauterine distention owing to the pressures of increased fetal mass.

However compelling may be the evidence that fetal growth is controlled by the maternal environment vis-à-vis the fetus itself, the crucial experiment to resolve this problem has yet to be carried out. Ideally, one would like to transfer fertilized ova to the uterus of another species of different size to see if prenatal development would conform to the dimensions of the fetal species or that of its foster mother. For reasons yet to be explained, interspecific ovum transfers do not develop normally. In some combinations (e.g., mouse and rat) development ceases when implantation fails to occur (40). In others, (e.g., sheep and goat) implantation may take place and development proceeds, but the fetuses seldom survive more than 50 days (Fig. 179) (43). Until

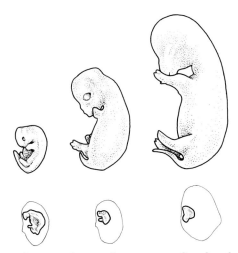

Fig. 179.   When preimplantation sheep embryos are transferred to the uterus of a goat they develop normally for 30, 40, and 45 days (left to right, above, respectively) before eventually dying *in utero*. Goat embryos (shown here with embryonic membranes intact) in the uterus of a ewe die prior to days 30, 40, or 45 of gestation (below). It is estimated that they survived only about 3 weeks (42).

the nature of these incompatibilities is discovered and corrected, such experiments in mammals must remain in the realm of the hypothetical. There may be brighter prospects in the case of the avian embryo, for it should be technically possible to exchange early blastodiscs between eggs of different sizes and species by grafting them onto yolks that are either too large or too small. If there are no interspecific incompatibilities in such recombinations, it would be interesting to note how the resultant hatching sizes and incubation periods might be affected.

DETERMINATE VERSUS INDETERMINATE SIZE

The postnatal growth of mammals and birds differs from that in many of the cold-blooded vertebrates by stopping at a predetermined size. Although it cannot be denied that the species-specific dimensions of such animals are genetically programmed, it is not known what the mechanism of this control might be. Either the overall body size is predetermined, its component organs and tissues growing to their appropriate dimensions in relation to the whole, or the various organs of the body are themselves subject to intrinsic constraints such that the overall somatic size is simply the sum of its parts. In higher vertebrates, certain organs, like the body as a whole, exhibit determinate growth. The long bones, for example, grow to lengths relatively independent of outside influences. The central nervous system, skeletal muscle, and heart are also limited in their potentials for growth owing to the premature cessation of mitotic activity in these tissues. In other organs growth is limited by the inability to augment the population of functional units at the histological level of organization. This applies to the number of nephrons per kidney, pulmonary alveoli per lung, seminiferous tubules per testis, and villi per intestine, to name a few. In each of these cases, there is an apparent species-specific predetermined number of such units which is produced early in development but which cannot be augmented later in life. Accordingly, their subsequent growth is limited to the enlargement of preexisting structures, a mode of growth that limits the ultimate size to which the organ can enlarge. Thus, one way in which birds and mammals insure the eventual cessation of somatic growth is to lose the capacity to multiply the cells or histological units of which some of their organs are composed before sexual maturity. In this way, the eventual size of the body is correlated with the growth limitations of certain key organs (13).

In contrast, animals of indeterminate size are not subject to such constraints. Some of the lower vertebrates, most notably fishes, as well as a variety of invertebrates, fall into this category. Those organs and tissues that are homologous with the ones in birds and mammals that have fixed numbers of functional units must keep pace with the sustained growth of the body as a whole. Accordingly, one finds a continuous increase in the population of nephrons, myocardial fibers, skeletal muscle fibers, nerve cells, and retinal photoreceptors. Hence, the developmental potentials of these creatures would appear to be limited largely by such exogenous factors as temperature and the availability of food.

Animals possessing the potential to grow throughout life must cope with the problem of how not to become too large. Even in those forms in which growth decelerates asymptotically, the body may reach considerable proportions given sufficient time. It has been claimed that aging may not be programmed into the life cycle so rigorously as it is in birds and mammals, the implications being that the predictable course of aging in higher vertebrates might have evolved in conjunction with the cessation of bodily growth at maturity in animals of determinate size (*14*). If aging is not the universal phenomenon it is often assumed to be, it may have evolved primarily in those animals that, in its absence, would be in danger of living indefinitely. Such would be the case among lower vertebrates were it not for the fact that many of them possess a lifelong capacity for continued growth which may itself insure the animal's demise through excessive increase in body mass. Thus, at the organismal level, as at that of the cell, the process of growth may exert a rejuvenating influence on living things, with aging being the consequence of the depreciations that set in when tissues fail to grow and renew themselves at a rate sufficient to offset the inexorable course of degradation which besets all protoplasm.

REFERENCES

1. Barr, M., Jr., Jensh, R.P., and Brent, R.L. Fetal weight and intrauterine position in rats. *Teratology* **2**, 241–246 (1969).
2. Beydoun, S.N., Haviland, M.E., and Abdul-Karim, R.W. The mechanism of placental hypertrophy in estrogen deprived pregnant rabbits. *Biol. Reprod.* **7**, 142–143 (1972).
3. Biggers, J.D., Curnow, R.N., Finn, C.A., and McLaren, A. Regulation of the gestation period in mice. *J. Reprod. Fertil.* **6**, 125–138 (1963).
4. Binns, W., James, L.F., and Shupe, J.L. Toxicosis of *Veratrum californicum* in ewes and its relationship to a congenital deformity in lambs. *Ann. N.Y. Acad. Sci.* **111**, 571–576 (1964).
5. Bruce, H.M., Renwick, A.G.C., and Finn, C.A. Effect of post-coital unilateral ovariectomy on implantation in mice. *Nature (London)* **219**, 733–734 (1968).
6. Cawley, R.H., McKeown, T., and Record, R.G. Parental stature and birth weight. *Am. J. Hum. Genet.* **6**, 448–456 (1954).
7. Csapo, A.I., and Fáy, E. Effect of hypophysectomy on ovariectomy-induced placental hypertrophy. *Acta Physiol. Acad. Sci. Hung.* **44**, 39–43 (1973).
8. Dickinson, A.G. Some genetic implications of maternal effects—an hypothesis of mammalian growth. *J. Agr. Sci.* **54**, 378–390 (1960).

9. Dickinson, A.G., Hancock, J.L., Hovell, G.J.R., Taylor, S.C.S., and Wiener, G. The size of lambs at birth—A study involving egg transfer. *Anim. Prod.* **4**, 64–79 (1962).

10. Drost, M., and Holm, L.W. Prolonged gestation in ewes after foetal adrenalectomy. *J. Endocrinol.* **40**, 293–296 (1968).

11. Eckstein, P., and McKeown, T. Effect of transection of one horn of the guinea-pig's uterus on foetal growth in the other horn. *J. Endocrinol.* **12**, 97–107 (1955).

12. Gammal, E.B. Effects of placental lesions on foetal growth in rats. *Experientia* **29**, 201–203 (1973).

13. Goss, R.J. Hypertrophy versus hyperplasia. *Science* **153**, 1615–1620 (1966).

14. Goss, R.J. Aging versus growth. *Perspect. Biol. Med.* **17**, 485–494 (1974).

15. Green, C.V. Size inheritance and growth in a mouse species cross (*Mus musculus* × *Mus bactrianus*). IV. Growth. *J. Exp. Zool.* **59**, 247–263 (1931).

16. Heggestad, C.B., and Wells, L.J. Experiments on the contribution of somatotrophin to prenatal growth in the rat. *Acta Anat.* **60**, 348–361 (1965).

17. Honnebier, W.J., and Swaab, D.F. The influence of anencephaly upon intrauterine growth of fetus and placenta and upon gestation length. *J. Obstet. Gynaecol. Br. Commonw.* **80**, 577–588 (1973).

18. Hsueh, A.M., Simonson, M., Hanson, H.M., and Chow, B.F. Protein supplementation to pregnant rats during third trimester and the growth and behavior of offspring. *Nutr. Rep. Int.* **9**, 31–45 (1974).

19. Hunter, G.L. The maternal influence on size in sheep. *J. Agr. Sci.* **48**, 36–60 (1956).

20. Jack, P.M.B., and Milner, R.D.G. Effect of decapitation and ACTH on somatic development of the rabbit fetus. *Biol. Neonate* **26**, 195–204 (1975).

21. Jost, A. La physiologie de l'hypophyse foetale. *Biol. Med.* **40**, 205–229 (1951).

22. Joubert, D.M., and Hammond, J. Maternal effect on birth weight in South Devon × Dexter cattle crosses. *Nature (London)* **174**, 647–648 (1954).

23. Liggins, G.C., Kennedy, P.C., and Holm, L.W. Failure of initiation of parturition after electrocoagulation of the pituitary of the fetal lamb. *Am. J. Obstet. Gynecol.* **98**, 1080–1086 (1967).

24. Mantalenakis, S.J., and Ketchel, M.M. Influence of pregnant parabiotic partners on time of parturition in rats. *J. Reprod. Fertil.* **11**, 313–316 (1966).

25. McKeown, T., Marshall, T., and Record, R.G. Influences on fetal growth. *J. Reprod. Fertil.* **47**, 167–181 (1976).

26. McKeown, T., and Record, R.G. Observations on foetal growth in multiple pregnancy in man. *J. Endocrinol.* **8**, 386–401 (1952).

27. McLaren, A. Effect of foetal mass on gestation period in mice. *J. Reprod. Fertil.* **13**, 349–351 (1967).

28. McLaren, A., and Michie, D. The spacing of implantations in the mouse uterus. *Mem. Soc. Endocrinol.* **6**, 65–75 (1959).

29. McLaren, A., and Michie, D. Control of pre-natal growth in mammals. *Nature (London)* **187**, 363–365 (1960).

30. Myers, R.E., Hill, D.E., Scott, R.E., Mellits, E.D., and Cheek, D.B. Fetal growth retardation produced by experimental placental insufficiency in the rhesus monkey. I. Body weight, organ size. *Biol. Neonate* **18**, 379–394 (1971).

31. New, D.A.T., Coppola, P.T., and Cockroft, D.L. Comparison of growth *in vitro* and *in vivo* of post-implantation rat embryos. *J. Embryol. Exp. Morphol.* **36**, 133–144 (1976).

32. Oh, W., and Guy, J.A. Cellular growth in experimental intrauterine growth retardation in rats. *J. Nutr.* **101**, 1631–1634 (1971).

33. Robb, R.C. On the nature of hereditary size limitations. I. Body growth in giant and pigmy rabbits. *Brit. J. Exp. Biol.* **6**, 293–310 (1929).

34. Roux, J.M., Jahchan, T., and Fulchignoni, M.C. Desoxyribonucleic acid and pyrimidine synthesis in the rat during intro-uterine growth retardation: Responsiveness of several organs. *Biol. Neonate* **27**, 129–140 (1975).

35. Sara, V.R., Lazarus, L., Stuart, M.C., and King, T. Fetal brain growth: Selective action by growth hormone. *Science* **186**, 446–447 (1974).

36. Smart, J.L., Adlard, B.P.F., and Dobbing, J. Further studies of body growth and brain development in "small-for-dates" rats. *Biol. Neonate* **25**, 135–150 (1974).

37. Stryker, J.L., and Dziuk, P.J. Effects of fetal decapitation on fetal development, parturition and lactation in pigs. *J. Anim. Sci.* **40**, 282–287 (1975).

38. Swaab, D.F., and Honnebier, W.J. The influence of removal of the fetal rat brain upon intrauterine growth of the fetus and the placenta and on gestation length. *J. Obstet. Gynaecol. Br. Commonw.* **80**, 589–597 (1973).

39. Swaab, D.F., Visser, M., and Tilders, F.J.H. Stimulation of intra-uterine growth in rat by α-melanocyte-stimulating hormone. *J. Endocrinol.* **70**, 445–455 (1976).

40. Tarkowski, A.K. Inter-specific transfer of eggs between rat and mouse. *J. Embryol. Exp. Morphol.* **10**, 476–495 (1962).

41. van Marthens, E., and Grauel, L. Ovarian, adrenal and litter restriction effects on fetal and placental development. *Nutr. Metab.* **17**, 198–204 (1974).

42. Walton, A., and Hammond, J. The maternal effects on growth and conformation in Shire horse-Shetland pony crosses. *Proc. Roy. Soc. London, Ser. B* **125**, 311–335 (1938).

43. Warwick, B.L., and Berry, R.O. Inter-generic and intra-specific embryo transfers in sheep and goats. *J. Hered.* **40**, 297–303 (1949).

44. Wigglesworth, J.S. Fetal growth retardation. Animal model: Uterine vessel ligation in the pregnant rat. *Am. J. Pathol.* **77**, 347–350 (1974).

45. Zeman, F.J., and Stanbrough, E.C. Effect of maternal protein deficiency on cellular development in the fetal rat. *J. Nutr.* **99**, 274–282 (1969).

Index

A

Acetylcholine, 164–171
  hypersensitivity, 164, 167, 169–171
  receptors, 164–167, 170, 171, 174, 176
Acinus
  pancreatic, 244, 246
  salivary, 236–241
Acromegaly, 68, 292
ACTH, 350, 356, 361–363, 422
Actinomycin D, 212, 246, 258, 304
Adaptation, physiological vs. genetic, 138, 312
Adenoma, 95, 287, 322
Adipocyte, 56–61
Adrenal cortex, 307, 355–363, 422
  cancer, 360
  enucleation, 362
  X-zone, 357, 359, 360
Adrenalectomy, 307, 354, 360–362, 422
Adrenaline, 18, 19, 143
Aflatoxin, 258
Aging, 8, 18, 30, 32, 33, 57, 65, 66, 79, 90, 140, 184, 213, 230, 231, 252, 256, 284, 300, 303, 425
Albumin, 252
Aldosterone (see also Deoxycorticosterone, Mineralocorticoid), 353, 354, 356, 357, 362
Allograft rejection, 86, 100, 148, 321, 403, 408
Allometry, 9, 253, 297
Alloxan, 262, 317, 351
Altitude (see also Hypoxia), 93, 145, 146, 291
Alveolus, pulmonary, 284–289, 291, 292, 424
  number, 284–289
  size, 284
Ameloblast, 31, 33

β-Aminopropionitrile, 36, 52
Amniotic fluid, 299
Amphibian, 286, 300, 328
  anuran, 91, 151, 158, 193, 201, 202, 286, 344, 377, 387, 390, 403
  larva, 91, 99, 170, 218, 272, 343
  urodele, 91, 99, 151, 158, 175, 187, 188, 190, 193, 194, 204, 215, 218, 220
Amylase, 235, 238–240, 244, 245, 248
Anemia, 93, 94, 96, 98–100, 144
Anencephaly, 421
Angiogenesis, 120–124
Angiotensin, 96, 314, 353, 362
Angiotensinogen, 96, 252, 314, 353
Antibody, 105, 176, 200, 211, 350, 381
  auto-antibody, 112
Antigen, 105, 106, 211, 212, 219, 408
Antilens serum, 211
Antineutrophil serum, 102, 104
Antiplatelet serum, 110
Antitemplate, 11
Antler, 24, 25, 126
Antrectomy, 268
Antrocolic transposition, 268, 269
Antrum, 268, 269
Aorta. 125, 126, 132
  ligation, 145, 315, 316
Aqueous humor, 213, 221, 223
Arteriole, afferent, 315
Arterio-venous fistula, 74, 75, 127–130, 144
Artery, 84, 120
  bronchial, 130
  hepatic, 260
  occlusion, 124, 125, 127–130, 145, 260, 291, 292, 408

429

Vallate papilla, 44
Vascular anastomosis, 291
Vas deferens, 375, 380, 381
Vasectomy, 380, 381
Vein, 120, 121
    hepatic portal, 252, 260, 388
    renal portal, 297
    transplantation, 126
Venom, 167, 170, 174, 235
Venous stasis, 75, 76
Ventricle, cardiac, 138, 139, 150, 151
    left hypertrophy of, 145, 150
    right hypertrophy of, 145–147
Veratrum californicum, 421
Villus, 272–275, 277–279, 424
    number, 273–275
Virus
    Friend, 97, 101, 110
    polycythemic, 97
    Rauscher, 97
Viscacha, 386, 389
Vitamin A, 28, 195, 198, 252, 330
Vitamin D$_2$, 344
Vitreous humor, 216, 218
Viviparity, 403–405, 414

W

Waltzing mouse, 291
Whale, 18, 33, 120, 138, 141, 284, 297
White blood cell (*see* Leukocyte)
Wilms tumor, 95, 322
Wolff's law, 73
Wound healing

in bladder, 330
contraction, 26, 36, 53, 54
dressing, 22, 24
in endothelium, 132, 133
in epidermis, 18, 22–24, 26, 27, 31
in intestine, 273, 274
in lens epithelium, 213–215
in liver, 253
in mesothelium, 55, 56
in renal tubule, 317
in salivary gland, 239
in stomach, 272
in uterus, 398

X

X-rays, 42, 54, 76, 78, 79, 100, 102–104, 110, 258
Xylocaine, 188
X-zone, adrenal, 357, 359, 360
X-zone degeneration, 359

Z

Zahnreihe, 39, 40
Zinc acetate, 377
Zinc deficiency, 377
Zollinger-Ellison syndrome, 268
Zona fasciculata, 356, 357, 360
Zona glomerulosa, 307, 314, 356, 357, 362
Zonal necrosis, 254
Zona reticularis, 356
Zymogen granules, 244, 245, 249, 349